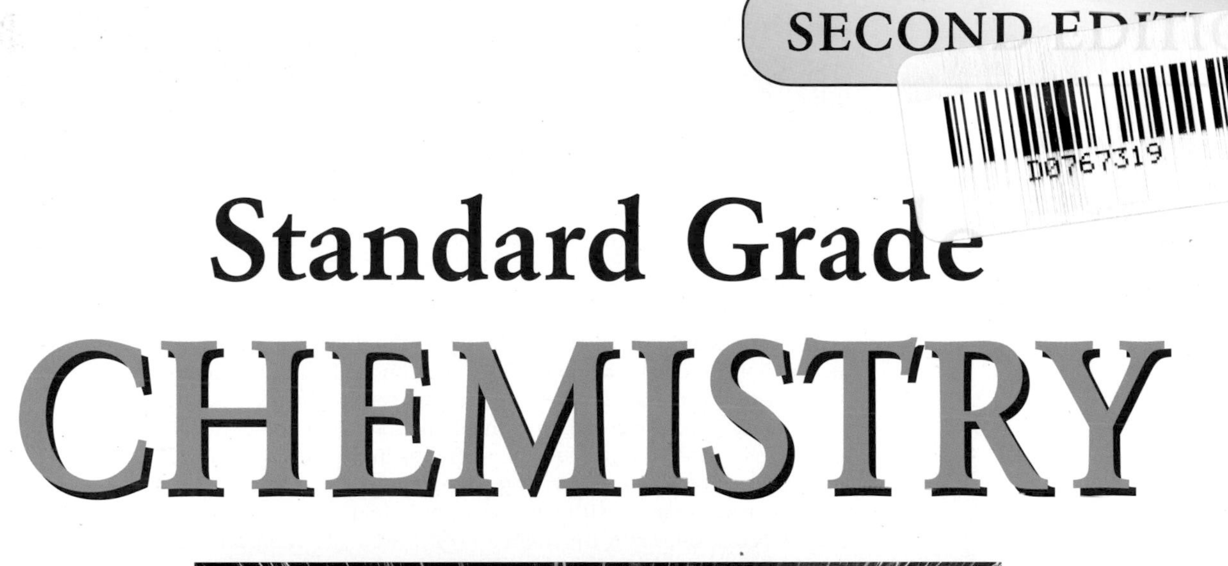

SECOND EDITION

D0767319

Standard Grade CHEMISTRY

Norman Conquest
Roddy Renfrew

Hodder Gibson
A MEMBER OF THE HODDER HEADLINE GROUP

To the Student

This book covers the Standard Grade Chemistry course. The shaded sections marked with a C contain Credit level material.

There are short questions throughout the text to check your understanding of what you have read. The answers to these questions are given in Chapter 24. At the end of each of the main chapters there is a selection of examination-type questions. Chapters 16–22 contain revision material.

We hope that you find this book useful.

Acknowledgements

The authors and publishers would like to thank the Scottish Qualifications Authority for permission to reproduce questions in this book. The SQA accept no responsibility whatsoever for the accuracy or method of working in the answers given.

The publishers would like to thank the following companies and individuals who have given permission to reproduce photographs in this book. Every effort has been made to trace and acknowledge ownership of copyright. The publishers will be glad to make suitable arrangements with any copyright holder whom it has not been possible to contact. Autoexpress (69, 120); BP/Amoco (48); Corbis (1, 4, 55 bottom, 78 bottom right, 84 top, 110 bottom, 119 top, 122, 134 bottom, 153); Empics (132 top); GSF Picture Library (73 all, 90 top, 117 bottom two, 138 left, 148 bottom); Hodder & Stoughton (87 bottom)/ NASA (1); Holt (137, 138 middle and right); Life File (6 bottom two, 19, 66, 89 top, 90 bottom, 98. 106 bottom, 107, 108, 110 top 117 top, 119 middle left, 125 both, 126, 127, 129, 149, 152 top); Norman Conquest (6 top right, 13, 15, 22 bottom, 33 all, 34 both, 40 all, 59 both, 64, 77, 78 bottom left, 80, 119 middle right, 128, 132 bottom, 148 top, 152 middle); Redferns (106 top); Ruth Hughes (147 right); Science Photo Library (21, 22 top, 25, 46, 52, 55 top, 61, 78 top, 79, 84 bottom, 89 bottom, 130, 134 top); Science and Society Picture Library (99); Scottish View Point (119 top); Still Moving Picture Company (147 left); Still Pictures (3, 87 top, 140 both).

Cover photo: Science Photo Library

Orders: please contact Bookpoint Ltd, 130 Milton Park, Abingdon, Oxon OX14 4SB. Telephone: (44) 01235 827720, Fax: (44) 01235 400454. Lines are open from 9.00 – 5.00, Monday to Saturday, with a 24 hour message answering service. You can also order through our website at www.hoddereducation.co.uk

British Library Cataloguing in Publication Data
A catalogue record for this title is available from The British Library

Published by Hodder Gibson, a member of the Hodder Headline Group, an Hachette Livre UK Company, 2a Christie Street, Paisley PA1 1NB. Tel: 0141 848 1609;
Fax: 0141 889 6315; email: hoddergibson@hodder.co.uk

ISBN-13: 978-0-34084718-3

First edition published 1995
This edition published 2002
Impression number 10 9 8 7 6
Year 2007

ISBN-13: 978-0-34084719-0

First edition published 1995
This edition published 2002
Impression number 10 9 8 7 6
Year 2007
With Answers

Copyright © 1995, 2002 Roddy Renfrew and Norman Conquest

All rights reserved. No part of this publication may be reproduced or transmitted in any form or by any means, electronic or mechanical, including photocopy, recording, or any information storage and retrieval system, without permission in writing from the publisher or under licence from the Copyright Licensing Agency Limited. Further details of such licences (for reprographic reproduction) may be obtained from the Copyright Licensing Agency Limited, of Saffron House, 6-10 Kirby Street London, EC 1N 8TS.

Printed in Italy for Hodder Gibson, 2a Christie Street, Paisley, PA1 1NB, Scotland, UK

CONTENTS

CHAPTER ONE

Introducing Chemistry

SECTION 1.1 A chemical world

Figure 1.1 Every second, millions of chemical reactions take place all over the world

It's a chemical world. Everything around you, all the materials you can touch, all the plants and animals on the planet, are made up of chemicals.

Chemists try to find out how chemicals behave and how they can be made into useful new substances. Think about the following: making toast; a fireworks display; a space shuttle taking off; a plant growing. They all have one thing in common – they all involve chemicals changing to produce new substances. These changes are called **chemical reactions**.

Chemical reactions occur almost everywhere. Inside your body, chemical reactions break down your food and give you the energy you need to live. In cooking, your food is changed to produce new chemicals giving new tastes and textures. In a gas fire, the burning gas joins with oxygen in the air to produce two different chemicals called water and carbon dioxide.

Questions

Q1 Look at the photograph in figure 1.2. Where do you think chemical reactions are taking place?

Figure 1.2 A fireworks display

With so many different chemical reactions going on in the world how can we begin to understand them? Fortunately, they all have some basic features in common.

All chemical reactions produce new substances

When sherbet is added to water it fizzes. The fizzing is caused by a gas produced by the reaction of sherbet with water. In fact, all chemical reactions produce new substances. You can tell if a chemical reaction has occurred because a new substance will have been formed. This can be shown in a simple experiment where copper and sulphur are heated together. A chemical reaction takes place and a new substance, called copper sulphide, is made (see figure 1.3).

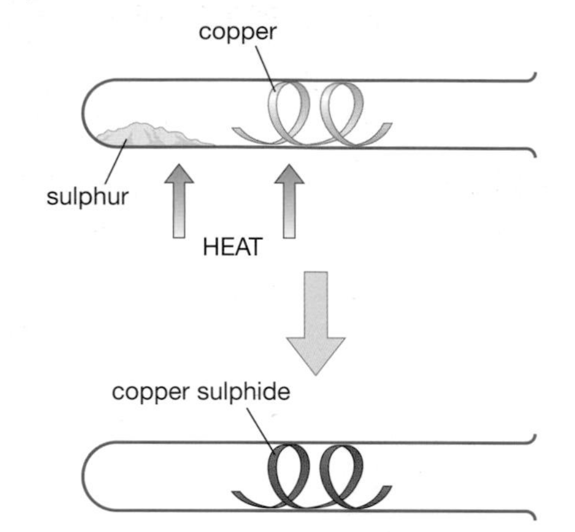

Figure 1.3 The reaction between copper and sulphur produces a new substance – copper sulphide

We can summarise what has happened by using a **word equation**, which tells us the names of the substances that react and the name of the new substance that is formed. In this case, the word equation would be:

copper + sulphur → copper sulphide

In the word equation, the + sign means 'and' and the arrow means 'react to produce'. For example, the following word equation is used for the reaction between calcium and water to produce calcium hydroxide and hydrogen:

calcium + water → calcium hydroxide + hydrogen

The substances on the left of the arrow, which react together, are called the **reactants**. The substances on the right of the arrow are called the **products**.

Chemical reactions can involve a change in appearance

You can see this in the copper and sulphur experiment, shown in figure 1.3. The new substance formed is a black powder quite unlike the yellow powdered sulphur and the brown copper that were in the test tube at the start of the experiment.

In many chemical reactions the new substance that is formed looks different from the substances at the start of the reaction. We say that there has been a 'change in appearance'.

Questions

Q2 The rusting of iron is a chemical reaction. Describe the change in appearance which takes place when an iron nail rusts.

Most chemical reactions involve energy changes

Some important types of energy	
heat	movement
light	chemical
sound	electrical

Table 1.1 Energy types

You will already know that energy exists in different forms. Table 1.1 gives the names of the main types of energy.

In a chemical reaction, the substances which react contain chemical energy. During the reaction this energy is changed into new forms of energy, usually heat and light.

When dynamite explodes, a chemical reaction occurs which gives out a huge amount of heat and movement energy. This is an example of an energy change.

Figure 1.4 What two other types of energy are given out in a dynamite explosion?

Questions

Q3 Name the types of energy given out in the following reactions:
a) wood burning,
b) a car engine running,
c) a firework rocket taking off.

Section 1.1 Summary

- *All* chemical reactions produce new substances.
- *Most* chemical reactions involve noticeable energy changes.
- *Many* chemical reactions involve a change in appearance.

SECTION 1.2 Elements, mixtures and compounds

Elements

Every chemical in the world is made up of **elements**. Just over one hundred elements have been discovered. Elements are substances that are made of only one type of atom. This is discussed in more detail in Chapter 3. You will be familiar with some elements already. Silver, used in jewellery, aluminium, used in cooking foil, and oxygen in the air you breathe are all elements.

All the known elements are listed in the periodic table, as shown on page 8 in the SQA Data Booklet. Every element has its own **symbol**. Symbols are very useful for representing elements. In most cases the symbol for an element is the first letter or the first two letters of the element's name. For example U is the symbol for uranium and Al is the symbol for aluminium.

In some cases the symbol for an element is based on its Latin name. The Latin word for gold is 'aurum' and its symbol is Au. In the past, most water pipes in houses were made from the element lead. The symbol for lead is Pb,

Questions

Q1 Use the periodic table on page 167 to find the symbols for the following elements:
a) iodine,
b) potassium,
c) mercury.

Q2 Try to find out why the symbol for iron is Fe.

which comes from the Latin word 'plumbum', meaning lead. So you can see how plumbers originally got their name.

Compounds

The section above dealt with elements on their own. However, elements can join together chemically – they do this when they take part in a chemical reaction.

When elements join together, they form more complicated substances called **compounds**. For example, the compound sodium chloride can be made by the chemical reaction between two elements – sodium and chlorine. This reaction is quite spectacular as it produces flashes of yellow flames (see figure 2.1). You will probably know that sodium chloride is the chemical name for common salt. It's interesting to think that something as harmless as this can be made from a violent reaction between a metal and a poisonous green gas. Elements, therefore, are the building blocks which join together to form compounds.

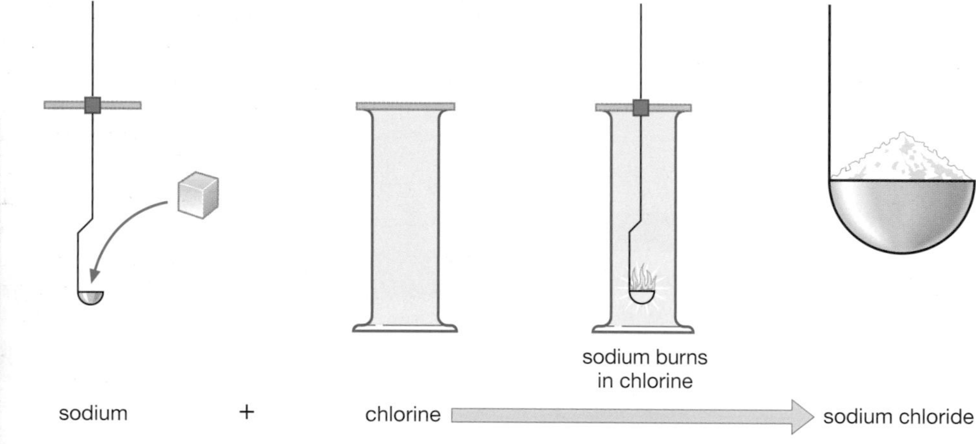

Figure 2.1 Sodium and chlorine react together to produce the compound sodium chloride

Figure 2.2 A few basic types of Lego brick can be used to build many different models

Surprisingly, every one of the millions of compounds that exist can be made from just over one hundred elements. We can think of elements as being like Lego pieces – with just a few basic types of Lego brick you can build hundreds of different models.

Naming compounds

Names often give you information. The name 'McDonald' means 'son of Donald'. 'Ben Aziz' means 'son of Aziz'. The names of compounds also tell you something about them.

The name of a compound containing two elements always ends in the letters **-ide** (see table 2.1).

Name of compound	Elements present
magnesium oxide	magnesium and oxygen
silver sulphide	silver and sulphur

Table 2.1 Compounds containing two elements

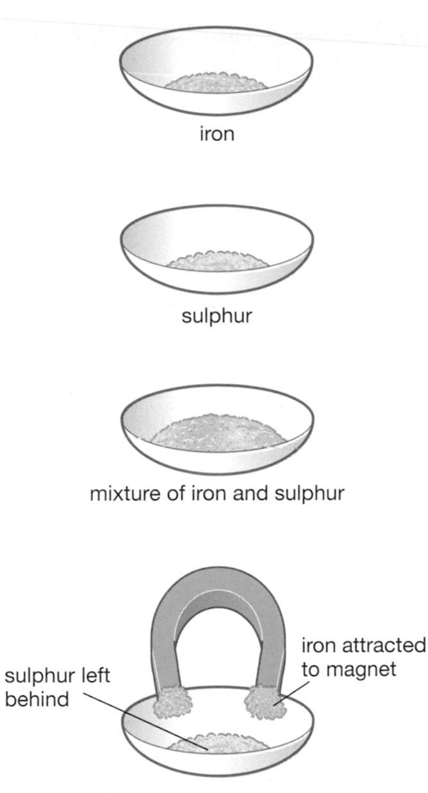

Figure 2.3 Separating a mixture of iron and sulphur

If a compound contains three elements and one of them is oxygen then its name usually ends in the letters **-ate** or **-ite** (see table 2.2).

Name of compound	Elements present
calcium carbonate	calcium, carbon and oxygen
lithium sulphite	lithium, sulphur and oxygen

Table 2.2 Compounds containing three elements

Questions

Q3 Give the names of the elements which are present in the following compounds: **a)** hydrogen sulphide, **b)** zinc nitrate, **c)** potassium chloride, **d)** barium nitride, **e)** magnesium chlorate, **f)** sodium nitrite.
Use the periodic table on page 8 in the SQA Data Booklet to help you.

Mixtures

Imagine some iron filings stirred together with some sulphur powder in a dish. This would be a **mixture** of iron and sulphur. A mixture of elements is different from a compound because in a mixture the elements are not chemically joined. Because of this it is often quite easy to separate the elements in a mixture. For example, the iron in the mixture of iron and sulphur can be separated by attracting the iron to a magnet (see figure 2.3).

The term 'mixture' is also used when different *compounds* are mixed together. They too can often be separated easily. For example, imagine a mixture of salt and grit used for putting on roads in winter. This mixture can be separated in two steps, as shown in figure 2.4. The salt is soluble in water and it dissolves when water is added. The grit can then be separated from the liquid by filtering. **Filtering** or **filtration** is a technique used to separate a soluble from an insoluble substance.

Questions

Q4 Can you think of a way of obtaining a sample of *dry* salt from the salt solution in figure 2.4?

Q5 In your own words, describe what is meant by the following terms: **a)** element, **b)** compound.

Q6 What is the most important difference between a *compound* made from iron and sulphur and a *mixture* of iron and sulphur?

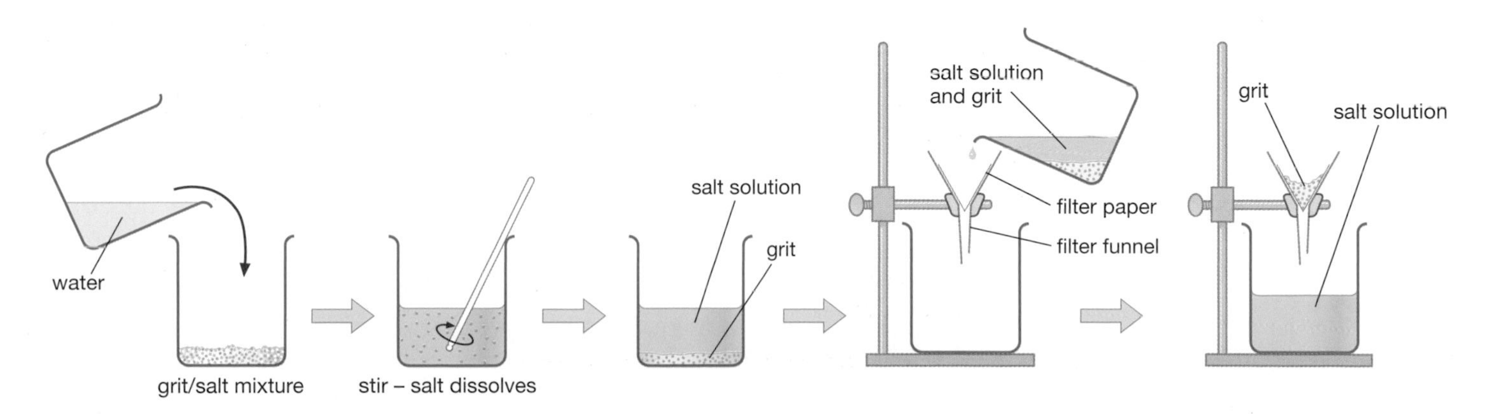

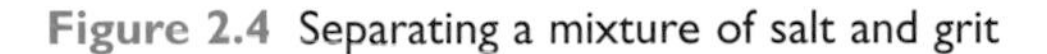
Figure 2.4 Separating a mixture of salt and grit

SECTION 1.3

Solutions and solubility

Figure 3.1 Adding sugar to tea in a tea cup

Solutions

What happens to the sugar that people put in their tea? Where does it go? The sugar dissolves in the water that makes up the tea. When sugar dissolves in water a **solution** is formed. Many substances will dissolve to form solutions. Solutions are special kinds of mixtures in which a **solute** has dissolved in a liquid which is called a **solvent**.

A solute is the name given to the substance that dissolves. Blood is a complex solution which carries many substances (solutes) around the body. Many soft drinks are solutions of sugar and flavourings in water.

Questions

Q1 Name the solutes and solvent in Irn Bru: use the information in figure 3.2 to help.

Figure 3.2

- **Dilute** solutions have only a small amount of solute dissolved in them compared with the amount of solvent.
- **Concentrated** solutions have a large amount of solute dissolved in them compared with the amount of solvent.
- **Saturated** solutions have as much solute as possible dissolved in the solvent.

Figure 3.3 A concentrated solution and a dilute solution

Questions

Q2 Explain the difference between concentrated and dilute apple juice using the words 'solute', 'solvent' and 'solution' in your answer.

Solubility

The amount of solute which is needed to make a saturated solution is called its **solubility**. The solubility of a solute can be measured by finding the maximum mass of solute which can be dissolved in a given mass of solvent at a fixed temperature.

Figure 3.4 shows that about 20 g of copper sulphate can be dissolved in 100 g of water at 20°C.

The change in solubility at different temperatures can be shown on a graph called a **solubility curve**. Solubility curves have been made for most solutes. They are used by chemists to find out how much solute can be dissolved in a solvent at different temperatures. The curves can also be used to predict what will happen when solutions are cooled down.

The curve in figure 3.4 shows that at 60°C it is possible to dissolve 40 g of copper sulphate in 100 g of water. But when the temperature drops to 40°C only 28 g can stay dissolved. The other 12 g of copper sulphate are pushed out of the solution and appear as crystals of copper sulphate.

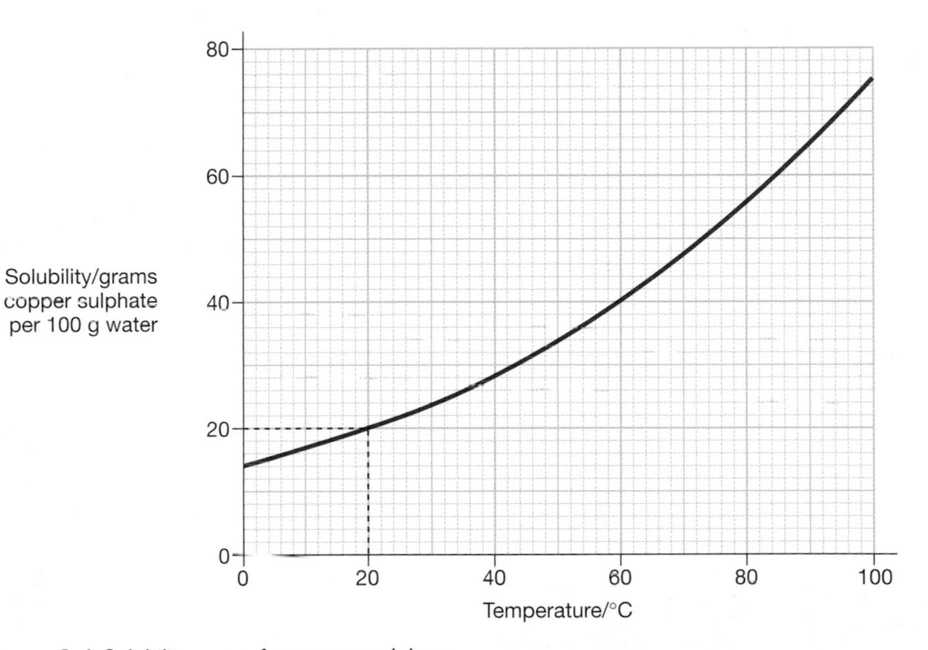

Figure 3.4 Solubility curve for copper sulphate

Questions

Q3 Use figure 3.4 to answer this question. What is the solubility of copper sulphate at 75°C?

Solutions and making sugar

The sugar which you buy in shops is made using a process that relies upon a knowledge of solutions and solubility curves.

In the sugar industry sugar crystals are produced by boiling up sugar beet or sugar cane and water to make a hot saturated solution. When the temperature drops, the saturated solution is no longer able to hold all of the sugar in solution, and crystals of sugar are formed.

By controlling the rate at which the sugar solution cools, different-sized crystals are produced. Caster sugar is made up of very small crystals, ordinary granulated sugar has larger crystals and preserving sugar is made of still larger crystals.

Chapter 1 Study Questions

1 On Andy's first day in chemistry, his teacher demonstrated an experiment to the class. Here is the report that Andy wrote in his notebook.

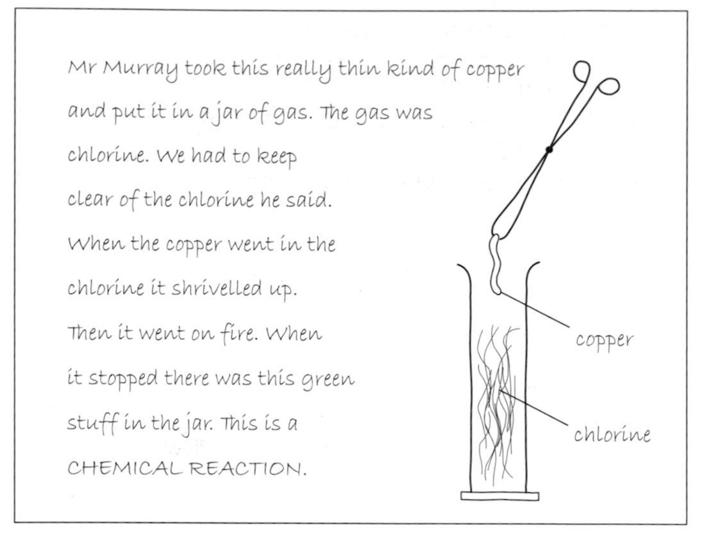

a) From Andy's report, give *two* pieces of evidence which suggest that a chemical reaction had taken place. (PS)

b) Write a word equation for this reaction.

SEB GENERAL (KU)

2 Which box (or boxes) shows a chemical change?

A gas burning in a cooker ring	**B** boiling water in an electric kettle	**C** dissolving sugar in a cup of tea
D an ice cube melting	**E** an epoxy resin glue setiing	**F** nail polish remover evaporating

CREDIT (KU)

3 Use the SQA Data Booklet to find which of the elements was discovered first.

A silicon	**B** aluminium
C nitrogen	**D** oxygen

GENERAL (PS)

4

A N	**B** Na	**C** Nb
D Ne	**E** Ni	**F** No

a) Identify the symbol for neon.

b) Identify the element that was discovered in 1751.

GENERAL (PS)

5 'Tincture of iodine' is an antiseptic produced by dissolving iodine in ethanol. Identify the term which describes the iodine used in making 'tincture of iodine'.

A solvent	**B** solute	**C** solution

GENERAL (KU)

6 The graph shows how the solubilities of three salts vary with temperature.

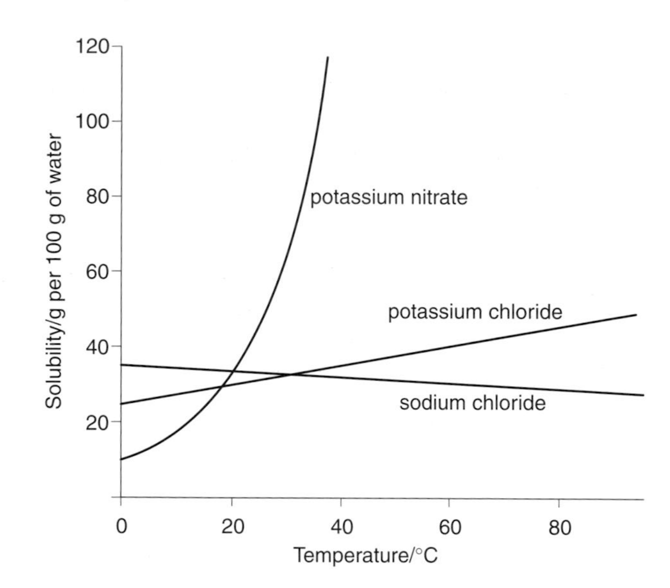

Identify the **true** statement(s).

A	At 10°C potassium nitrate is more soluble than sodium chloride.	A
B	At 20°C sodium chloride is more soluble than potassium chloride.	B
C	At 40°C potassium chloride is more soluble than potassium nitrate.	C
D	At 50°C sodium chloride is more soluble than potassium nitrate.	D

SEB CREDIT (PS)

7 A line graph showing how the solubility of potassium chlorate changes with temperature is shown below:

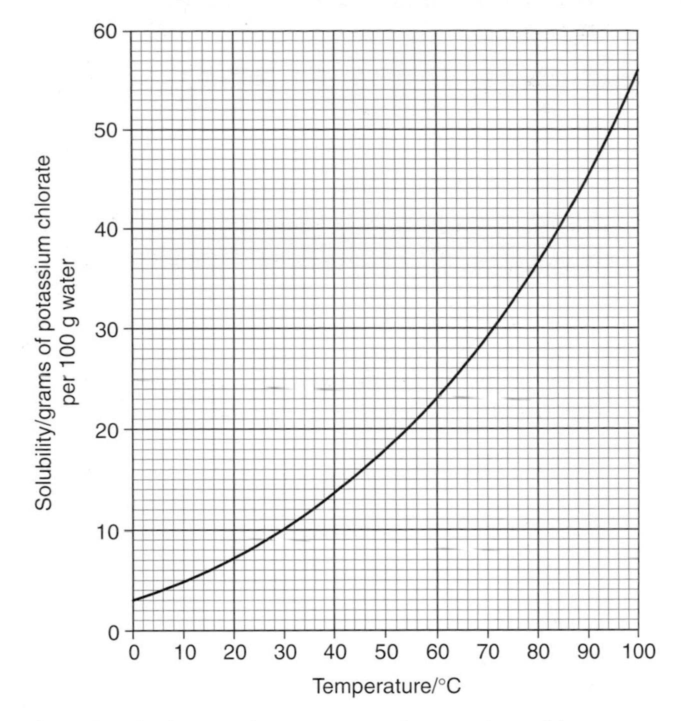

a) What is the maximum mass of potassium chlorate that would dissolve in 100 g of water at 62°C?

b) At what temperature is the solubility of potassium chlorate 15 g per 100 g of water? GENERAL (PS)

c) Calculate the mass of crystals that would form if a solution containing 80 g of potassium chlorate in 200 g of water were cooled from 90°C to 30°C. CREDIT (PS)

8 The table below gives the solubilities of sodium chloride and potassium nitrate at various temperatures. Each solubility is the mass in grams of solute that will dissolve in 100 g of water at the specified temperature.

Temperature/°C Solute	10	20	40	60
potassium nitrate sodium chloride	21.0 35.8	32.0 36.0	64.0 36.6	110.0 37.3

a) Plot solubility curves for the two solutes using the same set of axes.

b) Use the curves to estimate the temperature at which the two salts are equally soluble.

c) Use the curves to estimate the temperature at which the solubility of potassium nitrate is 70 g per 100 g of water. CREDIT (PS)

9 We live in a world of elements and compounds. We are surrounded by them all the time.

a) Name a *gaseous* element that is around us all the time.

b) Name a *solid* element you could find somewhere in your home.

c) Liquid elements are rare, but can you name a scientific instrument in which you might find a liquid metal element?

d) What do you think is the most common liquid *compound*?

e) Name a compound that is present in the air. GENERAL (KU)

10 Copy and complete the following tables:

a)

Compound	Elements present
potassium iodide _______ nitride _______ _______	_______ _______ lithium _______ calcium fluorine

b)

Compound	Elements present
_______ carbonate lead nitrate _______ _______	iron _______ _______ _______ _______ _______ copper sulphur oxygen

GENERAL (KU)

11 Some orange ammonium dichromate crystals were placed on a heat-resistant mat and touched with a hot spatula. Immediately, a green powder was thrown up into the air, as if from a volcano. The centre of the substance glowed red hot for about 20 seconds, after which all that was left was the green powder. Give *three* pieces of evidence which suggest that a chemical reaction has taken place. GENERAL (PS)

12 Rashid suggested that the burning of petrol or diesel in a car engine was an example of a chemical reaction. Explain how Rashid's claim can be justified by answering the following questions:

a) Is there a change in appearance as the reactants turn into products? Describe it as accurately as you can.

b) Does an energy change take place? If so, what forms of energy are released?

c) Are any new substances formed during the burning process? If there are, can you describe them, or perhaps name any of them? GENERAL (KU)

CHAPTER TWO

The Speed of Reactions

SECTION 2.1 Measuring the speed of reactions

Questions

Q1 Put the following reactions in order, starting with the slowest: glue setting, a car body rusting, a slice of apple turning brown, natural gas catching fire.

There have been several massive explosions in flour mills when flour dust has caught fire. These occur because the dust burns very quickly. An explosion is a very fast chemical reaction. Other reactions are much slower. For example, copper domes on the roofs of buildings will react with the air and turn green, but it can take many years for this to happen.

How to measure the speed of a reaction

The speed of a reaction is very important. In industry, chemical engineers need to know how long it will take for substances to react to make a product. Chemists have done a lot of work studying what makes a reaction fast or slow.

You know from Chapter 1 that all chemical reactions produce new substances. The speed of a reaction can be found by measuring how fast a new substance is produced.

The reaction between acid and marble chips is a useful one to investigate. When marble chips are added to dilute acid in a flask, carbon dioxide is produced. As the reaction goes on, the flask becomes lighter because the carbon dioxide gas escapes. The speed of the reaction can be measured by putting the flask on a balance and measuring the loss of mass at regular intervals (see figure 1.1). Some typical results are shown in table 1.1

Time/seconds	Mass of carbon dioxide produced/grams
0	0.0
10	1.2
20	2.0
30	2.5
40	3.0
50	3.3
60	3.6
70	3.8
80	3.9
90	4.0
100	4.0

Table 1.1

Figure 1.1 Measuring speed of reaction between marble chips and dilute hydrochloric acid

Once the results have been obtained, they can be plotted on a graph (see figure 1.2). If you put time on the horizontal axis, the slope of the graph tells you how fast the reaction is at different times.

- **Part A:** this is the steepest part of the curve, which means that the reaction is fastest here. Lots of carbon dioxide is given off each second.
- **Part B:** the slope is less steep here; the reaction is slowing down.

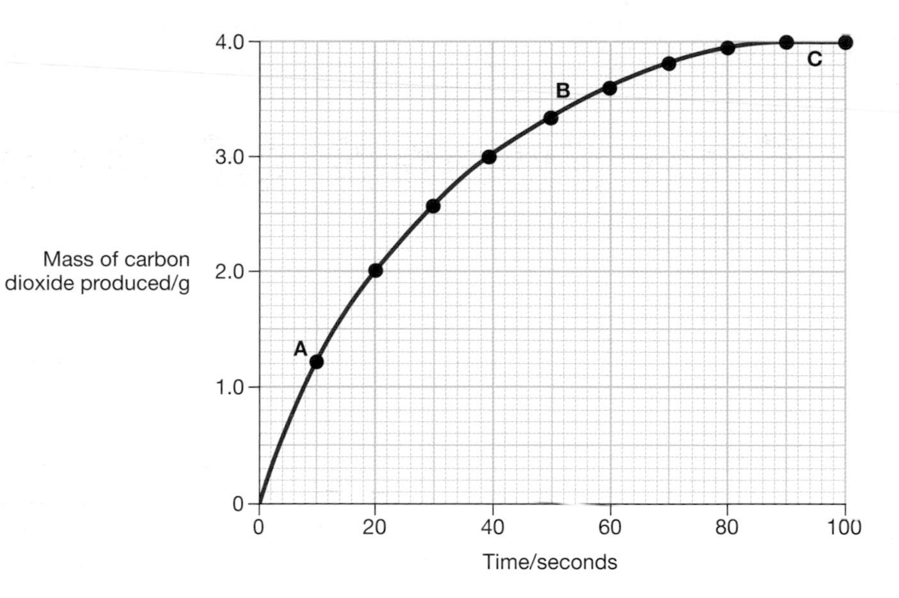

Figure 1.2

- **Part C:** here the graph has levelled off. This means that the reaction has stopped. No more carbon dioxide is being produced.

Questions

Q2 Table 1.2 gives the results of an experiment between marble chips and hydrochloric acid.
a) Plot a graph of these results.
b) Mark on your graph the part which shows the fastest part of the reaction.

Time/seconds	Mass of carbon dioxide produced/grams
0	0.0
10	1.0
20	1.8
30	2.4
40	2.8
50	3.0

Table 1.2

Speeding up reactions – making fair tests

Imagine you are an athlete and that you want to be able to run faster. You decide to change your diet, buy new running shoes and try out a weight-training programme. After doing all this, you find that there has been an improvement and you do run faster. However, because you changed all three things at the same time, you do not know which of them actually helped you to run faster. The only way to find out is to change one thing at a time and measure any increase in your speed.

The same problems can be found in chemistry. If you want to find out what makes a reaction go faster then you must change only one thing at a time. The things that can be changed in a chemical reaction are called **variables**.

Questions

Q3 Brand X washing-up liquid claims that it washes more dishes than brand Y. Describe the kind of experiment that you could do to find out if X really is better than Y. Mention all the things that would have to be kept the same in order to make it a fair test.

Q4 Some cooks say that water boils at a higher temperature when a little salt is added to it. Describe an experiment you could perform to see if this is true. Again, mention everything that would have to be kept the same.

Changing the speed of reactions

Particle size

Questions

Q1 In these experiments the concentration of the acid is kept the same in both flasks. What else has to be kept the same in order to make this a fair test?

What makes reactions go faster? Perhaps the size of the particles which are reacting has an effect on the speed. The reaction between marble chips and acid can be used to investigate the effect of particle size on the speed of a reaction.

Two experiments are set up as shown in figure 2.1. The first uses dilute acid and large marble chips. The second has the same concentration of acid and the same mass of marble, however now the marble is in the form of small chips.

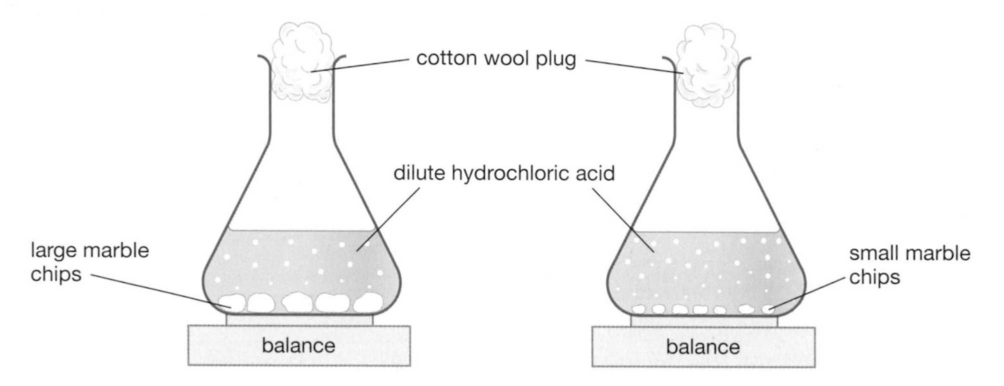

Figure 2.1 Investigating the effect of particle size on speed of reaction

In both cases the speed of the reaction is found by measuring the mass of the flasks every minute. The flasks become lighter as the carbon dioxide gas escapes. A simple subtraction gives the mass of carbon dioxide produced. The results are shown in table 2.1.

The results of the two experiments can be compared by plotting them on the same graph, as shown in figure 2.2.

The graphs show that the small marble chips react faster than the large marble chips. In other words, small particles react faster than large particles.

You probably expected the smaller marble chips to react faster; think of how fast sticks would burn compared with a solid block of wood. Cooking

Time/ minutes	Mass of carbon dioxide produced/grams: Small marble chips	Large marble chips
0	0.0	0.0
1	1.5	0.9
2	2.7	1.8
3	3.5	2.5
4	4.1	3.2
5	4.6	3.8
6	4.8	4.3
7	5.0	4.6
8	5.0	4.9
9	5.0	5.0
10	5.0	5.0

Table 2.1

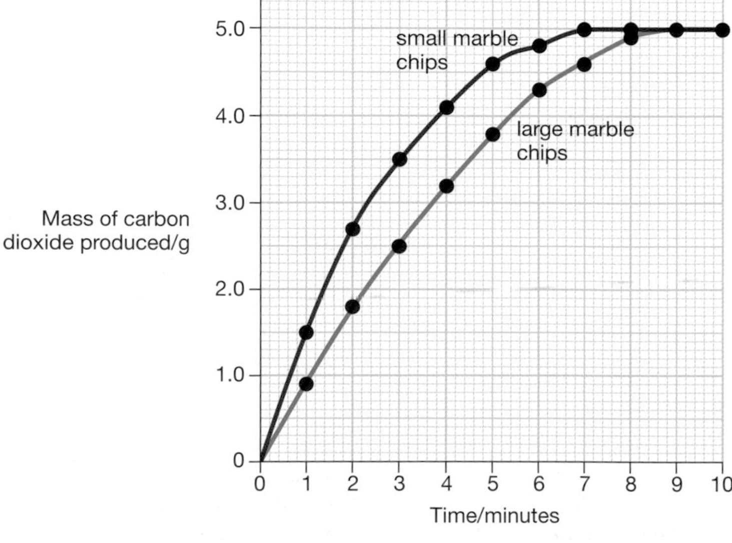

Figure 2.2

potatoes involves a series of chemical reactions – you may have noticed that small potatoes usually cook much faster than large ones.

Questions

Q2 This chapter began by mentioning explosions in flour mills. Why does the flour burn so quickly?

Concentration of solutions and the speed of reactions

The acid in the previous experiment is actually a solution. What happens if the concentration of the acid is increased?

As before, two flasks are set up, this time one has acid which is twice as concentrated as the other. To make the experiment fair, both flasks contain powdered marble. The results of the experiment are shown in figure 2.3.

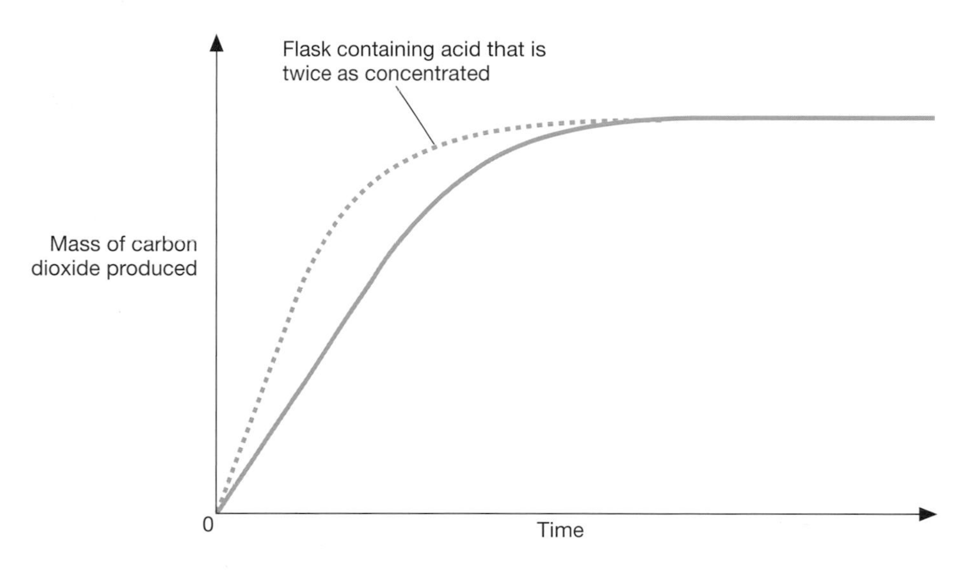

Figure 2.3 The effect of concentration of solution on speed of reaction

Questions

Q3 In gas cookers, you will often find that the metal first rusts near the burners. Why is this?

Notice that although the same amount of carbon dioxide is produced in both experiments, the more concentrated acid reacts faster than the other one. In general, the greater the concentration of the substance which is reacting, the faster the reaction. This is true for all solutions, not just acids. For example, cars corrode faster in cities where there is a high concentration of acid in the rain compared with country areas where the acid rain is less concentrated.

Figure 2.4 Why must we store milk at a low temperature?

Temperature and the speed of reactions

Do you think that reactions go faster when the temperature rises? The answer is yes – most reactions speed up when things get hotter. You can see this in the reaction between glucose and Benedict's solution. When these substances are warmed they react to produce an orange precipitate. The higher the temperature, the faster this precipitate is produced.

Fridges and freezers are used to keep food fresh. This is because the reactions which make food go bad occur more slowly at low temperatures.

Section 2.2 Summary

Slow reactions	Fast reactions
large particles	small particles
low concentrations	high concentrations
low temperatures	high temperatures

Catalysts

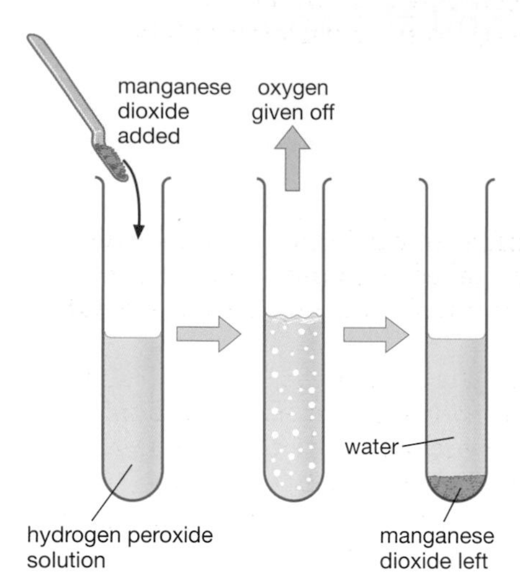

Figure 3.1 Using a catalyst to speed up a reaction

There is a way to speed up a reaction without increasing the temperature or the concentration of the reactants. That is, by adding a substance called a **catalyst**. Figure 3.1 shows the effect of adding manganese dioxide to hydrogen peroxide solution. Hydrogen peroxide solution is a liquid which decomposes slowly to give oxygen and water, as shown in the equation:

hydrogen peroxide → oxygen + water

Normally this reaction is so slow that you cannot see any signs of it taking place. But when manganese dioxide is added, bubbles of oxygen are produced very quickly. We can say that manganese dioxide is a *catalyst* for the decomposition of hydrogen peroxide.

Chemists have discovered many different catalysts. However, *all* catalysts:

- speed up some reactions
- are not used up during a reaction.

Questions

Q1 If you added exactly 1 g of manganese dioxide powder to a test tube of hydrogen peroxide how could you show, at the end of the reaction, that the manganese dioxide had not been used up?

Q2 Zinc reacts with dilute sulphuric acid to produce hydrogen gas. Copper is thought to be a catalyst for this reaction when in contact with the zinc. Given copper wire, granulated zinc and anything else you need, describe how you could find out whether or not the copper was a catalyst for the reaction.

Catalysts and industry

The margarine you might put in a sandwich starts off as a vegetable oil. In the food industry, hydrogen is used to react with the oil to turn it into solid margarine. This reaction has nickel metal as its catalyst. Table 3.1 shows some other industrial uses of catalysts.

Catalyst	Used for
iron	making ammonia
platinum	making nitric acid
vanadium oxide	making sulphuric acid

Table 3.1 Some catalysts used in industry

Catalysts are used in industry because they can speed up reactions. You may have heard the expression 'time is money'. In a factory, costs are reduced by making products more quickly. Catalysts also let some reactions take place at lower temperatures.

Questions

Q3 Give at least one reason why companies want to be able to carry out reactions at lower temperatures.

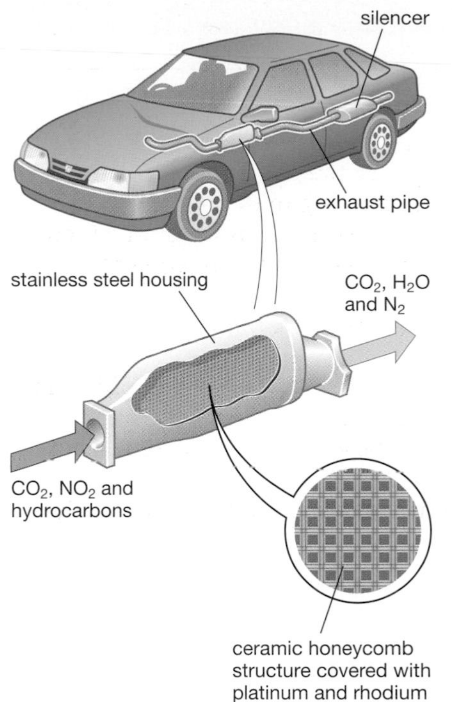

Figure 3.2

Catalysts and the environment

Figure 3.2 shows a car with a catalytic converter. This is a device that fits onto the exhausts of cars and helps to break down harmful oxides of nitrogen into substances that do less damage to the environment. The catalyst contains metals such as platinum.

Enzymes

The human body contains over 30 000 different catalysts. Catalysts which occur in nature, in plants and animals, are called **enzymes**. One of the human enzymes – called catalase – helps to break down hydrogen peroxide, which is a poison, into water and oxygen.

The liver contains a lot of catalase. The photographs in figure 3.3 show what happens when some liver is placed in hydrogen peroxide solution.

Enzymes are added to some washing powders because they are able to break down biological stains such as blood and fat. Enzymes are also used in the manufacture of cheese and yoghurt. Yeast provides a variety of enzymes which are used in making whisky and beer (see Chapter 15).

Figure 3.3 Liver contains catalase, which speeds up the breakdown of hydrogen peroxide into water and oxygen

Questions

Q4 Yeast is often used to provide enzymes to catalyse the reactions involved in making an everyday food. Can you name this food?

Q5 In your own words say what a catalyst is and give an example of one.

Chapter 2 Study Questions

1 Sea shells (mainly calcium carbonate, like marble) react with vinegar (an acid) to produce carbon dioxide gas. What would be the effect on the speed of reaction if:

a) the shells were crumbled
b) the vinegar was diluted with water
c) a catalyst was added
d) hot vinegar was used
e) the experiment was carried out in a refrigerator
f) the ordinary vinegar (5% acid) was replaced by another vinegar (7% acid)? GENERAL (KU)

2 Iain was asked by his teacher to find out whether globrite or maxlite candles give out more heat when burned.

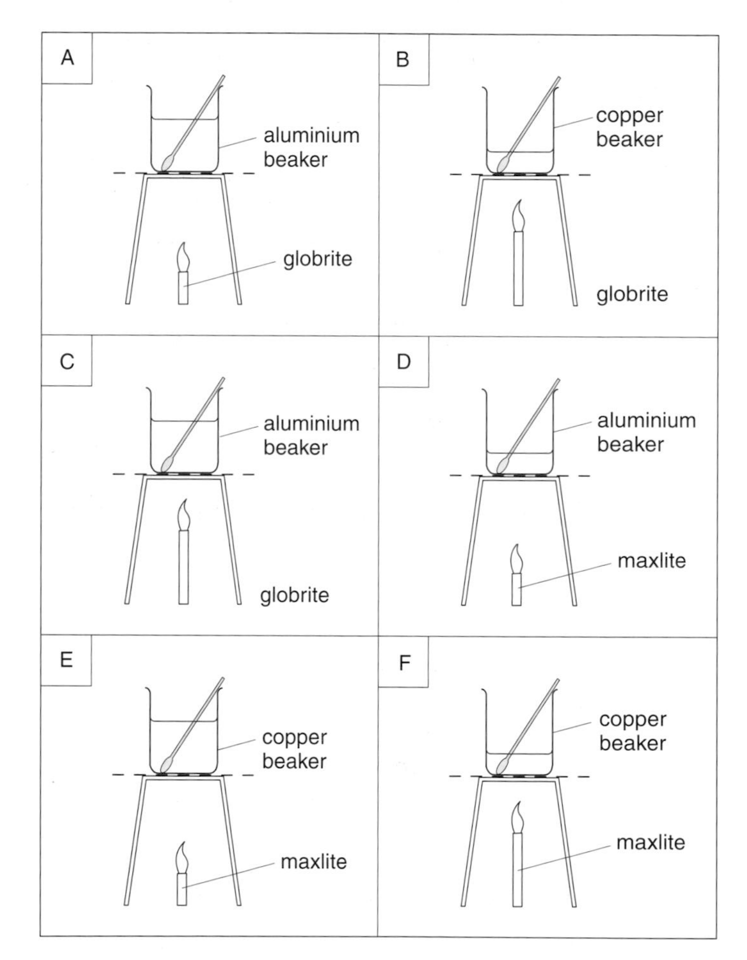

Which two arrangements could be used to make a fair comparison?

SEB GENERAL (PS)

3 Indigestion can be caused by too much acid in the stomach.

Six pupils set up experiments as part of an investigation. They were trying to find out which of two indigestion tablets dissolved in water more quickly. Three used Calmo tablets and three used Easo tablets. (See illustration top right.)

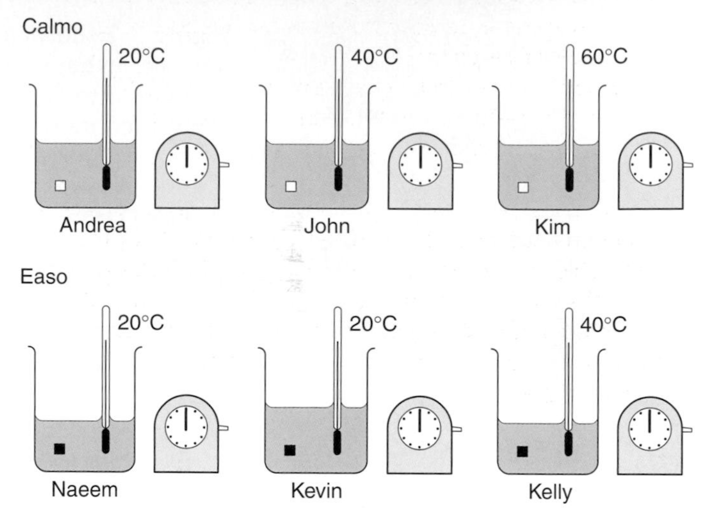

a) Why would Andrea and Kevin's experiments, taken together, provide the fairest way of comparing how quickly Calmo and Easo dissolved?
b) The results of the experiments carried out by Andrea, John and Kim can be compared. What would this comparison be designed to show?

SEB GENERAL (PS)

4 The reaction between sodium thiosulphate solution and a dilute acid causes the mixture to become cloudy in appearance due to the formation of sulphur. An indication of how fast the reaction proceeds can be obtained by observing the time taken for a pencilled 'cross', on a piece of paper beneath a beaker containing the reaction mixture, to disappear from view.

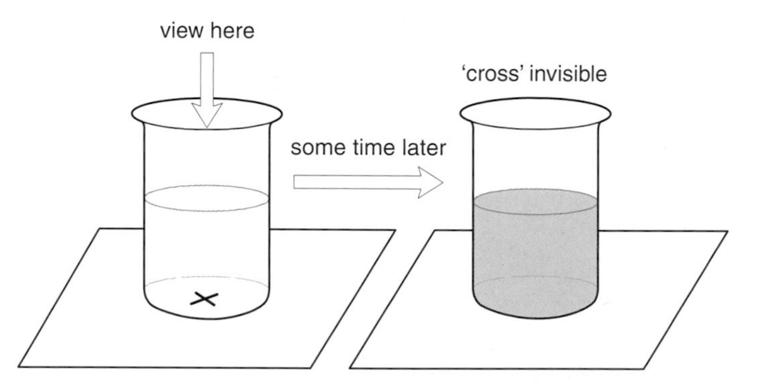

The reaction was carried out at three different temperatures, in each case using 20 cm^3 of sodium thiosulphate solution and one drop of acid. The following results were obtained:

Experiment	Temperature/°C	Time taken/s
1	10	208
2	20	104
3	30	52

a) At which temperature was the reaction *slowest*?
b) What conclusion can you reach regarding the effect of an increase in temperature on the speed of this particular reaction?
c) Based on the results given, predict the time taken for the disappearance of the pencilled 'cross' if the experiment were carried out at 40°C? GENERAL (PS)

5 Morag, Sadaf and Fraser were investigating the reaction between chalk and acid. They all used 100 cm^3 of the same hydrochloric acid, which in all cases was an excess. All of their chalk was used up. They each measured the rate at which carbon dioxide gas was given off from their reaction mixture. Morag used 5 g of chalk lumps and obtained the following graph:

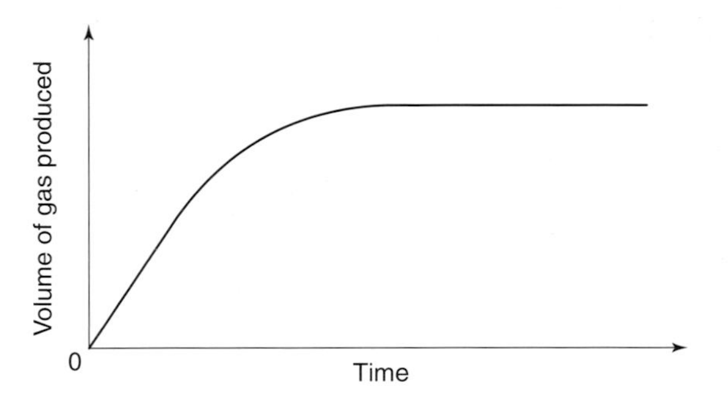

Copy the graph above and add to it the curves you would expect for the experiments carried out by Sadaf and Fraser, the details of which are:

a) Sadaf used 2.5 g of chalk lumps which were the same size as Morag's.
b) Fraser used 5 g of chalk lumps, but ground them down to a powder before mixing them with the acid.

Be sure to label each graph carefully. CREDIT (PS)

6 A group wanted to investigate the effect of changes in acid concentration on the reaction rate with zinc metal. They identified the following variables:

A the mass of zinc
B the temperature of the acid
C the size of the test tube
D the particle size of the zinc.

Which of the above variables need *not* be kept constant in the investigation? GENERAL (KU)

7 When hydrogen peroxide decomposes it releases oxygen gas. The reaction can be speeded up by the use of a catalyst.

Anne and Simon investigated the reaction using the apparatus shown top right.

Their experiments were the same, except that Anne used manganese dioxide as the catalyst, producing graph A, whereas Simon used copper oxide, which gave graph B.

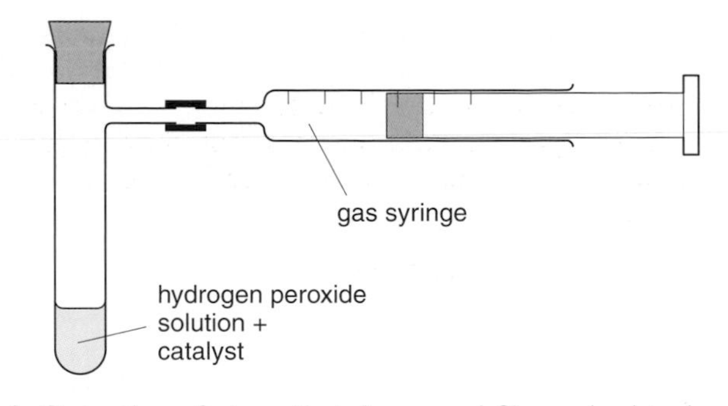

a) State *three* factors that Anne and Simon had to keep the same in their experiments so that the comparison of the catalysts was fair. (KU)
b) Which of the two oxides is the better catalyst? (PS)

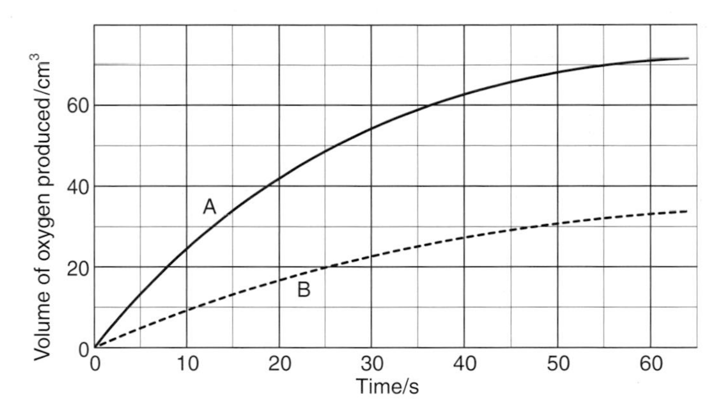

c) How long did it take to produce 25 cm^3 of gas in each experiment? CREDIT (PS)

8 The grid below shows factors which can affect the speed of a chemical reaction:

A increase in particle size	**B** decrease in particle size	**C** increase in concentration
D decrease in concentration	**E** increase in temperature	**F** decrease in temperature

Identify the factor that is responsible for the change in reaction rate in each of the following examples:

a) A carton of milk goes sour more slowly in a refrigerator than an identical carton left on a work surface in a kitchen.
b) 1 kg of wood shavings burns faster than a block of wood of equal mass.
c) Sulphuric acid causes iron nails to corrode. Battery acid and acid rain both contain sulphuric acid. Iron nails corrode more slowly in acid rain water than in battery acid.

GENERAL (PS)

9 Sam added manganese dioxide to hydrogen peroxide solution and measured the volume of oxygen produced.

Her results for two experiments at different temperatures are shown in this graph.

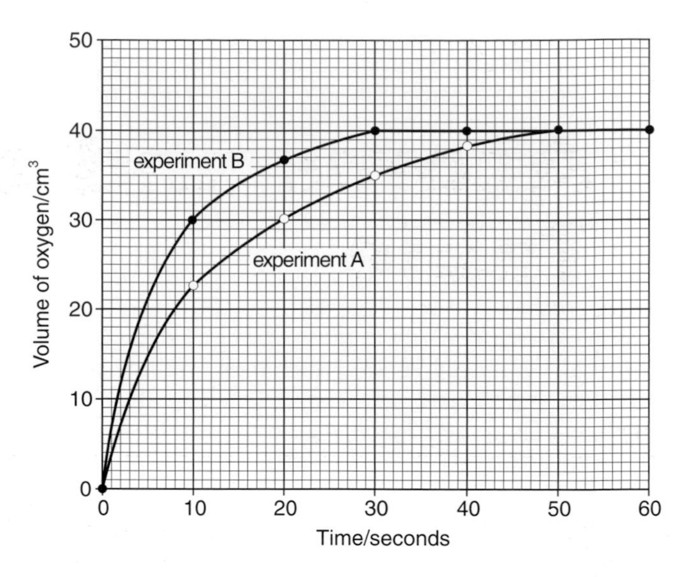

a) What volume of oxygen was collected in experiment **A**?

b) The same volume and concentration of hydrogen peroxide was used in both experiments **A** and **B**.
How can you tell this from the graph?

c) At the start, experiment **B** was faster than experiment **A**.
How can you tell this from the graph?

d) 0.2 g of the catalyst, manganese dioxide, was present at the start of experiment **A**.

What mass of catalyst would be left at the end of the experiment?

SQA GENERAL (KU)

CHAPTER THREE

Atoms and the Periodic Table

SECTION 3.1 The periodic table

Figure 1.1 The music in this store has been classified into different types

You know from Chapter 1 that there are just over one hundred different elements. Chemists have given a lot of thought to finding useful ways of putting the elements into groups. In most music stores, CDs and tapes are placed in groups such as chart music, dance, etc. This makes it easier to find the kind of music you are looking for.

Putting things into groups in this way is called **classifying**. We can classify the chemical elements in many different ways, for example according to their physical properties:

(a) *Are the elements solids, liquids or gases?*

This would give us three groups. Most of the elements are solids, a few are gases and only two are liquids. The two liquid elements are bromine and mercury.

(b) *Are the elements metals or non-metals?*

Table 1.1 shows some of the different physical properties of metal and non-metal elements. In general, non-metal elements do not conduct electricity. However carbon in the form of **graphite** is an exception. Graphite is a *good* conductor of electricity.

Properties of a typical metal	Properties of a typical non-metal
conducts electricity shiny	does not conduct electricity not shiny

Table 1.1 Comparing metals and non-metals

Questions

Q1 You have probably not used the element bismuth. It is a solid. Describe a simple experiment you could do to find out whether bismuth is a metal or a non-metal.

Q2 So far we have mentioned classifying elements as solids, liquids and gases and as metals and non-metals. Think of at least *two* other ways of classifying elements.

NON-METALS

METALS

H 1																	He 2
Li 3	Be 4											B 5	C 6	N 7	O 8	F 9	Ne 10
Na 11	Mg 12											Al 13	Si 14	P 15	S 16	Cl 17	Ar 18
K 19	Ca 20	Sc 21	Ti 22	V 23	Cr 24	Mn 25	Fe 26	Co 27	Ni 28	Cu 29	Zn 30	Ga 31	Ge 32	As 33	Se 34	Br 35	Kr 36
Rb 37	Sr 38	Y 39	Zr 40	Nb 41	Mo 42	Tc 43	Ru 44	Rh 45	Pd 46	Ag 47	Cd 48	In 49	Sn 50	Sb 51	Te 52	I 53	Xe 54
Cs 55	Be 56	*	Hf 72	Ta 73	W 74	Re 75	Os 76	Ir 77	Pt 78	Au 79	Hg 80	Tl 81	Pb 82	Bi 83	Po 84	At 85	Rn 86
Fr 87	Ra 88	**	Rf 104	Db 105	Sg 106	Bh 107	Hs 108	Mt 109									

*	La 57	Ce 58	Pr 59	Nd 60	Pm 61	Sm 62	Eu 63	Gd 64	Tb 65	Dy 66	Ho 67	Er 68	Tm 69	Yb 70	Lu 71
**	Ac 89	Th 90	Pa 91	U 92	Np 93	Pu 94	Am 95	Cm 96	Bk 97	Cf 98	Es 99	Fm 100	Md 101	No 102	Lr 103

Figure 1.2 A simplified periodic table, showing metals and non-metals

Questions

Q3 Using the periodic table on page 8 in the SQA Data Booklet, state whether the following are metals or non-metals:
a) scandium,
b) antimony,
c) silicon,
d) beryllium.

Q4 Which of the following is a naturally occurring element?
a) fermium,
b) neptunium,
c) plutonium,
d) polonium.

Q5 Give the names of all the elements that are gases at room temperature.

It has been found that the most useful way to classify elements is to arrange them in a table where the position of each element is based on its chemical properties. Over hundreds of years, chemists have worked on different ways to arrange the elements. The most successful of these is the 'Periodic Table of Elements', which was first developed by Russian chemist Dimitri Mendeleev in 1869. The modern form of the periodic table is shown in figure 1.2. A fuller version of the periodic table is given on page 8 in the SQA Data Booklet.

Look closely at the periodic table on page 8 in the SQA Data Booklet. Find the stepped line which separates metals from non-metals. Look for the symbol *. This is used to identify elements that are not found in nature, but which are made by scientists. None of the elements after number 92 occur naturally.

Groups within the periodic table

The main vertical columns within the periodic table are called **groups**. These are numbered from 1 to 7 with the last group on the right-hand side called Group 0. The groups are discussed more fully in sections 3.2 and 3.3.

SECTION 3.2 Some groups in the periodic table

Figure 2.1 Dimitri Mendeleev

As we mentioned in the last section, the periodic table was invented in 1869 by a Russian chemist called Dimitri Mendeleev (figure 2.1). He was a brilliant scientist (who, incidentally, had a haircut only once a year!).

Mendeleev looked at the ways in which each element reacted, and put all the elements that had similar types of reactions into the same groups. In other words, he classified the elements according to their chemical properties. The periodic table which we use today is based on Mendeleev's original classification. In this section we will look at the chemical properties of five important groups of elements.

Group 1 – the alkali metals

The alkali metals (figure 2.2) are all very reactive. You will probably have seen a piece of sodium added to water. It moves about the surface fizzing noisily as it produces hydrogen gas. The other product of the reaction is sodium hydroxide – a type of chemical called an **alkali**.

This reaction can be shown in a word equation:

sodium + water → sodium hydroxide + hydrogen

In fact, all the metals in Group 1 produce alkalis when added to water and this is why they are called the **alkali metals**.

The Group 1 metals also react readily with oxygen to form metal oxides. All the alkali metals are so reactive that they are stored under oil. This stops them from reacting with any oxygen or moisture in the air.

Another property of the Group 1 metals is that they are soft and most can be cut easily with a knife. When they are freshly cut they have a shiny surface, but this quickly turns dull as the metal reacts with oxygen in the air. The Group 1 metals and their compounds also give bright colours when placed in a bunsen flame – see the table on page 4 in the SQA Data Booklet.

Group 1 metals melt easily. One of them, sodium, is used in some types of nuclear reactor to remove heat from the centre of the reactor.

Group 1 Alkali Metals
Li lithium
Na sodium
K potassium
Rb rubidium
Cs caesium
Fr francium

Figure 2.2

Group 2 Alkaline Earth Metals
Be beryllium
Mg magnesium
Ca calcium
Sr strontium
Ba barium
Ra radium

Figure 2.3

Questions

Q1 Which substance do you think is formed when freshly cut potassium metal reacts with air? Write a word equation for the reaction.

Q2 Make a summary in your own words of the properties of the alkali metals.

Group 2 – the alkaline earth metals

These metals are similar to those found in Group 1. They also react with oxygen and with water. However, they are less reactive than the Group 1 metals and do not have to be stored under oil. Group 2 metals are also harder than the alkali metals and cannot be cut easily with a knife.

Magnesium is used to make special alloys for car wheels. Powdered magnesium and barium are both used in fireworks – they give off a shower of bright lights when they burn.

Group 7 Halogens
F fluorine
Cl chlorine
Br bromine
I iodine
At astatine

Figure 2.4 The elements in Group 7

Group 0 Noble Gases
He helium
Ne neon
Ar argon
Kr krypton
Xe xenon
Rn radon

Figure 2.5 The elements in Group 0

Group 7 – the halogens

These are all non-metals and they are all very reactive. For example, the gas chlorine reacts violently with sodium to produce sodium chloride:

sodium + chlorine → sodium chloride

As you go down the group, the physical states of the elements change; chlorine and fluorine are gases, bromine is a liquid and iodine is a solid.

The element chlorine is used in swimming pools and in disinfectants because it kills germs. Fluoride compounds are added to toothpaste to help prevent tooth decay.

Group 0 – the noble gases

These gases are the least reactive elements of all. They are called 'noble' to suggest that they do not join up easily with other elements. All the noble gases are found in very small amounts in the air.

Helium is lighter than air and so is used in airships and toy balloons. Neon is used in advertising signs because it glows bright red when an electric current is passed through it.

Figure 2.6 Because it is unreactive and less dense than air, helium is a safe gas to use in this airship

The transition metals

These metals are in a block in the centre of the periodic table (see figure 2.7). Transition metals have many very important uses. For example, the spatulas in your laboratory may be made of nickel. Chromium and nickel are added to steel to make stainless steel. When vanadium is added to steel it makes a very strong metal which is used for tools such as spanners. In Chapter 2 you saw that catalysts can be used to speed up chemical reactions. Many catalysts are transition metals, for example the ones used in catalytic converters in cars.

Questions

Q3 Which of the following are transition metals? Ba, Pb, Pt, La. Use the periodic table on page 8 in the SQA Data Booklet to help you.

TRANSITION METALS

Sc 21	Ti 22	V 23	Cr 24	Mn 25	Fe 26	Co 27	Ni 28	Cu 29	Zn 30
Y 39	Zr 40	Nb 41	Mo 42	Tc 43	Ru 44	Rh 45	Pd 46	Ag 47	Cd 48
La 57	Hf 72	Ta 73	W 74	Re 75	Os 76	Ir 77	Pt 78	Au 79	Hg 80
Ac 89	Rf 104	Db 105	Sg 106	Bh 107	Hs 108	Mt 109			

Figure 2.7 A simplified periodic table, showing the transition metals

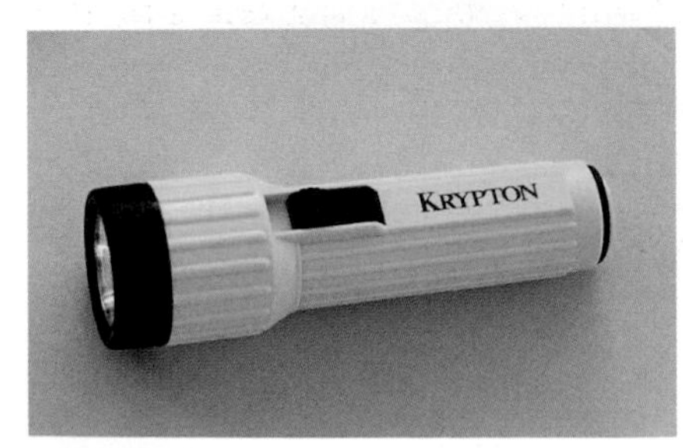

Figure 2.8 Why is krypton used in the bulb of this torch?

Inside the atom

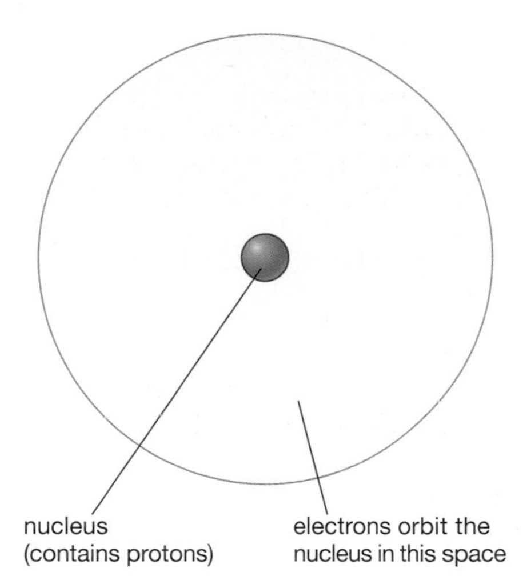

Figure 3.1 A simple diagram of an atom (not drawn to scale)

What are elements made from? The answer is very small particles called **atoms**. Each element is made from its own type of atom. For example, gold is made from gold atoms and helium from helium atoms. Although there are many different types of atoms they all have some important features in common:

1 Atoms have a small centre called a **nucleus**. The nucleus has a positive charge. This is because it contains positively charged particles called **protons**. Neutral particles called **neutrons** may also be present.

2 Atoms have particles called **electrons** which move around the nucleus. Electrons have a negative charge. Figure 3.1 shows a simplified drawing of an atom. A real atom is extremely small.

3 Atoms are electrically neutral. This is because the positive charge of the nucleus is 'cancelled out' by the negative charge of the electrons.

Atomic numbers and the periodic table

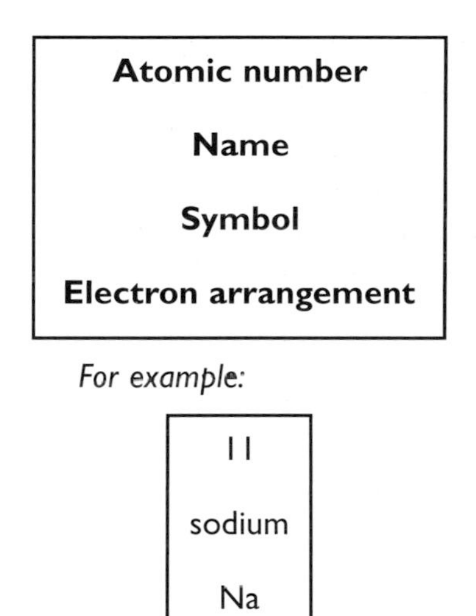

Figure 3.2 A typical entry in the periodic table

The periodic table is arranged so that it tells us not only about the properties of the elements but also about the number of protons and electrons in their atoms. We will now look at three important features of the periodic table.

1 **The atomic number**
You can see from figure 3.2 that each element has a special number given to it. This is called the **atomic number**. Notice that the atomic number increases by one unit as you move from one element to the next.

2 **The properties of the elements**
Remember from section 3.2 that the elements in the periodic table are arranged in groups and that the elements in each group all have similar properties. For example, all the elements in Group 0 are very unreactive gases.

3 **The number of outer electrons**
Why is it that all the elements in Group 1 are reactive metals? One reason is that they all have the same number of electrons on the outside of their atoms.

The electrons in an atom are arranged in layers, as shown in figure 3.3. The way an atom reacts depends on the number of electrons in its outer layer. Since all the elements in Group 1 have just one outer electron, they all react in similar ways. They have similar chemical properties.

The number of outer electrons can be shown in two ways. First there is a series of numbers called the electron arrangement (see section 3.4). The electron arrangement for sodium is 2,8,1. The number of outer electrons is the last number in the electron arrangement. This means that sodium has one outer electron.

The second way in which the number of outer electrons is given is by the group number. Every element in Group 1 has one outer electron, every element in Group 2 has two outer electrons and so on. Aluminium is in Group 3, therefore it has three outer electrons.

You should notice that Group 0 is different from the other groups in this respect. The first element, helium, has two outer electrons while the other elements have eight. This is explained further in section 3.4.

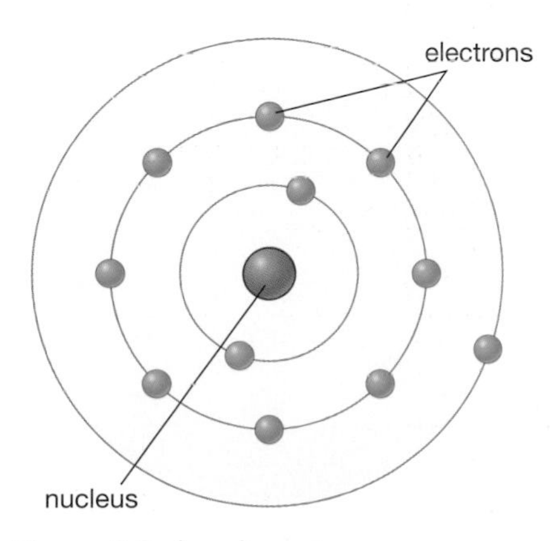

Figure 3.3 A sodium atom

Questions

Q1 How many outer electrons are there in the following?
  a) lithium,
  b) oxygen,
  c) neon.

Q2 How many outer electrons are there in the following groups?
  a) Group 7,
  b) Group 6,
  c) Group 4.

Q3 Group 0 (the noble gases) is called Group 8 in some older periodic tables. Why do you think this is?

Q4 An element has an electron arrangement of 2,8,3. Which group of the periodic table does it belong to?

Atomic mass

The mass of an atom is very small if measured in terms of grams. To measure the mass of an atom, we use a special unit called an **atomic mass unit (amu)**. Table 3.1 shows the masses of protons, neutrons and electrons.

Particle	Mass/atomic mass units
proton	1
neutron	1
electron	approximately zero

Table 3.1

The **mass number** of an atom is the total number of protons and neutrons it contains. This is explained more fully in section 3.4.

Section 3.3 Summary

- *Nucleus* the centre of an atom. It has a positive charge.
- *Neutrons* neutral particles found in the nucleus.
- *Protons* positively charged particles found in the nucleus.
- *Electrons* negatively charged particles which move round the nucleus.
- *Atomic number* the number of protons in an atom.
- *Mass number* the sum of the numbers of protons and neutrons in an atom.

More about atoms

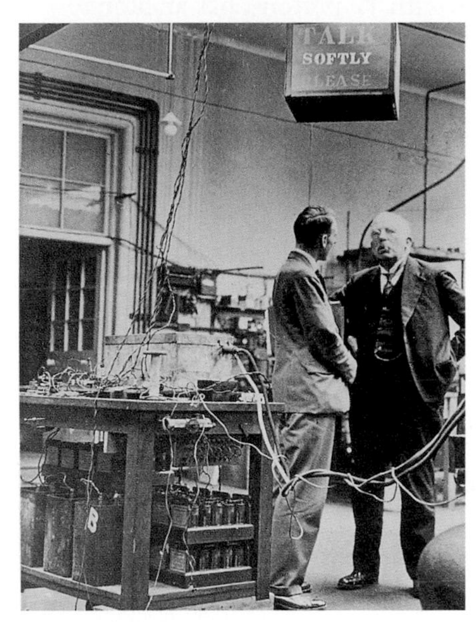

Figure 4.1 Ernest Rutherford in his Cambridge laboratory

Protons, neutrons and electrons

How could you find out what was inside a container lorry if all the doors were sealed? One way would be to smash it open – it would be drastic, but it would work! This was the kind of problem which faced scientists when they started to try to find out what was inside atoms. One ingenious solution was to use large amounts of energy to separate the particles which were inside. Most of this work was carried out at Cambridge University.

One of the first experiments, in 1897, used electricity to separate the electrons from a sample of chlorine gas. The electrons were attracted to a positive electrode – this is how it was discovered that electrons are negatively charged. Later it was shown that atoms also contain positive particles – these were called protons.

However, no one knew how the electrons and protons were arranged in an atom. In 1911, Ernest Rutherford, using his special talents for good scientific guesswork and high-powered mathematics, concluded that the atom must have the protons in a central nucleus with the electrons arranged round the outside. Despite this successful prediction, something was missing: the calculations for the mass of atoms did not add up – the protons and electrons alone did not make up the total mass of the atom.

Then, in 1932, a third particle was discovered – the neutron. It had the same mass as a proton but had no electrical charge – it was neutral.

From these and other discoveries, scientists were able to conclude that atoms consist of a central nucleus containing the protons and neutrons, with the electrons arranged outside the nucleus. Figure 4.2 shows a diagram of an atom of helium, showing the positions of the neutrons, protons and electrons.

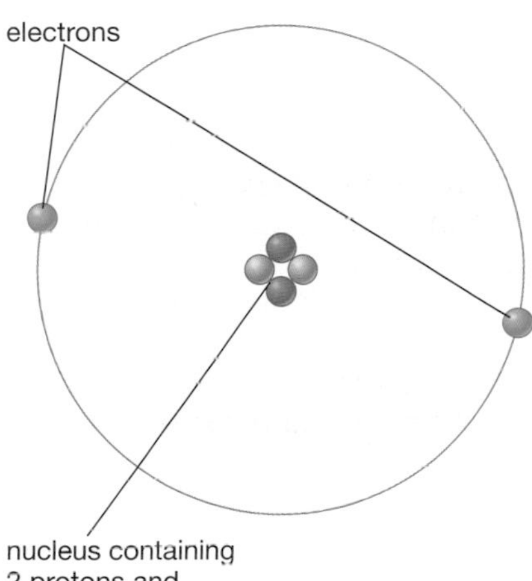

Figure 4.2 A helium atom

The number of neutrons in an atom can be found by subtracting the atomic number from the mass number.

Mass number

Table 4.1 summarises what you have learnt so far about atoms (atomic mass was discussed in section 3.3).

Particle	Charge	Mass/amu	Where found
proton	positive (+1)	1	nucleus
neutron	neutral (0)	1	nucleus
electron	negative (–1)	0 (approx.)	outside the nucleus

Table 4.1 Particles in an atom

Notice that the mass of an electron is almost zero compared to the mass other particles. This means that the mass of an atom is mostly made up from the protons and neutrons which it contains. Since each of these particles has a mass of 1 amu, the **mass number** of an atom is the total number of protons and neutrons it contains. For example, an atom of sodium contains 11 protons and 12 neutrons. Therefore its mass number is 11 + 12 = 23.

Questions

Q1 An atom of nitrogen contains 7 protons and 7 neutrons. What is its mass number?

Q2 An atom of potassium contains 19 protons and has a mass number of 39. How many neutrons does it contain?

Atomic number

This number (which was mentioned in section 3.3) is given by the number of protons in an atom, for example magnesium, with 12 protons, has an atomic number of 12. In an atom, the number of electrons is equal to the number of protons. Therefore, magnesium atoms must also have 12 electrons. The positive charge of each proton cancels out the negative charge of each electron.

To sum up, an atom's atomic number is equal to the number of protons in its nucleus, which is also equal to the number of electrons it contains.

Questions

Q3 Copy and complete the following

Element	Atomic number	Number of protons	Number of electrons
calcium	20	a)	b)
phosphorus	c)	d)	e)
f)	6	g)	6
h)	i)	17	j)

Electron arrangements

Energy level	Maximum number of electrons it can hold
1	2
2	8
3	18
4	32

Table 4.2

Electrons are arranged in an ordered way in atoms. It is helpful to picture them in layers built up around the nucleus. The layer nearest to the nucleus is the smallest, and the next layer has to be bigger to cover the first, rather like the way Russian dolls fit one inside the other.

These layers of electrons are called **energy levels**. They are numbered according to how close they are to the nucleus, the first energy level being the closest. Each energy level has a maximum number of electrons that it can hold. For example, the first energy level can hold up to two electrons. The number of electrons that can fit into each energy level is shown in table 4.2.

An atom of hydrogen has only one electron. This fits into the first energy level. A helium atom has two electrons and these both fit into the first energy level. However, a lithium atom has three electrons; the first two are placed in the first energy level, which is then full, and the remaining electron fits into the second energy level. The number of electrons fitting into the various energy levels of an atom is called the **electron arrangement**.

We can show the electron arrangement of an atom as follows. Lithium, with two electrons in the first level and one electron in the second level, has the electron arrangement 2,1. Silicon has two electrons in the first level, eight in the second and four in the third; its electron arrangement is 2,8,4 (figure 4.3).

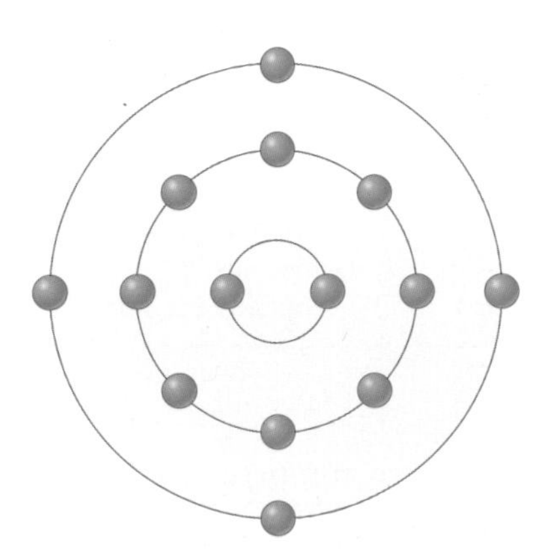

Figure 4.3 The electron arrangement of silicon is 2,8,4

Questions

Q4 Use the periodic table on page 167 to give the electron arrangements of the following elements:
- **a)** oxygen,
- **b)** boron,
- **c)** aluminium,
- **d)** argon.

Q5 Which atoms have the following electron arrangements?
- **a)** 2,5
- **b)** 2,7
- **c)** 2,8,1
- **d)** 2,8,7

Mass number

What gives an atom its mass? As mentioned in section 3.4, the mass of an atom is made up from the total number of protons and neutrons which it contains, called the mass number. For example, a sodium atom has 11 protons and 12 neutrons, therefore it has a mass number of 23.

Table 5.1 shows how the mass numbers of some other atoms are calculated.

Element	Number of protons	Number of neutrons	Mass number
helium	2	2	4
boron	5	6	11
lithium	3	4	7

Table 5.1

Using mass numbers and atomic numbers

If you know the atomic number and the mass number of an atom, then you can work out the number of protons, neutrons and electrons which make it up.

- The number of electrons and protons are the same as the atomic number.
- The number of neutrons is found by subtracting the atomic number from the mass number:

mass number	= protons + neutrons
atomic number	= protons
mass number – atomic number	= neutrons

For example, fluorine has a mass number of 19 and an atomic number of 9. Its atoms must therefore have 9 protons and 9 electrons. Subtracting 9 from 19 gives the number of neutrons (10).

Chemists use a special system for writing the atomic number, mass number and symbol for an element. In this system, the mass number is given on top with the atomic number underneath. For example:

lithium $^{7}_{3}\text{Li}$ fluorine $^{19}_{9}\text{F}$

Questions

Q1 Use the system described above to show the mass number, atomic number and symbol for:

a) a helium atom which contains 2 neutrons, 2 protons and 2 electrons,

b) a sulphur atom which contains 16 neutrons, 16 protons and 16 electrons.

$^{19}_{9}F^-$

Figure 5.1 The fluoride ion

Ions

During chemical reactions, atoms can lose electrons or have extra electrons added on. When this happens the atoms become electrically charged particles called **ions**. You remember that atoms are neutral and electrons have a negative charge. Therefore, if atoms *gain* electrons they become negative ions, and atoms become positive ions when they *lose* electrons. This is covered in more detail in Chapter 7.

We can show an extra electron by adding a negative sign to the symbol for an element. Each electron has a single negative charge, so two extra electrons are shown by 2^-, three extra electrons by 3^- and so on.

For example, the fluorine atom is shown as $^{19}_{9}F$. The fluoride ion is formed when the atom gains one electron. This ion is shown as $^{19}_{9}F^-$. Note that the atomic number does not change; it shows the number of protons – still 9 in this case. The fluoride ion therefore contains 9 protons, 10 neutrons and *10* electrons. Table 5.2 shows some further examples.

Ion	Number of protons	Number of electrons	Number of neutrons
$^{35}_{17}Cl^-$	17	18	18
$^{16}_{8}O^{2-}$	8	10	8
$^{23}_{11}Na^+$	11	10	12
$^{24}_{12}Mg^{2+}$	12	10	12

Table 5.2

Questions

Q2 Give the number of protons, neutrons and electrons in the following:

a) $^{32}_{16}S^{2-}$
b) $^{40}_{20}Ca^{2+}$
c) $^{15}_{7}N^{3-}$

Q3 Show the mass number, atomic number and symbol for

a) a sodium ion (Na^+), which contains 12 neutrons, 11 protons and 10 electrons,
b) an oxygen ion (O^{2-}), which contains 8 neutrons, 8 protons and 10 electrons.

Isotopes

In the 1930s, chemists discovered that there were two types of chlorine atom. Both had the same numbers of protons and electrons, the difference was in the numbers of *neutrons* – one type of chlorine atom had two more neutrons than the other. These two types of chlorine were called **isotopes**. The symbols are:

$^{35}_{17}Cl$	$^{37}_{17}Cl$
each of these atoms has 18 neutrons	each of these atoms has 20 neutrons

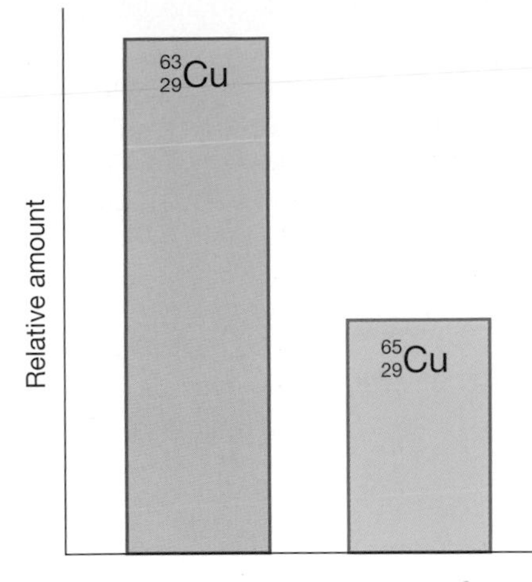

Figure 5.2 The relative amounts of the two isotopes of copper

In fact, most elements are made up of mixtures of isotopes, for example hydrogen consists of three isotopes:

$$^{1}_{1}H \quad ^{2}_{1}H \quad ^{3}_{1}H$$

Isotopes are atoms of the same element with a different number of neutrons. They have the same atomic number but a different mass number.

Relative atomic mass

Copper has two isotopes:

$$^{63}_{29}Cu \text{ and } ^{65}_{29}Cu$$

It has been found that 69 per cent of copper is made up of $^{63}_{29}Cu$ atoms and the remaining 31 per cent consists of $^{65}_{29}Cu$ atoms.

This means there are two different masses for copper atoms. It is often useful to have a figure for the *average* mass of a copper atom. This average value turns out to be 63.5 amu.

Similar calculations have been carried out for most other elements. The name given to the average mass of the isotopes of an element is the **relative atomic mass.** A table of relative atomic masses is given on page 4 of the SQA Data Booklet.

Questions

Q4 $^{16}_{8}O$ represents the most common isotope of the element oxygen. Give the symbol for the oxygen isotope which has 10 neutrons.

Q5 Give the numbers of neutrons in each of the following isotopes:

a) $^{24}_{12}Mg$

b) $^{11}_{5}B$

c) $^{14}_{6}C$

d) $^{13}_{7}N$

Section 3.5 Summary

◆ *Atomic number*	number of protons in an atom.
◆ *Mass number*	number of protons plus neutrons in an atom.
◆ *Isotopes*	atoms with the same atomic number but a different mass number.
◆ *Relative atomic mass*	the average mass of all the isotopes for a particular element.

Chapter 3 Study Questions

1 There are over one hundred known elements. Some of them are shown below.

A silicon	**B** potassium	**C** iodine
D barium	**E** lead	**F** helium

Use the SQA Data Booklet to help answer these questions:

a) Which metal element and non-metal element are in the same group of the periodic table?
b) Which element gives a green flame colour?
c) Which element was discovered most recently?

GENERAL (PS)

2 Refer to the SQA Data Booklet for information about the periodic table to help you answer the questions about these elements:

A mercury	**B** neon	**C** calcium
D bromine	**E** sodium	**F** silver

a) Which element is an alkali metal?
b) Which element is a very unreactive gas? GENERAL (KU)

3 The diagram below represents an atom. Copy and complete the two sentences beside it using words of your own choice.

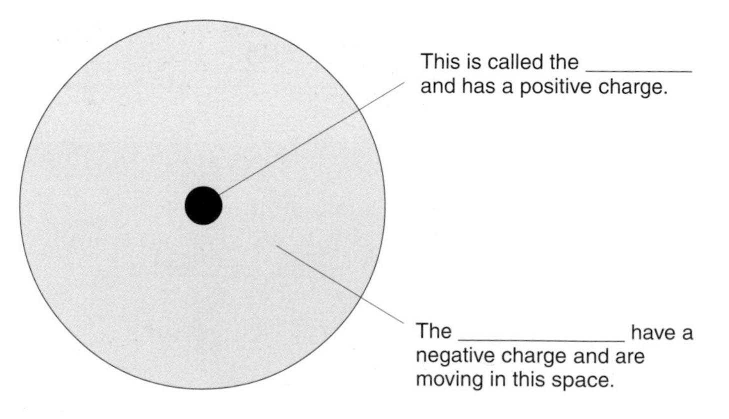

Explain why, despite the presence of charged particles inside them, atoms have no *overall* charge and are electrically neutral. GENERAL (KU)

4 Lithium and sodium are reactive metals with similar chemical properties. For example, both react with water to give the responding metal hydroxide and hydrogen.

a) Refer to the SQA Data Booklet and give the electron arrangements for these two elements. (PS)
b) What feature of their electron arrangements leads to lithium and sodium having similar chemical properties?

GENERAL (KU)

5 The outline below shows part of the periodic table. The positions of seven elements are shown by letters, but these do *not* represent their symbols.

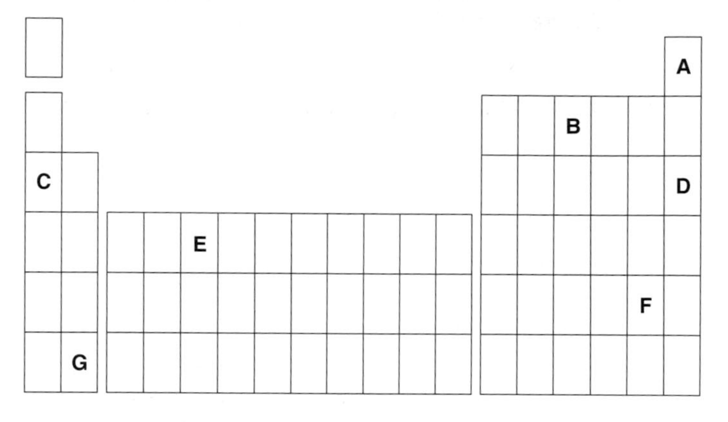

Which letter(s) represent:

a) halogen(s)
b) alkali metal(s)
c) noble gas(es)
d) transition metal(s)?
e) *two* elements that belong to the same group,

GENERAL (KU)

6 The diagram shows part of the periodic table.

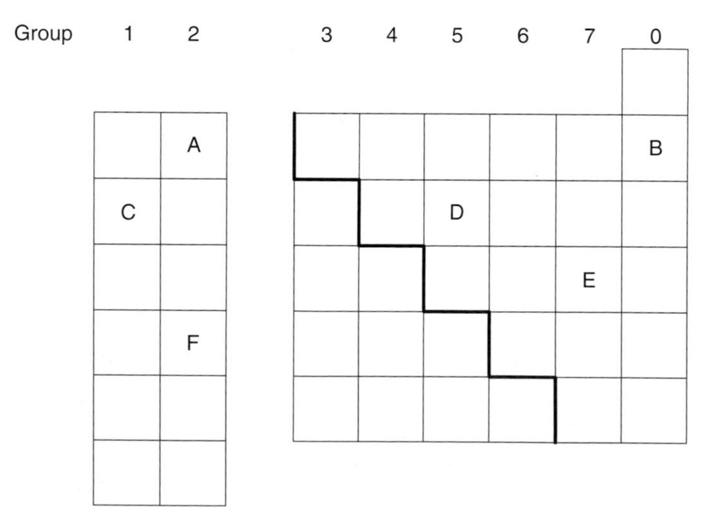

The letters do *not* represent the symbols for the elements.

a) Identify the element which is a halogen.
b) Identify the element which has the greatest number of electrons in its outer shell.
You may wish to use the SQA Data Booklet to help you.
c) Identify the element with the highest atomic number.

SQA GENERAL (PS)

7 Different numbers are used to give information about atoms.

A	Atomic number
B	Mass number
C	Number of outer electrons
D	Number of electron energy levels

a) Identify the number which is the number of protons plus neutrons in an atom.
b) Identify the number which is the same for sodium and chlorine atoms.

SQA CREDIT (PS)

8 Each box in the grid below refers to an element:

A the element with electron arrangement 2,8,3	**B** the element of atomic number 19	**C** Ar
D sodium	**E** the element which is a brown liquid at room temperature	**F** the element which has 6 electrons in each atom

Which box (or boxes) refers to:

a) a metal which does *not* react violently with water
b) a very unreactive element
c) elements in the same group of the periodic table
d) an element which is a gas at room temperature?

SEB GENERAL (KU)

9 Atoms are made up of smaller particles. The numbers of these particles can give information about the element.

A	The number of protons in the nucleus
B	The number of neutrons in the nucleus
C	The number of protons plus neutrons in the nucleus
D	The number of electrons outside the nucleus
E	The number of electrons in the outer energy level

a) Identify the number which is 7 for the halogens.
b) Identify the *two* numbers which are the same in all neutral atoms.
c) Identify the mass number of an element.

SEB CREDIT (KU)

10 The table gives the numbers of protons and neutrons in the nuclei of atoms of different elements.

a) The nucleus of an atom has a positive charge. Why is an atom neutral?
b) From the information in the table, write a statement linking the numbers of protons and neutrons in atoms.

Element	Number of protons	Number of neutrons
boron	5	6
phosphorus	15	16
zinc	30	35
zirconium	40	51
tin	50	69

c) Draw a bar graph showing the number of neutrons in the atoms of the elements listed in the table.

SQA GENERAL (PS)

11 There are three different types of silicon atom

Type of atom	Number of protons	Number of neutrons
$^{28}_{14}Si$		
$^{29}_{14}Si$		
$^{30}_{14}Si$		

a) Complete the table to show the number of protons and neutrons in each type of silicon atom.
b) What name is used to describe these different types of silicon atom? (KU)
c) A natural sample of silicon has an average atomic mass of 28.11.

What is the mass number of the most common type of atom in the sample of silicon?

SQA CREDIT (PS)

12 Heavy water is used in some nuclear reactors. It is like ordinary water except that the normal hydrogen atoms, $^{1}_{1}H$, known as **protium** atoms, are replaced by **deuterium** atoms, $^{2}_{1}H$.

a) Tritium, $^{3}_{1}H$, is another type of hydrogen atom. Copy and complete the table to show the number of protons, neutrons and electrons in a tritium atom.

	Number
protons	—
neutrons	—
electrons	—

b) What term is used to describe atoms like protium, deuterium and tritium?

SEB CREDIT (KU)

13 Use information in the SQA Data Booklet to find the missing terms in the following table. CREDIT (KU)

Element	Symbol	Atomic number	Electron arrangement
lithium	Li	3	2,1
boron	(a)	(b)	(c)
(d)	Si	(e)	(f)
(g)	(h)	20	(i)

14 Explain the meaning of the term 'relative atomic mass of an element' in terms of the mass numbers of the isotopes of which it is made up.

CREDIT (KU)

15 In a table of accurate values of relative atomic masses, magnesium is given a value of 24.31 but no magnesium isotope has this mass number. Explain.

CREDIT (PS)

16 Chlorine consists of two isotopes with mass numbers 35 and 37. Chlorine has a relative atomic mass of 35.5.

Explain which of the isotopes is present in the greater amount.

CREDIT (PS)

17 Silver consists of two isotopes, ^{107}Ag and ^{109}Ag. The relative atomic mass of silver is almost exactly 108.

What does this information tell you about the relative amounts of the two isotopes present in silver?

CREDIT (PS)

18 With the help of information contained within the SQA Data Booklet, find the missing terms in the table below.

Isotope	Mass number	Atomic number	Number of protons	Number of neutrons	Number of electrons
$^{9}_{4}Be$	(a)	(b)	(c)	(d)	(e)
(f)	19	(g)	9	(h)	(i)
(j)	(k)	17	(l)	20	(m)

CREDIT (KU)

19 With the help of information in the SQA Data Booklet, find the missing terms in the following table:

Ion	Number of protons	Number of neutrons	Electron arrangement
$^{23}_{11}Na^{+}$	(a)	(b)	(c)
$^{40}_{20}Ca^{2+}$	(d)	(e)	(f)
(g)	13	14	2,8

CREDIT (PS)

20 With the help of information in the SQA Data Booklet, find the missing terms in the following table:

Ion	Number of protons	Number of neutrons	Electron arrangement
$^{35}_{17}Cl^{-}$	(a)	(b)	(c)
$^{32}_{16}S^{2-}$	(d)	(e)	(f)
(g)	7	7	2,8

CREDIT (PS)

21 With the help of information in the SQA Data Booklet, find the missing terms in the following table:

Ion	Number of protons	Number of neutrons	Electron arrangement
$^{19}_{9}F^{-}$	(a)	(b)	(c)
$^{25}_{12}Mg^{2+}$	(d)	(e)	(f)

CREDIT (PS)

CHAPTER FOUR

How Atoms Combine

SECTION 4.1 Covalent bonds

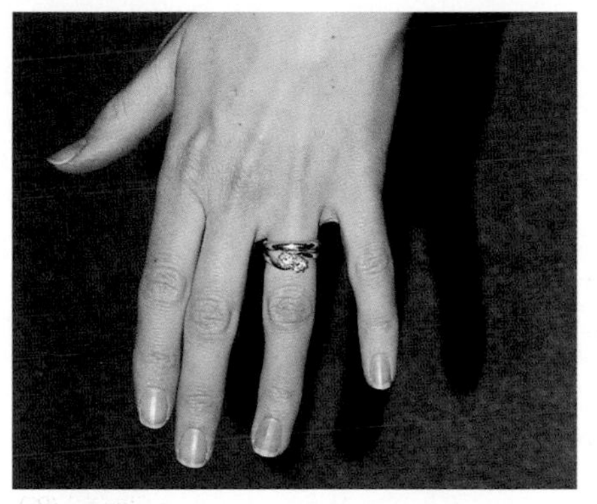

Figure 1.1 The atoms in this diamond are held together by covalent bonds

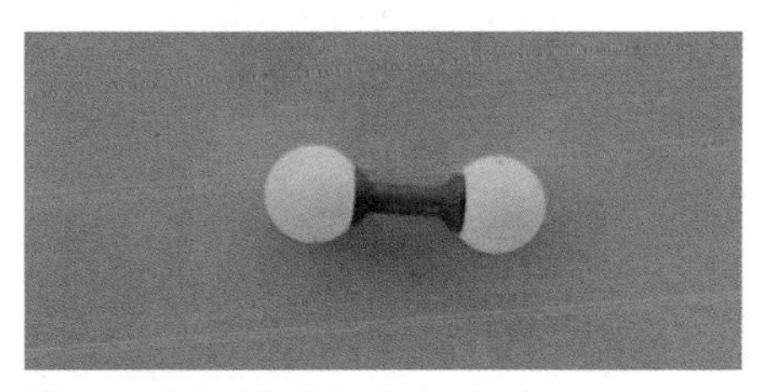

Figure 1.2 Model of two hydrogen atoms joined by a covalent bond

Figure 1.3 Model of a chlorine molecule

Figure 1.4 Model of an oxygen molecule, showing the double covalent bond

The diamond in figure 1.1 is made entirely of carbon atoms. These atoms are held together by chemical bonds. There are three main types of chemical bond. This section looks at one of them – **covalent bonds**.

Covalent bonds in elements

When two atoms are joined by a covalent bond they share one pair of electrons. You may have used molecular models to represent atoms joining. Figure 1.2 shows a molecular model of two hydrogen atoms joined by a covalent bond. In this case, the shared pair of electrons consists of one electron from each hydrogen atom.

Chemists use a special symbol to show the bonding between the hydrogen atoms. It is written as H–H. The line between the hydrogen atoms represents the bond.

When hydrogen atoms join they form a hydrogen molecule. A **molecule** is the name given to a group of two or more atoms joined by covalent bonds. The element chlorine occurs as chlorine molecules. Each molecule is made from two chlorine atoms joined by a covalent bond (see figure 1.3). A chlorine molecule is therefore shown as Cl–Cl.

The atoms in a chlorine molecule share only one pair of electrons. This gives a *single* covalent bond. In a hydrogen molecule, the atoms are also joined by single covalent bonds.

A molecule of oxygen, on the other hand, is formed by two oxygen atoms sharing *two* pairs of electrons. This gives a *double* covalent bond. Each oxygen atom has two electrons involved in the bond. This can be represented by O=O. Figure 1.4 shows a diagram of an oxygen molecule.

Nitrogen molecules are formed when two nitrogen atoms share *three* pairs of electrons. Each atom has three electrons involved in the bond, which is called a *triple* covalent bond (see figure 1.5). It can be written as N≡N.

Each of the molecules we have looked at so far has been made up of two atoms joined together. Molecules such as these are called **diatomic molecules**; 'diatomic' literally means 'two atoms'.

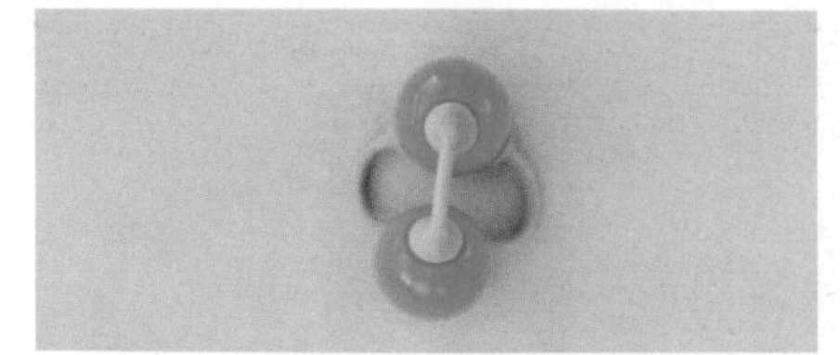

Figure 1.5 A nitrogen molecule

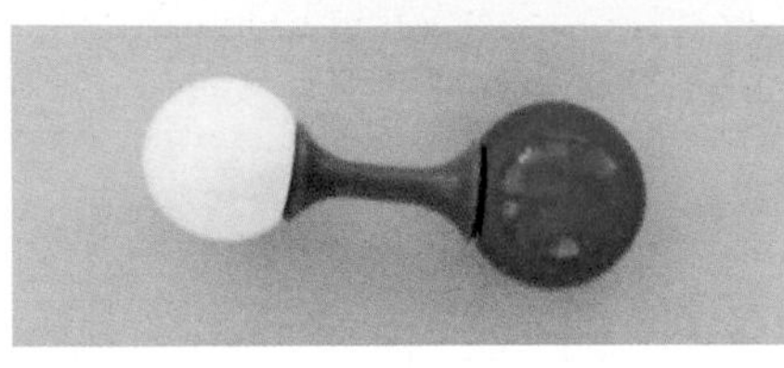

Figure 1.6 A hydrogen chloride molecule

Covalent bonding in compounds

Molecules can be formed between atoms of different elements. For example, a hydrogen and a chlorine atom will share one pair of electrons to form a diatomic molecule of hydrogen chloride. Figure 1.6 represents a hydrogen chloride molecule. This can be written as H–Cl.

As a general rule, if a compound is made up of non-metal elements only, then it will have covalent bonding.

Questions

Q1 Cl–Cl is the notation used for two chlorine atoms joined by a covalent bond. Use the same notation to show how the following are joined:

a) two fluorine atoms,

b) a hydrogen and a fluorine atom.

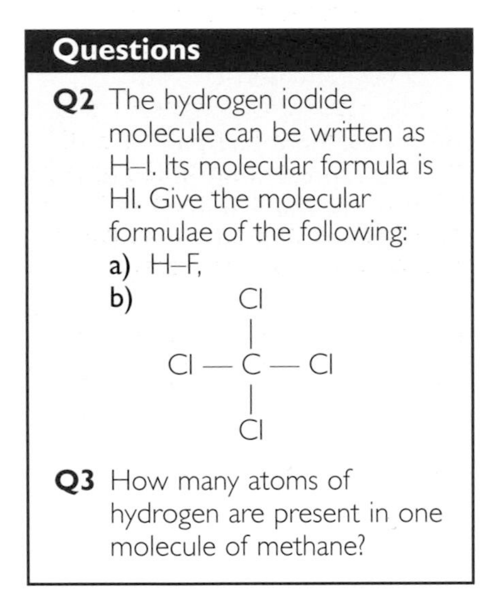

Questions

Q2 The hydrogen iodide molecule can be written as H–I. Its molecular formula is HI. Give the molecular formulae of the following:

a) H–F,

b)

```
       Cl
       |
 Cl —  C — Cl
       |
       Cl
```

Q3 How many atoms of hydrogen are present in one molecule of methane?

Chemical formulae

When a molecule of hydrogen chloride is formed there is always one hydrogen atom and one chlorine atom in the molecule – never two or three. This is true for all molecules; the number of atoms making up the molecule is always the same. For example, a molecule of nitrogen always contains two nitrogen atoms.

The **chemical formula** of a molecule shows the number of atoms of each element in the molecule. The formula consists of the symbols for the elements that make up the molecule, each followed by a number. An oxygen molecule, for example, contains two atoms, so the formula is O_2. The formula Br_2 for bromine tells you that there are two atoms of bromine in each molecule. If a molecule contains only one of a certain atom there is no need to put a '1' in the formula. For example, the formula for hydrogen chloride is HCl.

Formulae of this type are more correctly called **molecular formulae**. Note that we say formulae for more than one formula, not formulas.

Molecules can be formed from more than two atoms. For example, a water molecule is formed by two hydrogen atoms joining with one oxygen atom. The formula for water is H_2O.

Table 1.1 shows the formulae and the shapes of some important molecules.

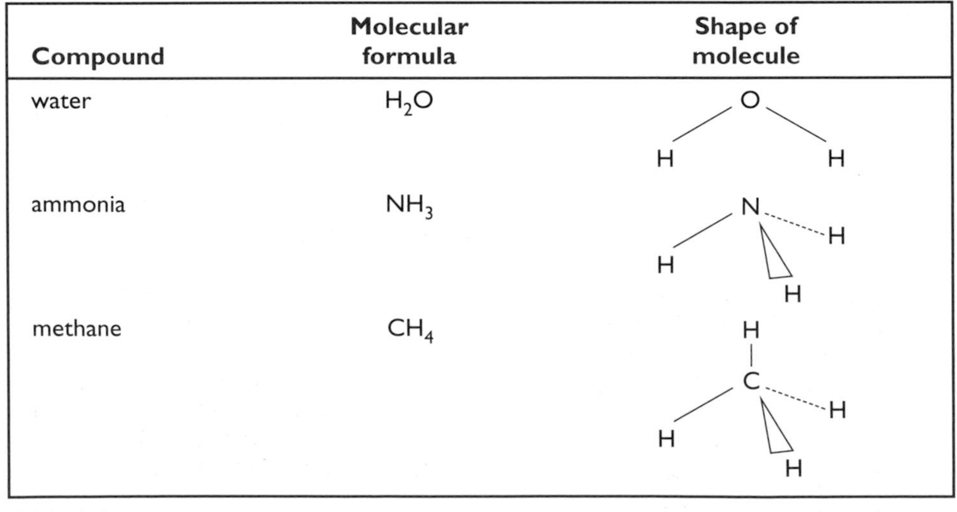

Compound	Molecular formula	Shape of molecule
water	H_2O	O, H, H
ammonia	NH_3	N, H, H, H
methane	CH_4	H, C, H, H, H

Table 1.1

Working out formulae

In the last section you saw that the formula for water is H_2O. But how do chemists know this? For most compounds, the formula has been found by carrying out experiments to measure the proportions of elements that are present. However, there is also a method using **valency numbers**, which allows you to predict the likely formula for a compound. Valency means 'combining power'. If you know the valency of a particular atom you can work out the number of other atoms that it will combine with.

Valency method

The valency of an atom depends on how many of its electrons are involved in forming chemical bonds. This in turn depends on the number of outer electrons that the atom has. Look at the electron arrangements shown in the periodic table on page 167. This shows that in each group, most of the elements have the same number of outer electrons. This means that they all have the same valency. For Groups 1–4 the valency number is the same as the group number. For Groups 5–7 the valency is found by subtracting the group number from 8. For example the elements in Group 5 have a valency of 8 – 5 = 3. Table 2.1 shows the valencies for Groups 1–0. Note that the noble gases in Group 0 have valencies of zero, that is, they have no combining power.

Group number	1	2	3	4	5	6	7	0
Valency	1	2	3	4	3	2	1	0

Table 2.1

Valencies can be used to work out the formulae for compounds containing two elements using a 5-step method.

Example 1
The formula for the compound formed between silicon (Group 4) and oxygen (Group 6).

Step 1	write down the symbols for the element	Si	O
Step 2	put in the valencies	4	2
Step 3	cross over the valencies	Si_2	O_4
Step 4	cancel out any common factor	Si_1	O_2
Step 5	omit '1' if present	Si	O_2

The formula for this compound is SiO_2

Example 2
The formula for the compound formed between phosphorus (Group 5) and chlorine (Group 7).

Step 1	write down the symbols	P	C1
Step 2	put in the valencies	3	1
Step 3	cross over the valencies	P_1	$C1_3$
Step 4	cancel out any common factor (not required in this case)		
Step 5	omit '1' if it is present	P	$C1_3$

$PC1_3$ is the formula for this compound.

Questions

Q1 Use valencies to find the formulae for the following compounds:
a) boron fluoride,
b) carbon sulphide,
c) hydrogen iodide.

Prefix	Meaning
mono	one
di	two
tri	three
tetra	four
penta	five
hexa	six

Table 2.2

Questions

Q2 Use the names of the following compounds to work out their formulae:
a) sulphur dioxide,
b) nitrogen monoxide,
c) uranium hexafluoride,
d) diphosphorus pentoxide.

Questions

Q3 Use the valency method to work out the formulae for the following:
a) copper(I) oxide,
b) tin(IV) chloride,
c) iron(II) oxide,
d) iron(III) sulphide.

Using prefixes

The valency method does not work for every compound. However, in some cases the compound's name helps you to find the formula. The information is contained in the prefixes mono, di, tri, etc. Table 2.2 shows what these prefixes stand for.

Examples

- Carbon monoxide (mono = 1)
  The formula is CO (one carbon atom joined to one oxygen atom).
- Carbon dioxide (di = 2)
  The formula is CO_2 (one carbon atom joined to two oxygen atoms).
- Phosphorus pentachloride (penta = 5)
  The formula is PCl_5.
- Dinitrogen tetroxide
  Here there are two nitrogen atoms for every four oxygen atoms. The formula is N_2O_4.

Elements with more than one valency

Some elements, such as the transition metals, can have more than one valency. Chemists use Roman numbers in the names of compounds to show which valency is to be used. For example, copper can have a valency of 1 or 2. In a compound of copper chloride, where the copper has a valency of 1, the compound's name is written as copper(I) chloride, which has a formula of CuCl. In copper(II) chloride, the valency of the copper is 2. The formula for this compound is $CuCl_2$. These formulae can be worked out as follows:

(a) Copper(I) chloride

Step 1	write down the symbols	Cu	Cl
Step 2	put in the valencies	1	1
Step 3	cross over the valencies	Cu_1	Cl_1
Step 4	cancel out any common factor (not required in this case)		
Step 5	omit '1' if it is present	Cu	Cl

CuCl is the formula for this compound.

(b) Copper(II) chloride

Step 1	write down the symbols	Cu	Cl
Step 2	put in the valencies	2	1
Step 3	cross over the valencies	Cu_1	Cl_2
Step 4	cancel out any common factor (not required in this case)		
Step 5	omit '1' if it is present	Cu	Cl_2

$CuCl_2$ is the formula for this compound.

A closer look at bonding

In section 3.4 you looked at the electron arrangements of atoms. This section looks at how these change when atoms form covalent bonds. If two atoms share a pair of electrons, then each one gains a share in an extra outer electron. For example, the electron arrangement for chlorine is 2,8,7 – it has seven outer electrons. When two chlorine atoms form a covalent bond, each one has a share in an extra electron. This gives each chlorine atom a new electron arrangement of 2,8,8. Argon, one of the noble gases in Group 0, also has an electron arrangement of 2,8,8.

Chemists have found that when atoms share electrons, they usually end up with the same electron arrangement as the noble gas nearest to them in the periodic table. Note that the 'nearest noble gas' is the one with an atomic number closest to that of the atom which is forming a bond. For example, the nearest noble gas to nitrogen (atomic number 7) is neon (atomic number 10). Table 3.1 shows the electron arrangements of the noble gases.

Questions

Q1 Give the electron arrangement for:
a) a fluorine atom,
b) a fluorine atom which has formed a covalent bond.

Q2 An oxygen atom will share two electrons when it takes part in covalent bonding. Give the electron arrangement of an oxygen atom plus two electrons, and state the name of the noble gas which has the same electron arrangement as this.

Noble gas	Electron arrangement
helium	2
neon	2,8
argon	2,8,8
krypton	2,8,18,8
xenon	2,8,18,18,8
radon	2,8,18,32,18,8

Table 3.1 Noble gas electron arrangements

There are small amounts of noble gases in the air around us. They exist as single atoms, not joined to any other atoms. The noble gases in the air are **monatomic**.

Bonding diagrams

The following system is used to show a simplified picture of an atom's outer electrons. It allows us to show what happens when atoms share electrons. In this system, two different shapes are used for the arrangement of the outer electrons. For an atom with one or two outer electrons, a circle is used. For example, hydrogen, which has one outer electron, and helium, which has two outer electrons, would be drawn as in figure 3.1.

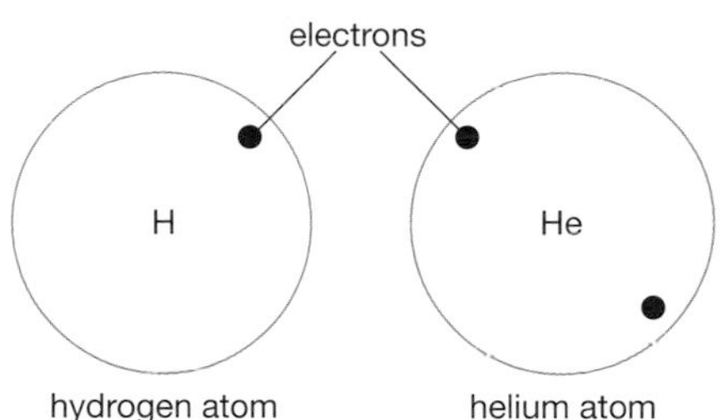

Figure 3.1 Atoms with 1 or 2 outer electrons

For all other atoms, a four-lobed shape is used, rather like four petals on a flower. Each lobe can hold up to two electrons. Some examples are shown in figure 3.2.

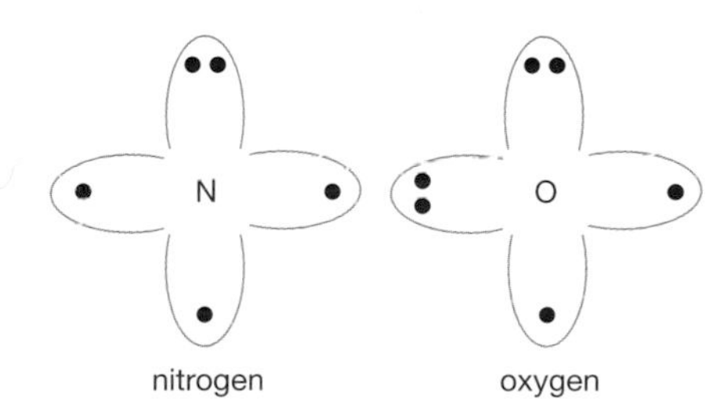

Figure 3.2 Atoms with more than 2 outer electrons

Notice that in the nitrogen and oxygen atom diagrams, each of the four lobes contains at least one electron. When drawing your own diagrams, you should add the electrons one-by-one to each lobe before putting two electrons side-by-side in the same lobe.

Remember, these are simplified diagrams – real atoms are much more complicated.

Questions

Q3 Draw diagrams to show the outer electrons in atoms of:
a) sulphur, b) bromine, c) neon.

Hydrogen (H–H)

A hydrogen atom has only one electron in the first energy level, which can hold two electrons. By sharing their electrons, two hydrogen atoms can each obtain the electron arrangement of the noble gas helium. This is shown in figure 3.3. To tell the difference between the electrons of the two atoms, one is shown as a dot, the other as a cross.

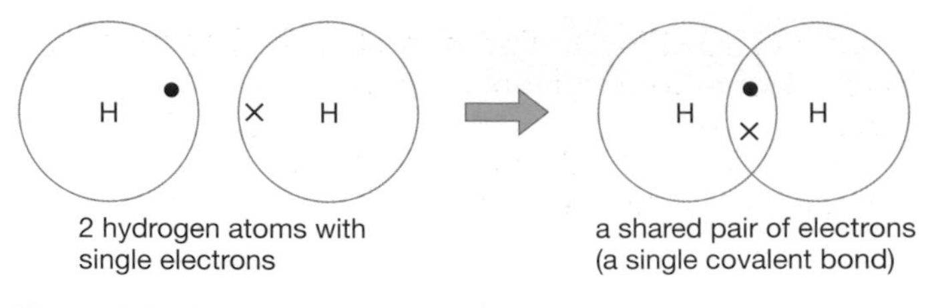

Figure 3.3 Two hydrogen atoms combining

Chlorine (Cl–Cl)

In a chlorine atom, one of the four outer lobes is only half filled. However, two chlorine atoms can achieve a noble gas arrangement by sharing a pair of electrons (see figure 3.4).

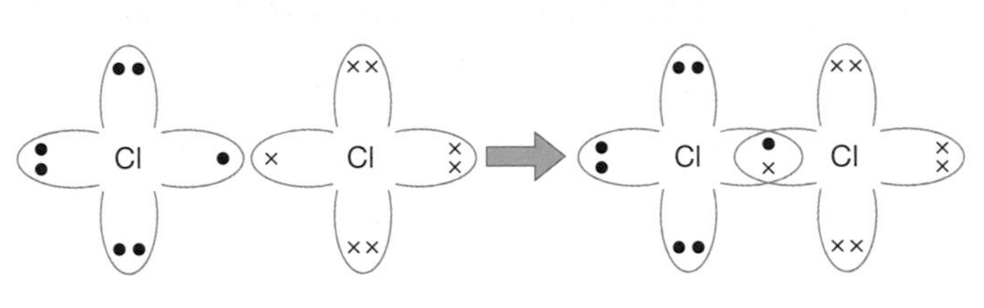

Figure 3.4 Two chlorine atoms combining

Oxygen (O=O)

Oxygen is in Group 6 and has six outer electrons. In an oxygen atom there are two filled and two half-filled lobes. The atoms of oxygen can achieve an electron arrangement of 2,8 by sharing two pairs of electrons (see figure 3.5).

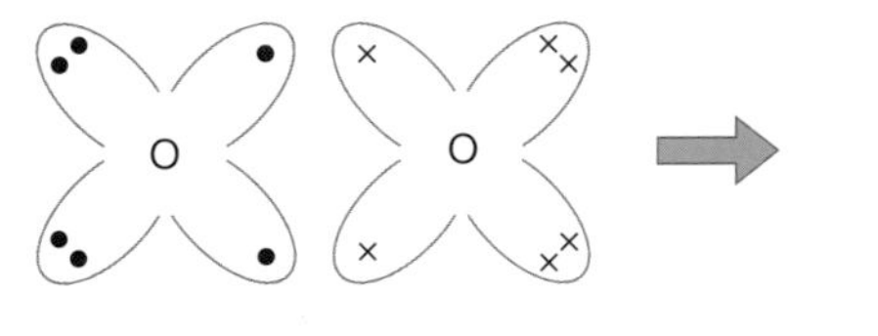

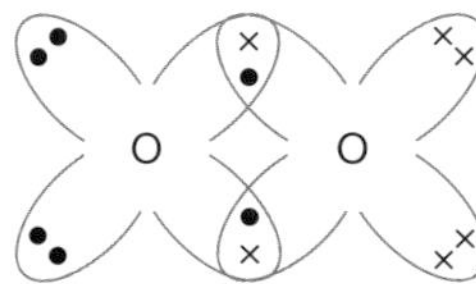

Figure 3.5 Two oxygen atoms combining

Nitrogen (N≡N)

A nitrogen atom has three half-filled lobes. Two nitrogen atoms will combine as shown in figure 3.6 by sharing three pairs of electrons to make a triple covalent bond.

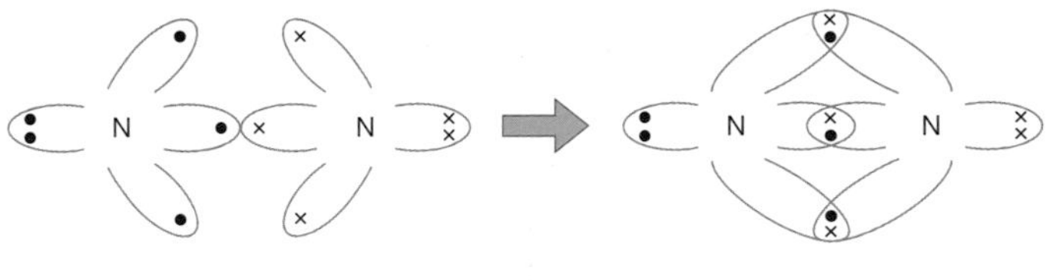

Figure 3.6 Two nitrogen atoms combining

Compounds

The way atoms of different elements join together can be shown using the same system of bonding diagrams as above.

Hydrogen chloride (H–Cl)

Here, the single electron from the hydrogen atom is shared with one electron from the chlorine atom (see figure 3.7).

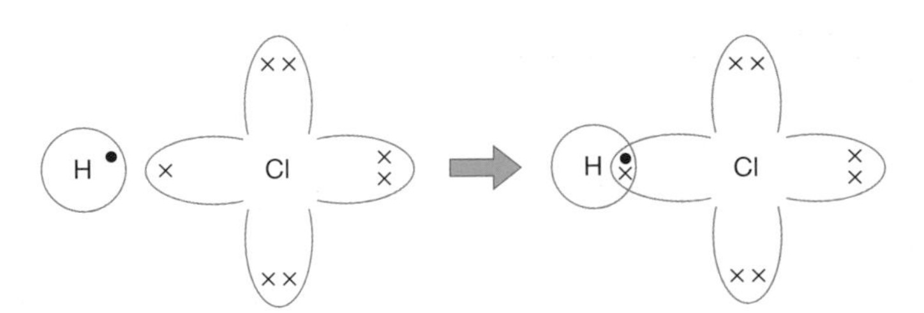

Figure 3.7 Hydrogen and chlorine atoms combining

Water (H_2O)

Oxygen has two half-filled lobes. This means two electrons are needed to give oxygen the same electron arrangement as a noble gas. Each hydrogen atom can only supply one electron for sharing. Therefore two hydrogen atoms are needed to join with one oxygen atom to make a molecule (figure 3.8).

Ammonia (NH_3)

Three hydrogen atoms supply electrons to fill the three half-filled lobes in the nitrogen atom (figure 3.9).

Methane (CH_4)

This compound consists of molecules formed between carbon and hydrogen atoms. Each carbon atom has four outer electrons. This gives each carbon

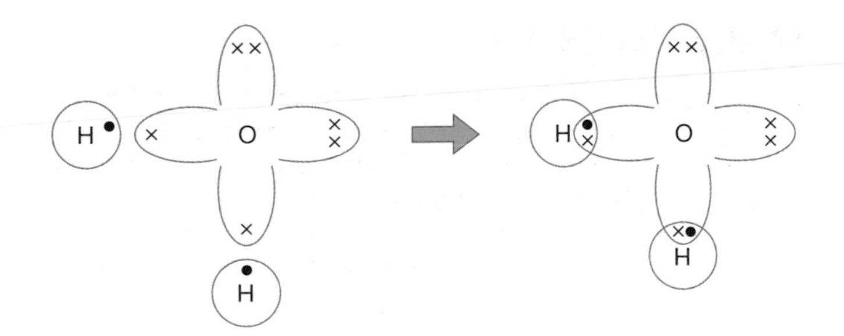

Figure 3.8 Hydrogen and oxygen atoms combining

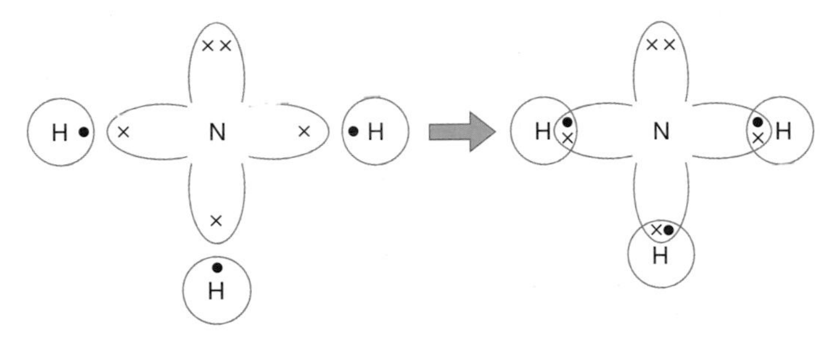

Figure 3.9 Hydrogen and nitrogen atoms combining

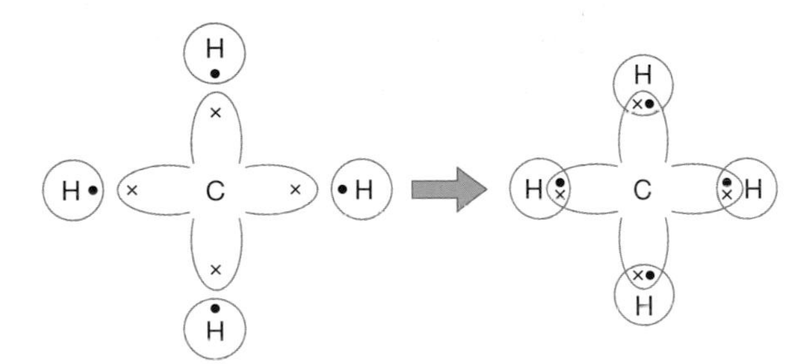

Figure 3.10 Hydrogen and carbon atoms combining

four half-filled lobes. Therefore, four hydrogen atoms are required to supply four electrons for sharing (figure 3.10).

Remember, the diagrams used here are all simplified. Section 4.4 discusses the *real* shapes of molecules.

Simpler electron diagrams

Drawing the shapes for the orbitals in atoms and molecules can take a lot of time.

You don't always have to do this. You can draw simpler electron diagrams by just showing the outer electrons. A molecule of hydrogen chloride could be drawn in two ways, as shown in figure 3.11.

Questions

Q4 Draw diagrams to show how the atoms combine to form the following compounds:
a) hydrogen fluoride (HF), **b)** silicon hydride (SiH_4), **c)** chloromethane (CH_3Cl).

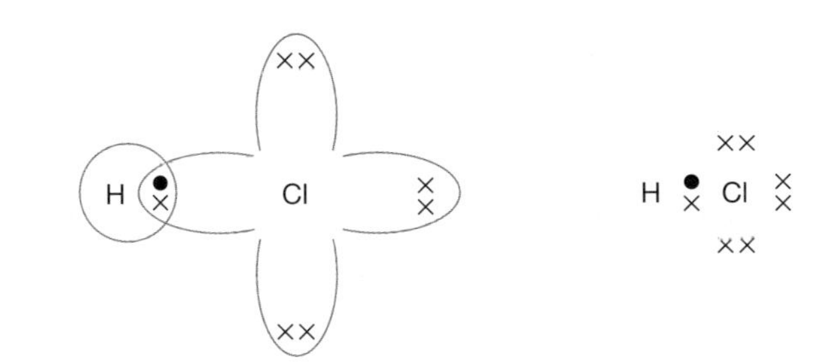

Figure 3.11 A simpler electron diagram for hydrogen chloride

Table 3.2 shows some simpler electron diagrams.

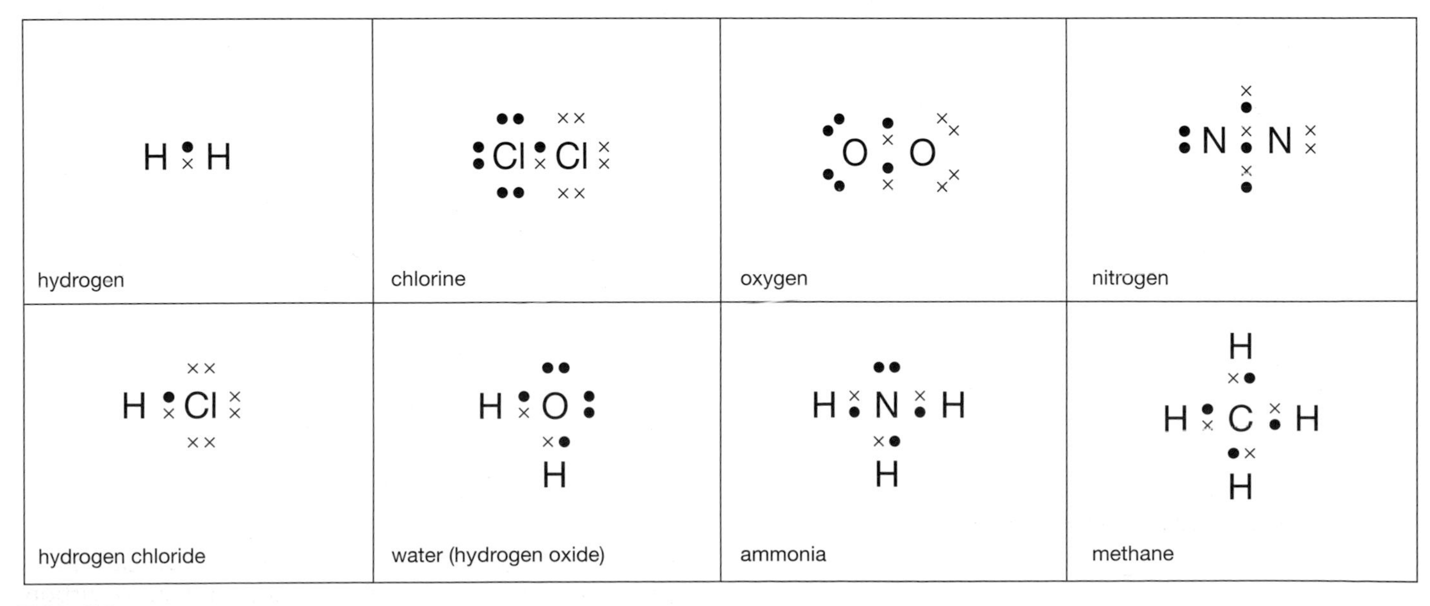

Table 3.2

SECTION 4.4 The shapes of molecules

Figure 4.1 The tetrahedral shape

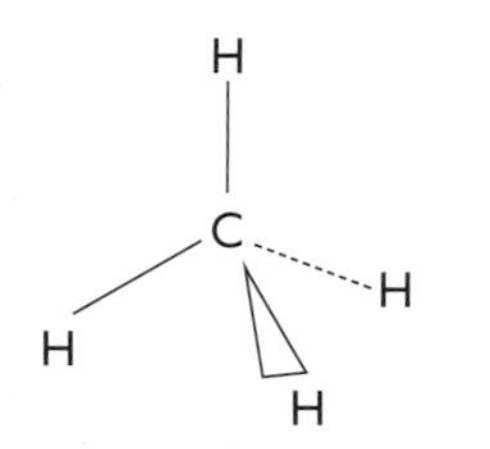

Figure 4.2 The three-dimensional shape of methane

The diagrams used so far are simplifications; they do not show the true shapes of molecules. For example, the electron in a hydrogen atom is contained in a three-dimensional sphere and not in a flat circle. The shape of the four lobes is also more complicated. They form a tetrahedral shape where each lobe can be imagined as pointing to the corners of a tetrahedron, as shown in figure 4.1.

Knowing these simple shapes helps us predict the shape of the actual molecules. The true three-dimensional shape of methane (the simplest compound formed between hydrogen and carbon) is shown in figure 4.2.

It is difficult to draw three-dimensional molecules and so chemists use special conventions: a solid line shows a bond in the same plane as the paper; a dotted line shows a bond behind the paper; and a wedge shape shows a bond pointing outwards. This particular method of showing the shape of a molecule is known as the **perspective formula**.

Table 4.1 shows the different ways of representing a selection of compounds containing two elements.

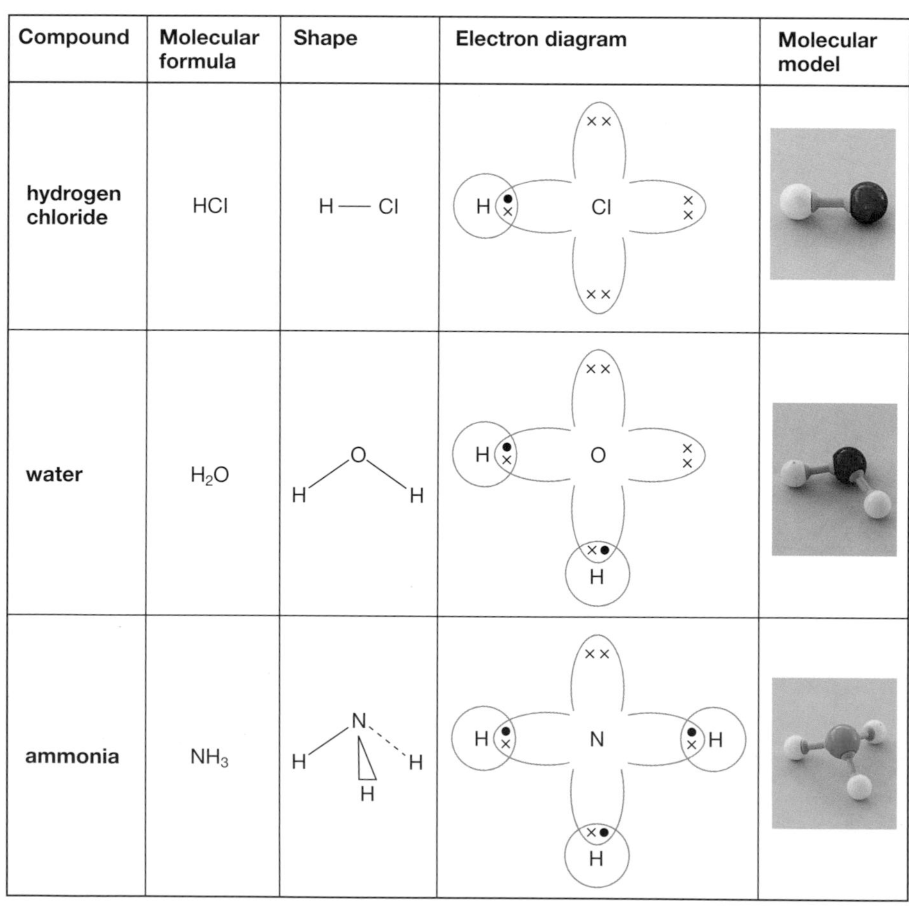

Compound	Molecular formula	Shape	Electron diagram	Molecular model
hydrogen chloride	HCl	H—Cl	H Cl	
water	H_2O	H O H	H O H	
ammonia	NH_3	H N H H	H N H H	

Table 4.1

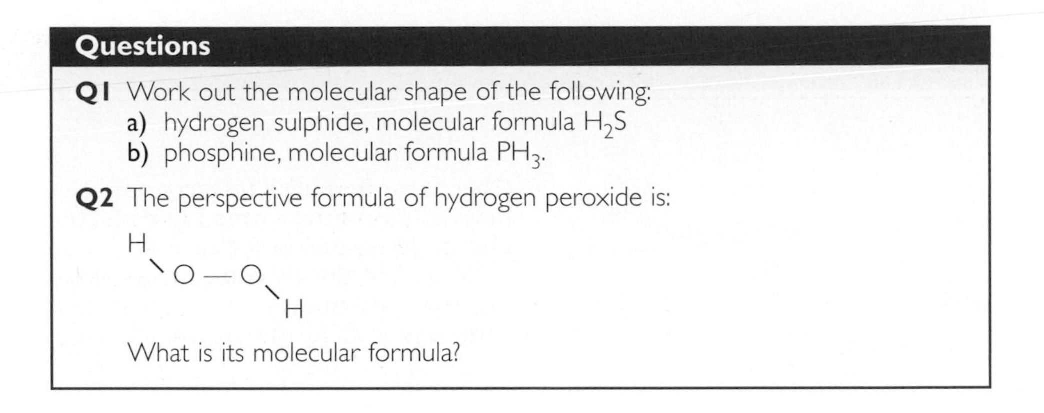

Questions

Q1 Work out the molecular shape of the following:

a) hydrogen sulphide, molecular formula H_2S

b) phosphine, molecular formula PH_3.

Q2 The perspective formula of hydrogen peroxide is:

$$H\diagdown O—O\diagdown H$$

What is its molecular formula?

What is a covalent bond?

Why do atoms stay joined together when they share a pair of electrons? The answer lies in the attraction that exists between negatively charged electrons and the positively charged nucleus of an atom. You probably know that a positive charge and a negative charge will attract each other. Look at the two hydrogen atoms bonded in figure 4.3. The atoms move closer together until the nuclei find a 'position of balance'. Any closer, and repulsion of the positively charged nuclei pushes the atoms apart. Any further apart, and the attraction between the nuclei and the shared electrons pulls the atoms back together again. In this way, the two hydrogen atoms stay together, as do all atoms that form covalent bonds.

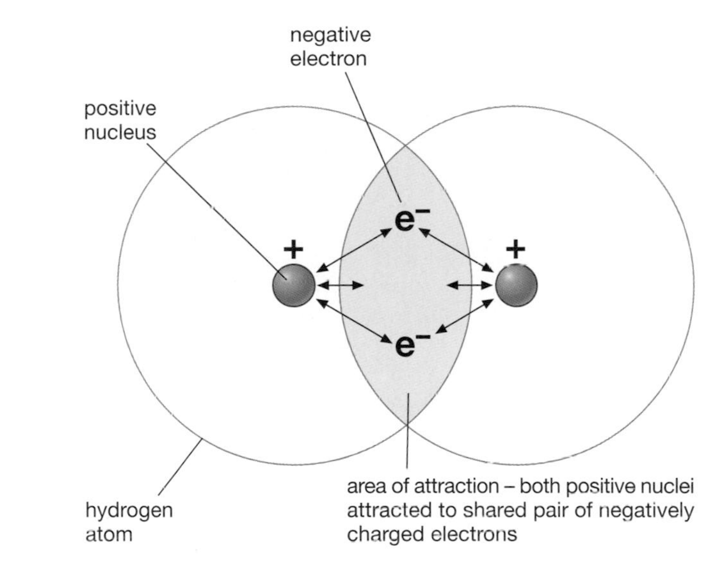

Figure 4.3 Two hydrogen atoms sharing a pair of electrons

Balanced equations

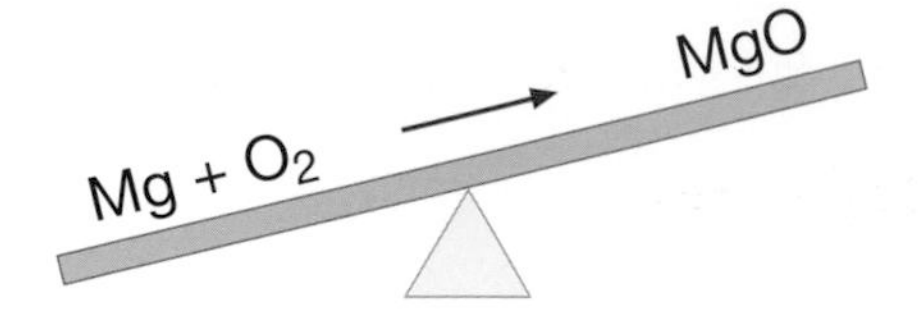

Figure 5.1 Two oxygen atoms on the left, but only one on the right

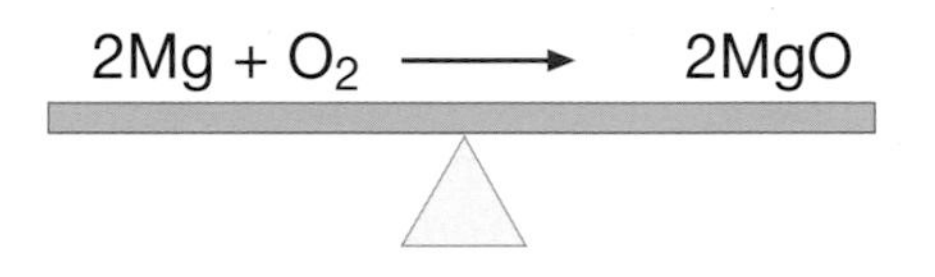

Figure 5.2 Same number of atoms of oxygen and magnesium on both sides of the equation

Chemists often have to carry out calculations, for example to find the mass of a product formed in a reaction. To do this, they need a **balanced chemical equation** for each reaction.

You have already come across word equations: these give the names of the reactants and products in a reaction. For example, magnesium combines with oxygen to form magnesium oxide. The word equation for this is:

magnesium + oxygen → magnesium oxide

To write this as a balanced equation we must first write the formulae for each substance:

$$Mg + O_2 \rightarrow MgO$$

This is called an **unbalanced equation.** On the left-hand side of the arrow there are two atoms of oxygen (the formula O_2 tells you this). However, on the right-hand side there is only one oxygen atom shown. The equation is therefore not balanced (figure 5.1).

To balance an equation the numbers of each type of atom on both sides of the arrow must be equal. Balancing is done by putting numbers *in front* of the formulae for the reactants and products, as required. The equation above can be balanced by putting a '2' in front of the Mg and also in front of the MgO (figure 5.2):

$$2Mg + O_2 \rightarrow 2MgO$$

The following rules are useful for writing balanced equations:

- Almost all elements are represented by a single symbol, for example copper is shown as Cu, sulphur as S. However, **diatomic elements**, for example, are written with a subscript '2', as in table 5.1.

Element	Formula
hydrogen	H_2
oxygen	O_2
nitrogen	N_2
fluorine	F_2
chlorine	Cl_2
bromine	Br_2
iodine	I_2

Table 5.1 Diatomic elements

- You should be able to use the valency method described in section 4.2 to work out the formulae for many compounds.
- You cannot balance an equation by changing the formula of a reactant or product, for example changing HCl to HCl_2 to give an extra chlorine is not allowed.

Examples:

1 Magnesium and sulphur combine when heated to produce magnesium sulphide. Notice that in this case no numbers have to be put in to balance the equation.

$$Mg + S \rightarrow MgS$$

2 Hydrogen and chlorine react explosively in bright light to produce hydrogen chloride:

$$H_2 + Cl_2 \rightarrow 2HCl$$

3 Ammonia (NH_3) burns in oxygen to produce nitrogen and water:

$$4NH_3 + 3O_2 \rightarrow 2N_2 + 6H_2O$$

Questions

Q1 For each of the following, write a word equation and then a balanced equation:

a) Carbon monoxide (CO) burning in oxygen with a blue flame to give carbon dioxide.

b) Sodium metal reacting with sulphur when heated to give sodium sulphide.

c) Magnesium metal reacting with dilute hydrochloric acid (HCl) to produce magnesium chloride and hydrogen gas.

State symbols

The formula for water is H_2O. However, this does not tell you if this is liquid water, steam or solid ice. It is often important to know the physical state of a substance, that is, if it is a solid, a liquid, a gas or even a solid which has been dissolved in water. This information is given by special symbols called **state symbols** which can be added after the formula for a substance (see table 5.2).

(s)	solid
(l)	liquid
(g)	gas
(aq)	aqueous (dissolved in water)

Table 5.2 State symbols

Using these symbols, liquid water is $H_2O(l)$, solid ice is $H_2O(s)$ and steam is $H_2O(g)$. Solid sodium chloride is NaCl(s) but when it is dissolved in water to make sodium chloride solution it becomes NaCl(aq).

Questions

Q2 Give the formulae with state symbols for the following:

a) ammonia gas,

b) liquid ammonia,

c) liquid oxygen (remember: oxygen is a diatomic element),

d) carbon dioxide gas,

e) solid carbon dioxide.

Q3 Ammonia freezes at –78°C and boils at –33°C. Give the formula with state symbols for ammonia

a) at –90°C

b) at –65°C.

Chapter 4 Study Questions

1 The grid shows the arrangement of atoms and molecules in pure substances and in mixtures.

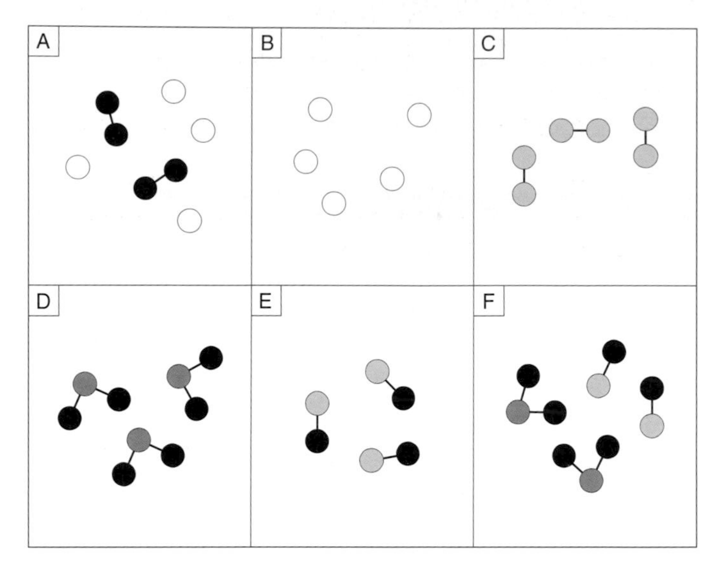

a) Identify the element which is made up of molecules.
b) Identify the mixture which is made up of compounds.

SQA GENERAL (PS)

2 a) The elements in Group 7 exist as **diatomic** molecules.
  (i) What is meant by diatomic?
  (ii) What type of bonding is present in a diatomic molecule? (KU)

b) Information on Group 7 elements is shown in the table.

Name	Atomic number	Boiling point/°C
fluorine	9	−188
chlorine	17	−35
bromine	35	59
iodine	53	184
astatine	85	

  (i) What happens to the boiling point as the atomic number increases?
  (ii) Predict the boiling point of astatine.

SQA GENERAL (PS)

3 The grid top right shows the arrangement of atoms and molecules in pure substances and in mixtures.

a) Identify the **compound** which is made up of **diatomic** molecules.
b) Identify the **mixture** of **monatomic** elements.
c) Identify the compound which could be hydrogen fluoride.

SQA GENERAL (PS)

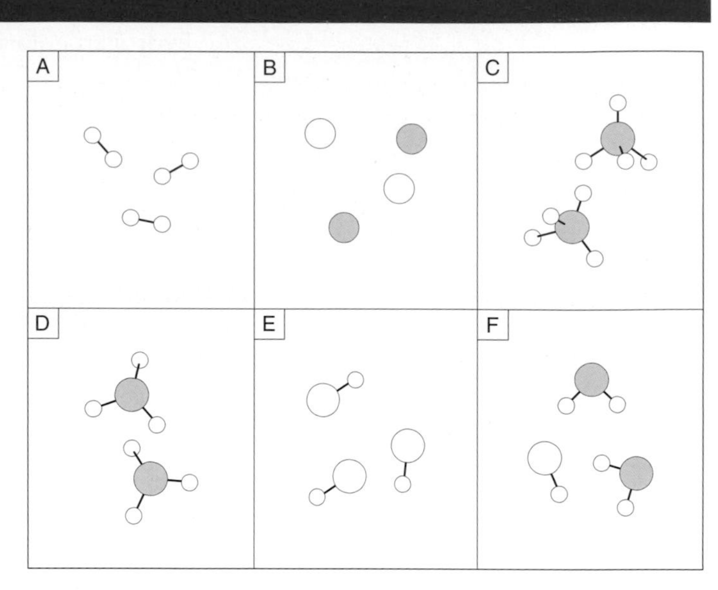

4 The grid contains the names of some elements and compounds together with full structural formulae for their molecules.

Identify the element(s).

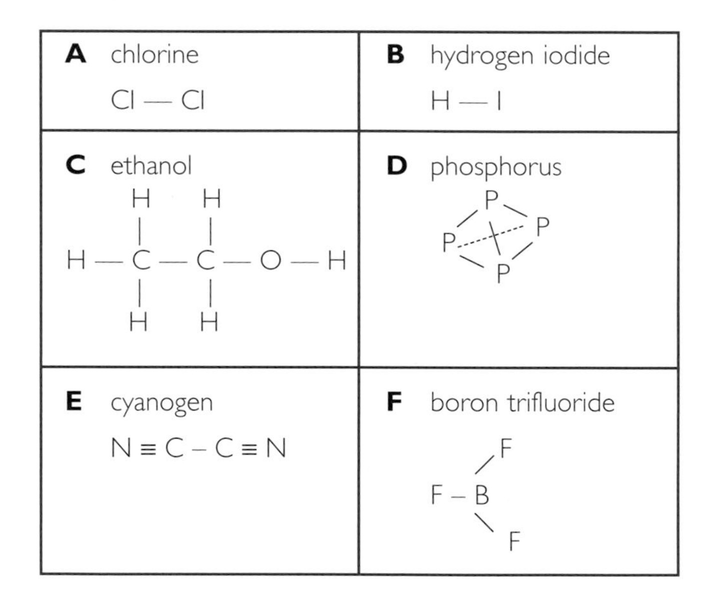

A chlorine Cl — Cl	**B** hydrogen iodide H — I
C ethanol H—C(H)(H)—C(H)(H)—O—H	**D** phosphorus P_4 (P, P, P, P)
E cyanogen N ≡ C – C ≡ N	**F** boron trifluoride F – B(F)(F)

GENERAL (KU)

5 A molecule of ethane has the formula C_2H_6. Copy and complete the following sentences:

a) The elements present in ethane are . . .
b) The number '2' in the formula means that . . .
c) The atoms in an ethane molecule are held together by . . .

SEB GENERAL (KU)

6 Two non-metallic elements X and Y exist as diatomic molecules. In a reaction, it was found that one molecule of X reacted with two molecules of Y to produce two molecules of a compound of X and Y. No other product was made by the reaction.

Give the formula for the compound of X and Y which was produced during the reaction. GENERAL (PS)

7 Formulae for compounds containing only two elements can be predicted with the help of the periodic table, if the elements concerned are 'main group elements'.

Use the information on page 1 in the SQA Data Booklet to help you work out the formulae for the following compounds:

a) boron sulphide
b) silicon fluoride
c) sulphur chloride
d) phosphorus iodide
e) iodine bromide. GENERAL (KU)

8 Prefixes can sometimes be used as a guide to formulae. The formula for nitrogen dioxide is NO_2, while that for diphosphorus pentoxide is P_2O_5.

Make use of the prefixes in the following compounds to help you write formulae for them.

a) nitrogen monoxide
b) sulphur trioxide
c) uranium hexafluoride
d) difluorine monoxide
e) dinitrogen tetroxide. GENERAL (KU)

9 The grid below contains six molecules made using a 'molecular models' kit:

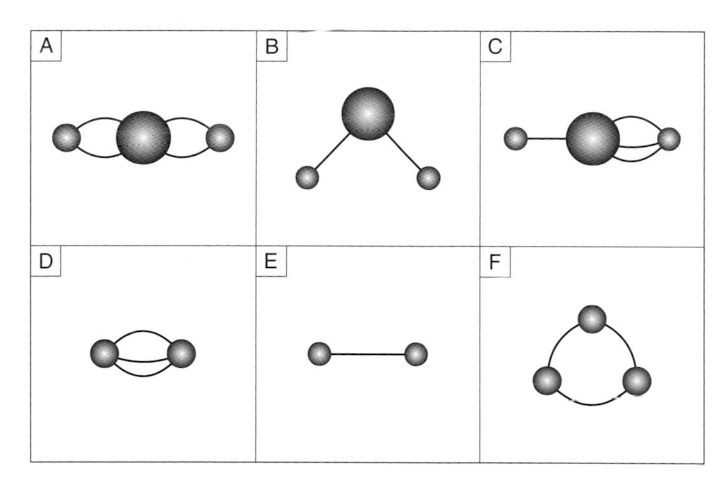

Identify the molecule which could represent:

a) hydrogen
b) water
c) carbon dioxide
d) nitrogen. GENERAL (KU)

10 A hydrogen molecule may be represented by the following formula:

H – H

In this formula, H represents a hydrogen atom and – represents a covalent bond.

Explain the role of the nuclei in the formation of this covalent bond. CREDIT (KU)

11 By means of diagrams showing outer electrons, show how covalent bonds form in the following molecules:

a) bromine, Br_2
b) hydrogen fluoride, HF
c) oxygen, O_2
d) water, H_2O
e) phosphine, PH_3
f) silane, SiH_4. CREDIT (KU)

12 The shape of a carbon dioxide molecule can be shown using symbols to represent atoms and lines to represent covalent bonds, as follows:

O=C=O

Use similar diagrams to show the shapes of the following molecules:

a) water, H_2O
b) hydrogen bromide, HBr
c) ammonia, NH_3
d) methane, CH_4
e) dibromomethane, CH_2Br_2 CREDIT (KU)

13 In a hydrogen chloride molecule, a **covalent bond** holds a hydrogen atom and a chlorine atom together. Explain the meaning of the term which is in bold. GENERAL (KU)

14 Hydrazine is a covalent compound which has been used as a rocket fuel. The formula for a molecule of hydrazine is N_2H_4. Explain what information the formula gives you about molecules of hydrazine. GENERAL (KU)

15 A main group element Z is known to bond covalently with chlorine to form a compound with the formula ZCl_3. In this compound, both element Z and chlorine have the stable electron arrangements of noble gases by sharing outer electrons.

a) To which main group of the periodic table is Z likely to belong?
b) By means of a diagram showing outer electrons, show how covalent bonds form in a molecule of this compound.
c) By means of a diagram, using lines to represent covalent bonds, show the expected *shape* of a molecule of this compound. CREDIT (PS)

16 Copy out the following equations and balance them by inserting numbers in the spaces provided if required:

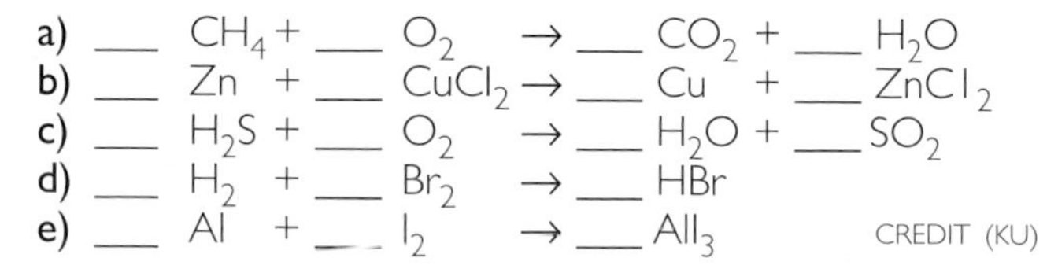

a) ___ CH_4 + ___ O_2 → ___ CO_2 + ___ H_2O
b) ___ Zn + ___ $CuCl_2$ → ___ Cu + ___ $ZnCl_2$
c) ___ H_2S + ___ O_2 → ___ H_2O + ___ SO_2
d) ___ H_2 + ___ Br_2 → ___ HBr
e) ___ Al + ___ I_2 → ___ AlI_3 CREDIT (KU)

CHAPTER FIVE
Fuels

SECTION 5.1 Fossil fuels

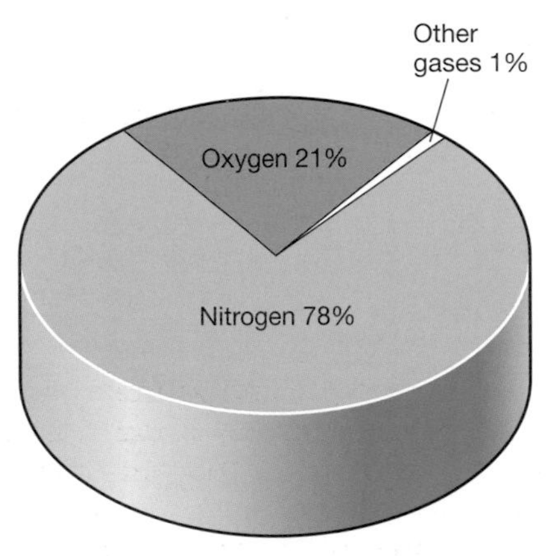

Figure 1.2 Composition of air

What do coal, petrol and wood all have in common? They can all be used as fuels. A fuel is a substance which gives out energy when it burns. For example, when wood is burnt in a fire, heat energy is given out.

Fuels supply heat for homes and for cooking. Coal and natural gas are used in many power stations to make electricity.

Figure 1.1 Fuels give energy for heat, light and transport

Questions

Q1 The safest way to put out a chip-pan fire is to put a damp cloth over the top of the pan. Explain how this is able to put out the fire.

Fuels give out energy when they burn. In order to burn they need oxygen. Scientists use the term 'combustion' to describe the process of burning. During combustion, fuels use the oxygen in the air.

Air contains oxygen and other gases. The pie chart in figure 1.2 shows that air contains 78% nitrogen, 21% oxygen plus a small amount of other gases. As a rough guide, we can say that air is made up of one-fifth oxygen and four-fifths nitrogen.

Oxygen is the gas which is involved in combustion. When a glowing splint is placed in a test tube of oxygen it re-lights and burns brightly (see figure 1.3). Oxygen is the only common gas that can do this. This is therefore used as the chemical test for oxygen.

The combustion of fuels produces heat energy. A reaction which gives out heat is called an **exothermic reaction**. 'Exo' means *out* and 'thermic' means *heat*.

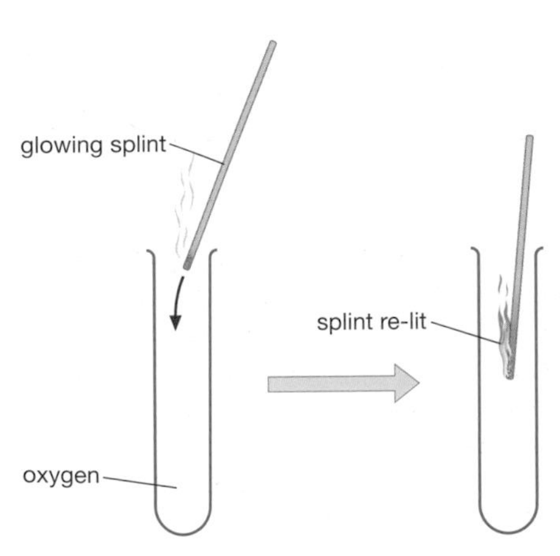

Figure 1.3 Oxygen will re-light a glowing splint

What is natural gas?

The gas fuel we use in cookers and gas fires is called **natural gas**. It is a fossil fuel. It is often found along with crude oil deposits.

Natural gas contains mostly methane. Before it is supplied to homes and industry other hydrocarbon gases, such as ethane, propane and butane, are removed. In Scotland this is carried out at the Mossmorran refinery in Fife. The final product is around 95% methane. A small amount of an unpleasant

Figure 1.4 The three main fossil fuels: coal, oil and natural gas.

smelling gas is added to the final product. This is done for safety reasons, to make it easier for people to smell any gas leaks. More information on methane and hydrocarbons is given in Chapter 6.

How fossil fuels were formed

The three main fuels which are used in the UK are coal, oil and natural gas. They are called **fossil fuels** because they were formed from once-living things.

The formation of coal started about 300 million years ago. At that time the Earth contained large areas of swamp and plant-life. When plants die, they break down into substances which return to the soil. However, when coal was being formed many dead plants lay in swampy areas and did not decay in the usual way. Instead, they were covered with layers of mud and earth. As the layers built up, the lower ones were compressed more and more and eventually turned into coal (see figure 1.5).

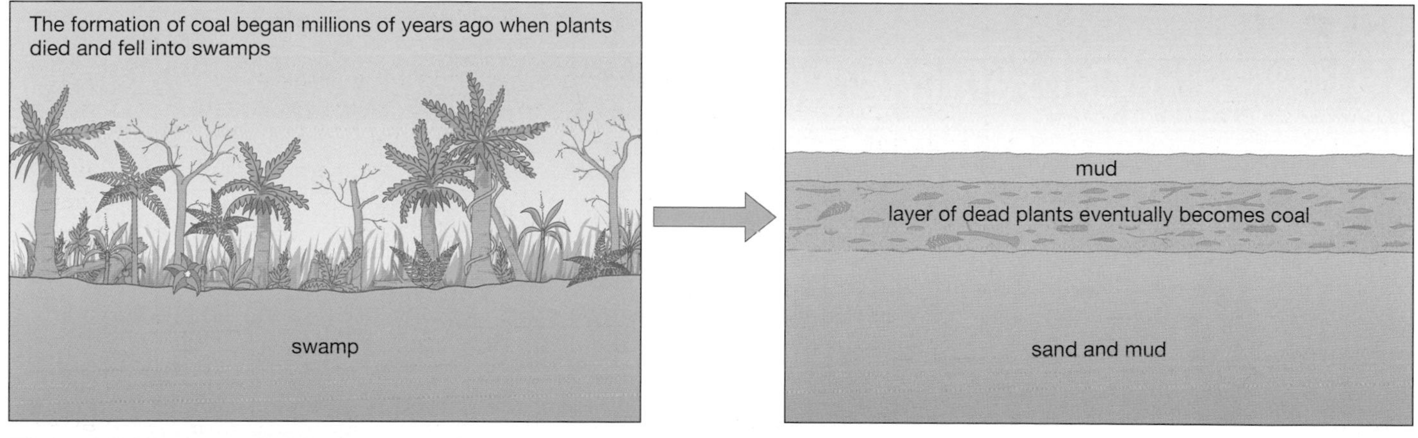

Figure 1.5 Formation of coal

Questions

Q2 Give three ways in which the formation of coal and oil are similar.

Oil and gas were formed millions of years ago in the oceans (see figure 1.6). As microscopic plants and animals died and sank to the bottom they were covered with layers of sand and other materials. Eventually the pressure of these layers above helped to turn them into oil and natural gas. Much of the oil and gas escaped but the rest was trapped in layers of porous rock sandwiched between other layers of non-porous rock, which sealed in the oil and gas. Later, movements of the Earth changed the positions of the oceans leaving some oil and gas deposits in land areas such as Nigeria and Texas, USA.

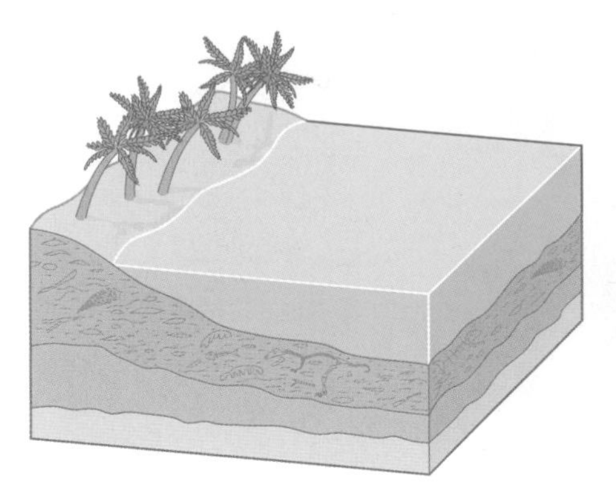

Millions of years ago, when plants and animals in the sea died, they fell to the sea bed.

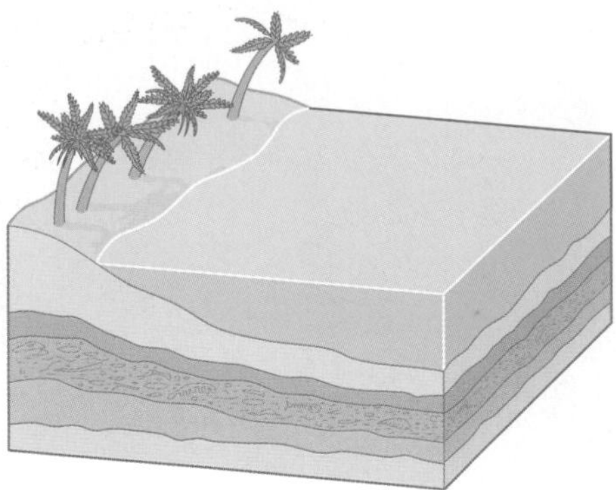

They were covered by layers of mud and rocks. Over a long period of time, more and more of these layers built up.

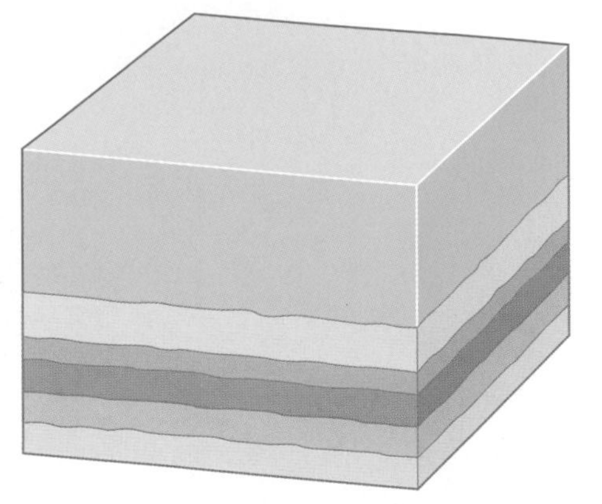

Enormous pressures were created as the layers sank deeper and deeper. The pressure changed the dead plants and animals into oil or gas.

Figure 1.6 Formation of oil and natural gas

Part of the world	Percentage
Western Europe	0.7
Asia	20.0
India	56.0
Africa	60.0

Table 1.1 Percentage of energy obtained from wood burning

There are large amounts of oil, natural gas and coal in the world today. However, one day they will run out. For this reason they are called **finite resources** because supplies are limited.

It is very difficult to estimate just how long fossil fuels will last. It is likely that there are supplies of coal to last around 300 more years. Oil and gas, however, will probably run out in 50–80 years time. Oil will last longer than gas. Of course, if we can save energy then the fuels will last longer.

Fossil fuels are the main sources of energy for many countries. However, in developing countries it is estimated that around 2000 million people use firewood to supply energy for cooking and heating (see table 1.1). Unfortunately, trees are being cut down for wood faster than they can be replaced with new trees. As a result, many people are facing a growing shortage of fuel.

SECTION 5.2 Oil refining

Figure 2.1 A fractional distillation tower at the BP/Amoco refinery, Grangemouth

The oil that comes out of most oil wells is a black, sticky liquid with a very unpleasant smell. It is called crude oil and it contains a rich mixture of valuable chemicals called **hydrocarbons**. These are compounds which contain only carbon and hydrogen atoms – no other elements are present. The problem for chemists is how to separate the different hydrocarbons in crude oil.

To understand how the problem is solved you need to know that different chemicals melt and boil at different temperatures – they have different melting points and boiling points. These changes of state are shown in figure 2.2.

Each of the hydrocarbons in crude oil has its own boiling point. If the oil is heated slowly, then each will boil and turn into a gas at different temperatures. The different gases can be collected separately. When the gases cool they will condense and turn back into liquids. This process is called **distillation**. Each separate liquid is called a **fraction**.

The name for the complete process of separating the different fractions in crude oil is **fractional distillation** (figure 2.3). The fractions that are obtained first consist of gases, then come clear, light-coloured liquids. After that the fractions gradually become darker in colour and thicker. The final fraction which is left is known as bitumen residue, and is the sticky material used in making roads.

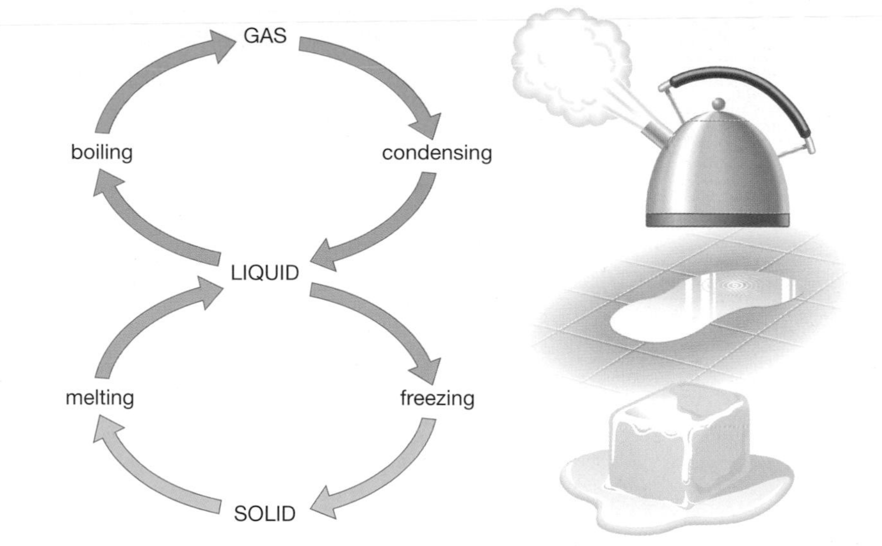

Figure 2.2 Changes of state

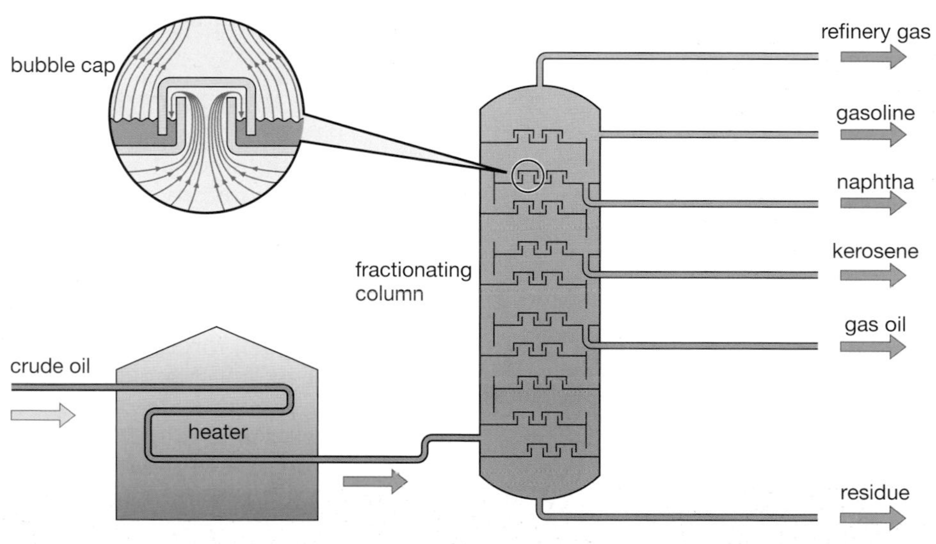

Figure 2.3 Fractional distillation of crude oil

Table 2.1 summarises the differences between the fractions. Table 2.2 shows the names of the fractions that are obtained from crude oil, along with their uses.

Note:
Flammability means how easily a substance will catch fire.
Viscosity means how thick a liquid is.
For example, syrup has a higher viscosity than water.

Fractions	Boiling points	Flammability	Viscosity	Speed of evaporation
first	low boiling points	catch fire easily	low – gases and runny liquids	fast
last	high boiling points	do not catch fire easily	highly viscous	slow

Table 2.1 Summary of differences between fractions of crude oil

Questions

Q1 A mixture of ethanol (boiling point 79°C) and water (boiling point 100°C) is distilled. Which liquid will boil first?

Crude oil is converted into fractions in **oil refineries**. The biggest refinery in Scotland is operated by BP at Grangemouth. This is a very large site which carries out not only fractional distillation, but also a range of other processes concerned with turning crude oil into useful products. For example, almost all the petrol on sale in Scotland comes from Grangemouth, no matter what the brand.

Fraction	Boiling range/°C	Carbon atoms per molecule	End uses
refinery gas	<20	1–4	fuel gas
gasoline	20–65	5–6	petrol and petrochemicals
naphtha	65–180	6–11	petrol and petrochemicals
kerosene	180–250	9–15	jet fuel/heating
gas oil	250–350	15–25	diesel
residue	>350	>25	heavy fuel oil/wax, etc

Table 2.2 Fractions obtained from crude oil and their uses

Properties of the fractions

The fractions obtained from crude oil have different properties because they contain hydrocarbons of different sizes. The simplest way to think about the size of a hydrocarbon is to count the number of carbon atoms in its molecule. There are between one and four carbon atoms in the hydrocarbon molecules of the first fractions.

Table 2.2 shows the number of carbon atoms in molecules of six of the fractions, along with the temperature ranges in which they boil.

Boiling point ranges

In liquids, small forces of attraction exist between the molecules. These forces hold the liquid together. When they boil energy is needed to separate the molecules and turn them into a gas.

The first fractions that are distilled have low boiling point ranges. This is because they consist of small molecules. Boiling point ranges of the different fractions vary because of differences in the size of the molecules.

There are small forces of attraction that exist between molecules. These forces are bigger and stronger for large molecules. In other words for bigger molecules there is more attraction between them. Therefore, it takes more energy to separate larger molecules. This means that the boiling point range increases as the size of the molecules increases. You can see in table 2.2 the increase in boiling point ranges for the different fractions.

Viscosity

Fraction	North Sea oil	Venezuelan oil
refinery gas	1	0
gasoline/naphtha	19	1
kerosene/paraffin	16	5
gas oil (diesel)	23	25
residue (bitumen, waxes)	41	69

Table 2.3 Composition of two crude oils (percentage by mass)

The viscosity of a liquid is a measure of how 'thick' it is. Syrup is a liquid with a high viscosity. As the forces of attraction between the molecules increase then liquids become less runny and more viscous. Therefore, the fractions with the bigger molecules have a high viscosity. They are also more viscous because they are made of long chains of atoms which tend to get tangled up and therefore 'clump' together.

Flammability

The small hydrocarbons are more flammable than the larger molecules because they can react more quickly with the oxygen in the air. They are also less viscous because of the smaller forces holding them together. The large hydrocarbons have greater forces acting between the molecules.

Crude oil from different parts of the world contains different amounts of the various fractions. Oil from the North Sea is quite runny because it contains a lot of small molecules, whereas oil from Venezuela is very thick (see table 2.3). There is an oil refinery based in Dundee which requires oil with a high proportion of large molecules, so Venezuelan oil is imported; it is brought to Scotland in heated tankers.

Note:
Gasoline and some of the naphtha are further processed to give petrol, which has 5–10 carbon atoms in its molecules.

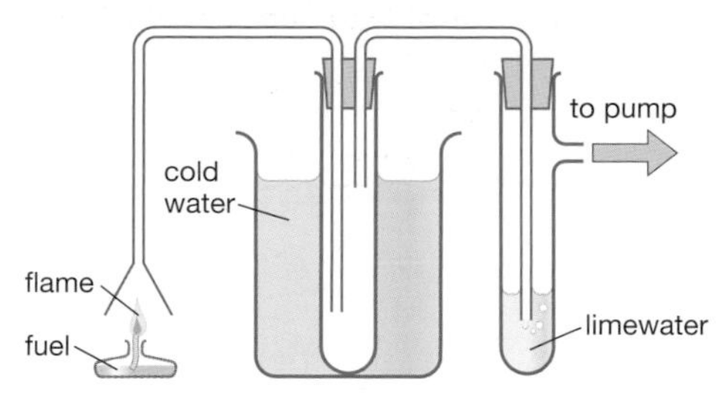

Figure 3.1 Combustion of a fuel to give carbon dioxide and water

All fossil fuels contain carbon. When they are burned they give out energy. They also all give out carbon dioxide, which can be a **pollutant** – a substance which damages the environment. The carbon in fuels reacts with oxygen in the air to produce carbon dioxide.

Most fuels also contain hydrogen. This produces water (H_2O) when it combines with oxygen. This can be shown in a laboratory experiment (see figure 3.1).

The burning fuel produces gases which are drawn through the apparatus by the pump. One of these gases is carbon dioxide, which turns the limewater in the test tube milky. This is the test for carbon dioxide. The drops of condensation on the other test tube are water. You can show that a liquid is water if it boils at 100°C and freezes at 0°C.

Experiments such as the one in figure 3.1 show that *when fossil fuels burn they produce carbon dioxide and water*. For example, the unbalanced equation for the combustion of methane is:

$$CH_4 + O_2 \rightarrow CO_2 + H_2O$$

Questions

Q1 Balance the equation on the right for the combustion of methane.

The greenhouse effect

How does carbon dioxide damage the environment? The Earth is surrounded by its atmosphere – a mixture of gases. Heat from the sun passes through these gases to warm up the Earth. Heat is also lost from the Earth into space (see figure 3.2).

There are gases known as 'greenhouse gases'; they let the heat from the sun in but do not let heat back out again. This is similar to what happens in a greenhouse, which stays warm because it traps heat from the sun through its glass. We need the greenhouse gases to keep the Earth warm.

Carbon dioxide is one of the greenhouse gases. The amount of carbon dioxide we produce is increasing every year. No one is sure what the effects of this will be, but it is thought that some parts of the world will warm up by around 4°C over the next few decades. Other parts may even cool down. There could be an increase in the levels of rivers and seas – this may well cause flooding in certain areas. Burning fuels for heating, for cars and other vehicles and for making electricity in power stations all produces carbon dioxide and adds to the greenhouse effect.

The experiment in figure 3.1 tells us two things about the atoms present in the fuel. First, if carbon dioxide is formed then the carbon must come from somewhere. There is very little carbon in the air (only 0.03 per cent of the air is carbon dioxide) therefore the carbon must be present in the fuel. Second, the hydrogen in the water that was formed must also come from the fuel. This experiment therefore proves that fuels contain hydrogen and carbon.

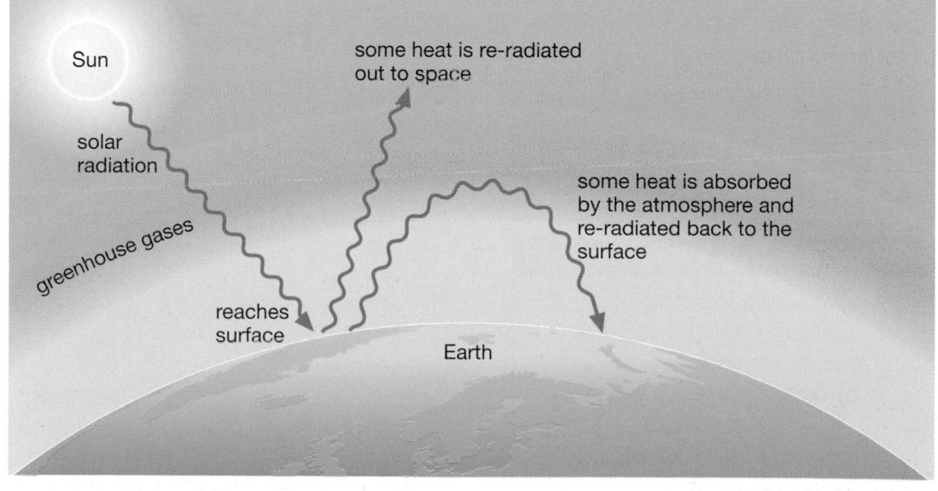

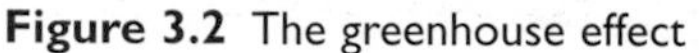
Figure 3.2 The greenhouse effect

Questions

Q2 Carbon dioxide and water both contain oxygen. Why does the experiment in figure 3.1 give no evidence to suggest that there is oxygen present in the petrol?

Questions

Q3 The warming of the Earth caused by the greenhouse effect is likely to cause several problems. Give at least three unwanted results of a rise in the temperature of the Earth.

Carbon monoxide

The carbon in fuels usually burns to make carbon dioxide, CO_2. To do this each carbon atom has to join with two oxygen atoms. Often when a fuel burns, there is not enough oxygen to produce carbon dioxide. Instead, the gas carbon monoxide is formed. This has the formula CO.

Carbon monoxide is a poisonous gas. In large enough amounts it can kill by destroying the blood's ability to carry oxygen.

Sulphur dioxide and acid rain

Most fossil fuels contain small amounts of sulphur. When the fuel is burnt the sulphur joins with oxygen to produce the gas sulphur dioxide. This is a poisonous gas which, when carried by the wind, can travel large distances. The sulphur dioxide can dissolve in rain water to produce an acid. This is one of the components of acid rain.

The greatest amounts of sulphur dioxide are produced by coal- and oil-fired power stations, which burn large amounts of fuel. New technology is now being used which can remove the sulphur dioxide from the gases leaving the power stations.

Figure 3.3 Erosion caused by acid rain

Car engines and pollution

Inside a car engine petrol is burnt to give energy. It is this energy which eventually makes the car move. To get the petrol to burn it is mixed with air and ignited by an electric spark produced in the engine (see figure 3.4).

However, nitrogen and oxygen in the air next to the spark also reacts with the air. Nitrogen is a very unreactive gas, but the spark supplies enough energy to make the nitrogen react with oxygen to form nitrogen dioxide. This is a poisonous reddish brown gas and, like colourless sulphur dioxide, dissolves in rain water to produce an acid solution.

In 'lean burn' engines the amount of air which is mixed with the petrol is increased. This allows the petrol to be burnt more efficiently and so produces less carbon monoxide and unburnt hydrocarbons.

Catalytic converters also cut down on some of the pollutants produced by car engines. However, they have no beneficial effect on carbon dioxide levels (see section 2.3). Of course, one way of cutting down on pollution from cars is to encourage people to use cars less. However, this is likely to be difficult and unpopular.

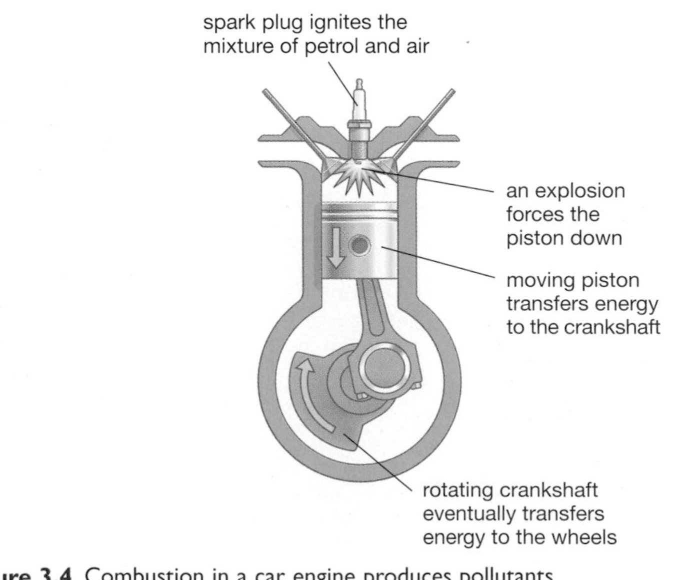

Figure 3.4 Combustion in a car engine produces pollutants

Questions

Q4 a) Make a list of the different types of pollution caused by cars.
b) Suggest ways of reducing the types of pollution mentioned in your answer to part a).

Chapter 5 Study Questions

1 The following table shows the percentage of energy consumed in the world from various sources in 1987.

Nuclear	Hydro-electric	Gas	Coal	Oil
5	7	19	29	40

a) Draw a **bar graph** to show this information.
b) Use the data to complete the table below.

World energy consumption from various sources for 1987 (expressed as a percentage)		
Fossil fuels	**Other sources**	**Total**
		100

SEB GENERAL (PS)

2 Distillation of crude oil produces several fractions.

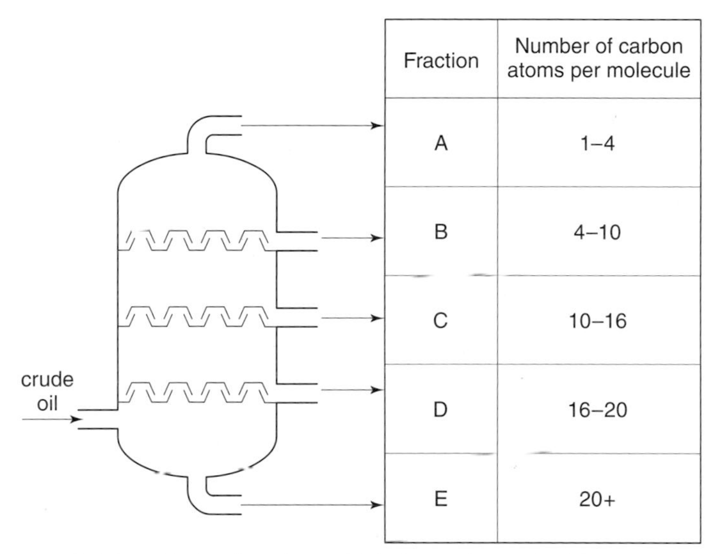

Fraction	Number of carbon atoms per molecule
A	1–4
B	4–10
C	10–16
D	16–20
E	20+

a) Identify the fraction which is used as a fuel for jet aircraft.
b) Identify the fraction with the lowest boiling point.

SQA CREDIT (KU)

3 At Longannet Power Station finely powdered coal is burned in four large furnaces.

a) Which element in the air is needed for coal to burn?
b) Describe how coal was formed. GENERAL (KU)

4 a) Describe how crude oil was formed.
b) Crude oil contains many different hydrocarbons. What is a hydrocarbon? GENERAL (KU)

5 The first lamps were not electric. A gas called acetylene was burned to produce the light.

The acetylene gas was made by adding water to a solid called calcium carbide. Acetylene gas is insoluble in water.

Draw a diagram of the apparatus you would use to prepare and collect a gas jar of acetylene in your chemistry laboratory. In the diagram, label the chemicals used.

SES GENERAL (PS)

6 Crude oil is one of our most important raw materials. It has transformed our way of life over the past fifty years.

In oil refineries, the following groups of hydrocarbons are obtained. Their approximate boiling ranges are also shown:

A light gas oil 250–300°C	**B** kerosene 180–250°C	**C** naphtha 40–180°C
D heavy gas oil 300–350°C	**E** residue >350°C	**F** petroleum gas <40°C

a) Which group (or groups) contains hydrocarbons which are more flammable than those in kerosene? (PS)
b) Which group (or groups) contains larger hydrocarbon molecules than those in light gas oil?

SEB CREDIT (KU)

7 When a fuel such as coal burns, oxygen reacts with the coal in an exothermic reaction.

a) What is meant by the term 'exothermic'? (KU)
b) Maureen was given two stoppered test tubes. One contained nitrogen and the other oxygen. Describe a test that Maureen could carry out which would show the presence of oxygen in one of the test tubes. (KU)
c) Oxygen and nitrogen are the main components of air in the proportions of 1:4. Draw a labelled pie chart to illustrate this. (You should ignore all of the other gases of the air.) GENERAL (PS)

8 Coal, oil and natural gas are finite resources. Various estimates of how long these fossil fuels will last have been made. One such estimate is as follows:

Resource	Lifetime/years
coal	110
natural gas	30
oil	25

a) Present this information on graph paper in a suitable form, labelling the horizontal axis 'time/years'. (PS)
b) Explain what is meant by the term 'finite resource'. (KU)
c) Explain what is meant by the term 'fossil fuel'.

GENERAL (KU)

9 A liquid fuel was thought to produce water vapour and carbon dioxide when it burned. Jean set up the apparatus shown below to investigate this:

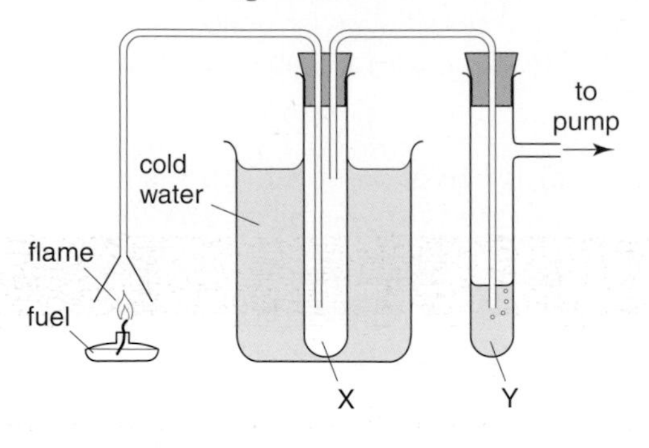

a) How should Jean test the liquid that formed in test tube X in order to prove that it was water? (KU)

b) Name the liquid which should be present in test tube Y in order to show that carbon dioxide was formed when the fuel burned. (KU)

c) Jean's teacher suggested that it would be a good idea to repeat the experiment without the burning fuel and with air drawn through the apparatus for the same period of time. Explain why this extra experiment should be carried out.

GENERAL (PS)

10 James investigated the burning of a fuel and obtained the following results:

(i) Water was a product of combustion.
(ii) Carbon dioxide was *not* a product of combustion.

What conclusion was James able to come to regarding the elements present in the fuel:

a) based on result (i) alone?

b) based on result (ii) alone? CREDIT (KU)

11 Several poisonous gases can be produced as a result of burning fuels.

a) Name a poisonous gas which can be formed as a result of incomplete combustion of petrol in a car engine.

b) Name a poisonous gas which is a major cause of acid rain and is produced by the burning of coal (and, to a lesser extent, fuels derived from oil).

GENERAL (KU)

12 Attempts are being made to reduce the pollution present in the exhaust gases of cars.

(i) All new cars are now fitted with catalytic converters as part of their exhaust system.
(ii) Car engines have been developed which are described as 'lean burn'. In these there is a lower than usual fuel to air ratio.

a) State what happens to pollutant gases as they pass through a (hot) catalytic converter.

b) Explain how a 'lean burn' engine reduces pollutants in exhaust gases. CREDIT (KU)

CHAPTER SIX

Hydrocarbons

SECTION 6.1 Alkanes

Figure 1.1 Aeroplanes and cars both use alkanes as fuels

Hydrocarbons are compounds which contain only hydrogen and carbon. There is a vast number of different hydrocarbons and so chemists have grouped them into 'sub-sets'. The **alkanes** make up the simplest sub-set of all the hydrocarbons.

Uses of alkanes

Alkanes are very important substances. Most of the fuels which we use are made from alkanes. Table 1.1 shows some alkanes and their uses. Notice that alkanes all have the same name ending: **-ane**

In addition, alkanes are used as the starting materials in the manufacture of a huge range of other substances such as plastics and drugs.

Alkane	Uses
methane	(natural gas) for cooking, heating
propane	used in gas cylinders for homes
butane	sold in blue cylinders as 'camping gas'?
octane	a component of petrol

Table 1.1 Some alkanes and their uses

The structure of alkanes

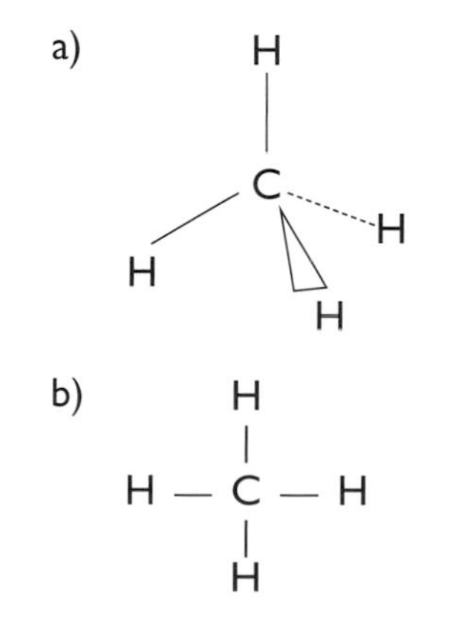

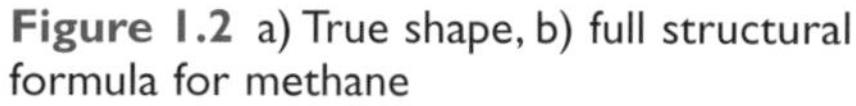

Figure 1.2 a) True shape, b) full structural formula for methane

Alkanes contain covalent bonds. Each carbon atom shares four outer electrons with other atoms. What makes alkanes, and in fact all hydrocarbons, different from other compounds is that they are able to form long, chain-like molecules containing many carbon atoms joined together.

In Chapter 4 you saw how methane forms a molecule with a tetrahedral shape. Figure 1.2 shows a diagram of a methane molecule. This is a difficult shape to draw and so chemists use a special type of formula called the **full structural formula** to give a simplified picture of the structure of the molecule. For the alkanes after methane it is sometimes more convenient to use **shortened structural formulae**. These simply show how many CH_3 and CH_2 groups are present, but still give some idea of molecular structure e.g. $CH_3CH_2CH_3$ for propane.

The simplest type of formula used for alkanes and other hydrocarbons is the **molecular formula**. This shows only the number of hydrogen and carbon atoms present. Table 1.2 shows the names and different types of formulae for the first eight alkanes.

Name	Molecular formula	Full structural formula	Physical state at 25°C
methane	CH_4	H H–C–H H	gas
ethane	C_2H_6	H H H–C–C–H H H	gas
propane	C_3H_8	H H H H–C–C–C–H H H H	gas
butane	C_4H_{10}	H H H H H–C–C–C–C–H H H H H	gas
pentane	C_5H_{12}	H H H H H H–C–C–C–C–C–H H H H H H	liquid
hexane	C_6H_{14}	H H H H H H H–C–C–C–C–C–C–H H H H H H H	liquid
heptane	C_7H_{16}	H H H H H H H H–C–C–C–C–C–C–C–H H H H H H H H	liquid
octane	C_8H_{18}	H H H H H H H H H–C–C–C–C–C–C–C–C–H H H H H H H H H	liquid

Table 1.2 Names and formulae of some alkanes

You can use the following phrase for remembering the names of the first five alkanes in their correct order:

Must Every PRefect Be PErfect

This gives the first letter or letters for methane, ethane, propane, butane and pentane. From hexane onwards, the names are based on Greek prefixes such as *hexa* for 6 etc.

The general formula for the alkanes

If you look at the full structural formulae in table 1.2 you can see that there is a pattern. Each carbon atom is joined to two hydrogen atoms, and there are two extra hydrogen atoms – one at each end of the molecule. In mathematical terms, this gives a general formula for the alkanes as:

$$C_nH_{2n+2}$$

where n is the number of carbon atoms in the molecule.

For example, if $n = 1$, then there will be 1 carbon atom and $(2 \times 1) + 2 = 4$ hydrogen atoms, giving a molecular formula of CH_4. In hexane, where $n = 6$, the formula will be C_6H_{14}.

Questions

Q1 What are butane and methane used for?

Q2 Write names for the following alkanes:
a) C_2H_6
b) C_4H_{10}
c) C_7H_{16}

Q3 Give full structural formulae for the following alkanes:
a) propane,
b) pentane,
c) octane.

Q4 Use the general formula to give the molecular formulae for:
a) decane, the alkane with 10 carbon atoms,
b) the alkane with 50 carbon atoms.

Alkenes

Alkenes are similar to alkanes in many ways. They are both groups of hydrocarbons containing covalent bonds. The main difference between them is that each alkene contains a special bond called a **carbon-to-carbon double bond,** which can be represented by C=C. As you will see, the presence of this bond affects the properties of alkenes.

Table 2.1 shows the names and formulae for the first four alkenes. There are four points to note:

- The first part of the alkene names are the same as those of the alkanes.
- All the alkene names end in **-ene**.
- There is no alkene with one carbon atom (no 'methene') – this is because each alkene must contain two carbon atoms to form a C=C bond.
- Each alkene contains only one C=C bond.

Name	Molecular formula	Full structural formula	Shortened structural formula
ethene	C_2H_4	H H \ / C=C / \ H H	CH_2=CH_2

propene	C_3H_6	H H H
		  | | /
		H–C–C=C
		  | \
		  H H
			CH_3CH=CH_2

butene	C_4H_8	H H H H
		  | | | /
		H–C–C–C=C
		  | | \
		  H H H
			CH_3CH_2CH=CH_2

pentene	C_5H_{10}	H H H H H
		  | | | | /
		H–C–C–C–C=C
		  | | | \
		  H H H H
			$CH_3CH_2CH_2$CH=CH_2

Table 2.1 Names and formulae of some alkenes

Uses of alkenes

The C=C bond is very reactive, much more so than the C–C single bond present in alkanes. Because of this, the alkenes are reactive molecules which can be changed readily into new products. For example, small alkenes such as ethene and propene are used in the plastics industry to make polythene and polypropene (see Chapter 13).

There is truth in the expression 'one bad apple spoils the whole barrel' – over-ripe fruit gives off the gas ethene, which ripens other fruits close by.

Oranges from Israel are a greenish colour when they are picked in February. Before they reach the shops, ethene gas is passed over them to help them ripen and to turn the skins orange. You can ripen an orange yourself by putting it in a bag with a ripe banana.

Questions

Q1 Give the names of the alkenes with the following full structural formulae:

a)
```
H     H H H
 \    | | |
  C=C-C-C-H
 /    | |
H     H H
```

b)
```
H     H H H H
 \    | | | |
  C=C-C-C-C-H
 /    | | |
H     H H H
```

Q2 a) Draw the full structural formulae of the following alkenes: (i) ethene, (ii) propene.
b) Draw the shortened structural formulae of (i) ethene, and (ii) propene.

Homologous series

A **homologous series** is a group of chemicals with similar chemical properties and which can be represented by a general formula. The alkanes form a homologous series because they share the following characteristics:

- They all share the general formula C_nH_{2n+2}
- They all have similar chemical properties, for example they all burn.
- They show a gradual change in some physical properties as the molecules get bigger, for example the boiling points of the alkanes increase gradually as the molecules increase in size.

When a liquid boils its molecules have to move further apart. It requires more heat energy to separate large alkane molecules than small ones. Therefore large alkane molecules boil at a higher temperature than smaller alkanes.

The alkenes also form a homologous series. They have the general formula C_nH_{2n}, share similar chemical properties and show a gradual change in certain physical properties such as boiling point.

Questions

Q3 Which type of chemical bond is present in alkenes but not in alkanes?

Cycloalkanes

The alkanes show that carbon atoms can join to form chains, but they can also form rings of carbon atoms. This gives another homologous series of hydrocarbons called **cycloalkanes**. The first four members are shown in table 2.2.

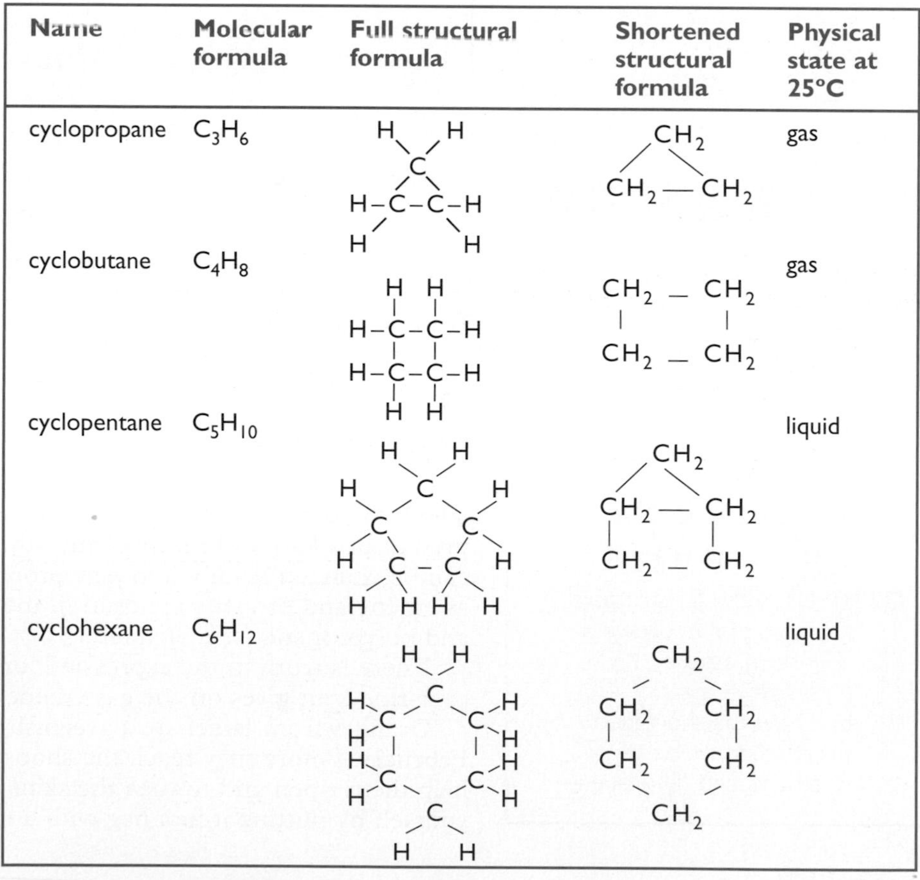

Name	Molecular formula	Full structural formula	Shortened structural formula	Physical state at 25°C
cyclopropane	C_3H_6			gas
cyclobutane	C_4H_8			gas
cyclopentane	C_5H_{10}			liquid
cyclohexane	C_6H_{12}			liquid

Table 2.2 Some cycloalkanes

All cycloalkanes burn to give carbon dioxide and water. Cyclohexane (C_6H_{12}) is used to manufacture nylon and other plastics.

Questions

Q4 The boiling point of alkenes increases gradually as the molecules get bigger. Name two other properties which you would expect to change gradually as you move from small to larger alkene molecules.

Q5 Work out the general formula for the cycloalkanes.

Q6 What evidence is there in the section on cycloalkanes to suggest that they form a homologous series like the alkanes?

SECTION 6.3 Isomers

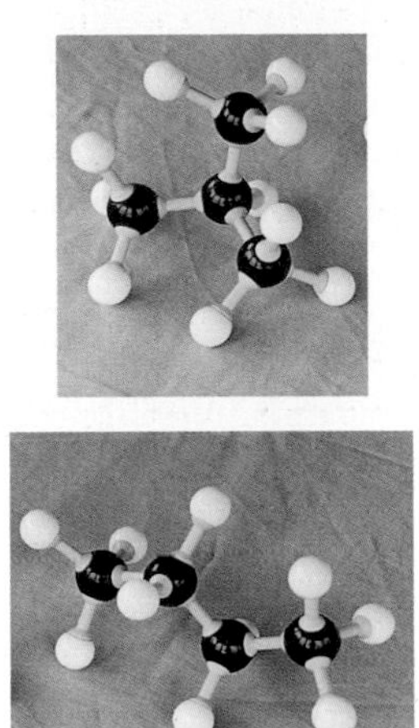

Figure 3.2 Molecular models of the two isomers of butane

There are two forms of butane. Both have the same molecular formula, C_4H_{10}. Both will burn in the same way but they both have slightly different boiling points. This puzzled chemists in the past, until they discovered that the two butanes had different structures. The two molecules are called **isomers** of butane. They are shown in figure 3.1.

Isomers are defined as molecules which have the same molecular formula but different structural formulae. In other words, isomers contain the same numbers of each kind of atom, but they are arranged differently.

There are no isomers of methane, ethane or propane, but after that the number of possible isomers increases rapidly. Pentane has three isomers (figure 3.3), hexane five and octane eighteen.

In certain cases, such as in drugs, identifying the correct isomer can be extremely important. The drug thalidomide, which was given to pregnant women in the 1960s to prevent morning sickness, produced severe defects in the babies which were later born. However, it was later discovered that an isomer of the drug would have been perfectly safe and would have produced no side-effects.

```
                     H H H
   H H H H         H-C-C-C-H
   | | | |           |   |
 H-C-C-C-C-H         H | H
   | | | |           H-C-H
   H H H H             |
                       H
```

Figure 3.1 Two isomers of butane

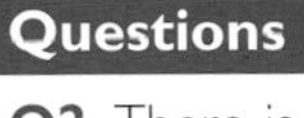

Questions

Q1 Draw full structural formulae for three of the isomers of hexane (C_6H_{14}).

Questions

Q2 There is one alkene which does not have a cycloalkane as an isomer. Which alkene is it?

```
   H H H H H
   | | | | |
 H-C-C-C-C-C-H
   | | | | |
   H H H H H
```

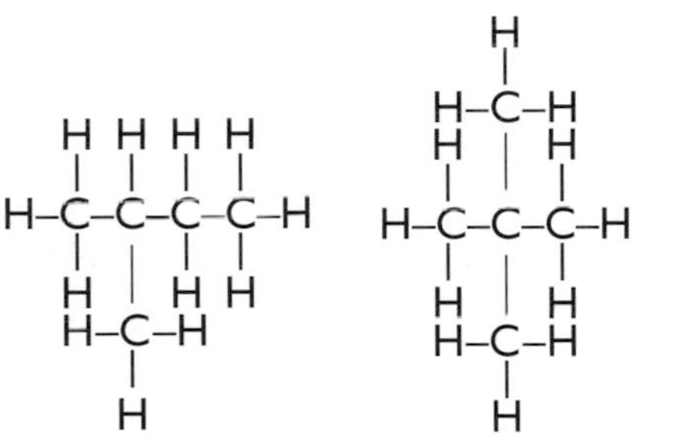

Figure 3.3 Three isomers of pentane

Alkenes also have isomers. Figure 3.4 shows two isomers of hexene. You have seen that both alkenes and cycloalkanes have the same general formula of C_nH_{2n}. This means that for each cycloalkane there is an alkene with the same molecular formula but with a different structural formula. In other words, each cycloalkane is an isomer of an alkene. Two such isomers are shown in figure 3.5. Note that only the alkene has a C=C double bond.

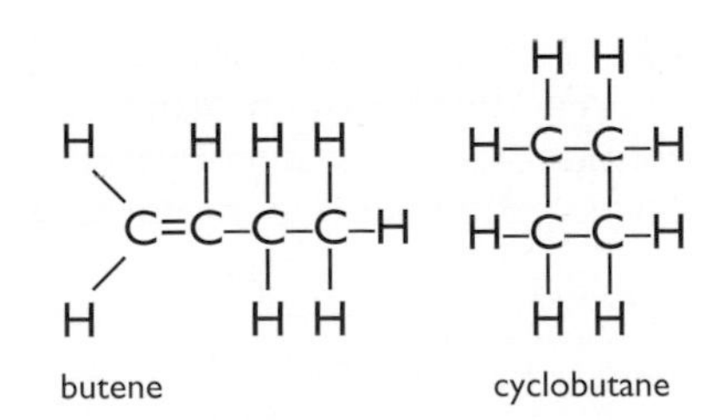

Figure 3.5 Butene and cyclobutane are isomers

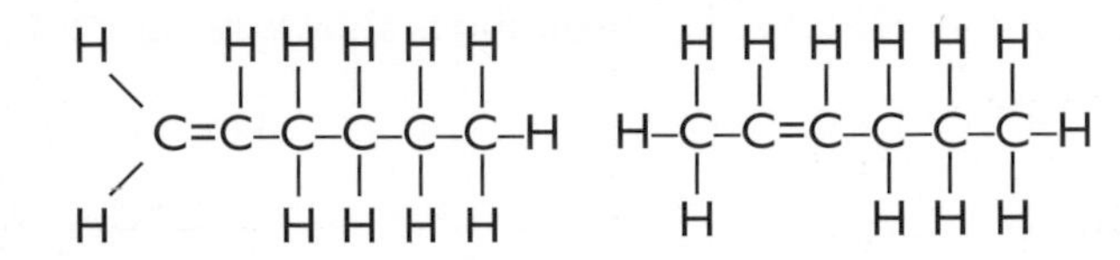

Figure 3.4 Two isomers of hexene

Saturation and unsaturation

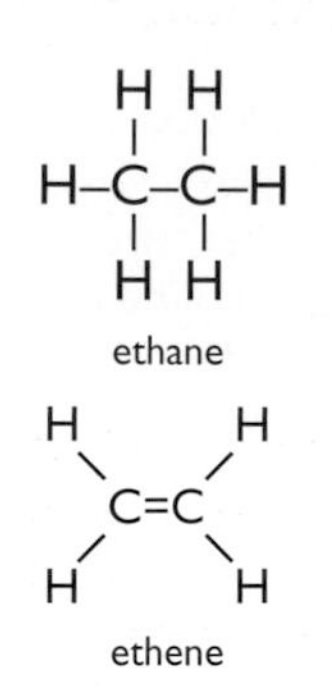

Figure 4.1 Structures of ethane and ethene

What are the differences between alkanes and alkenes? Figure 4.1 shows that they have different *structures*.

Also, the C=C bond in alkenes allows them to react in ways which alkanes cannot. The C=C bond is a reactive bond and will break easily. This can be shown in the reaction with hydrogen, where two hydrogen atoms 'add on' to the C=C bond in an alkene (figure 4.2).

The type of reaction shown in figure 4.2 is called an **addition reaction.** This is one in which molecules join together to produce a larger molecule. No other product is formed.

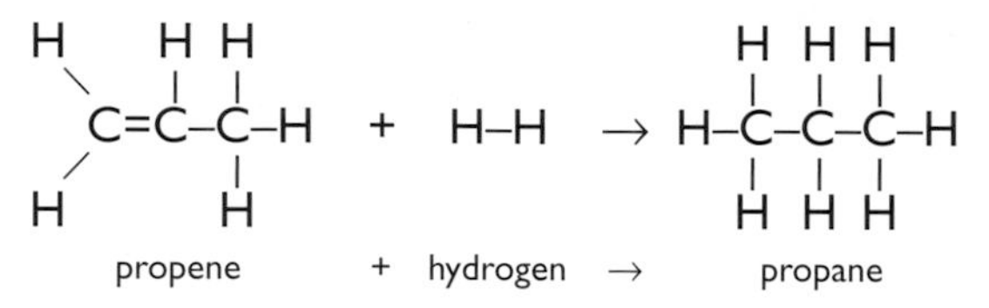

Figure 4.2 Addition reaction between propene and hydrogen

Notice that the alkene propene has turned into an alkane. For the reaction to work, the gases propene and hydrogen have to be heated with a catalyst. A similar reaction cannot take place between an alkane and hydrogen because the alkane molecules are 'full up' or **saturated** with hydrogen atoms – no more hydrogen atoms can be 'added on'. Alkanes are said to be **saturated molecules** because they cannot add on any extra hydrogen atoms. Alkenes, however, *can* add on extra hydrogen atoms. They are therefore said to be **unsaturated molecules.**

In the above reaction:

(unsaturated) alkene + hydrogen → (saturated) alkane

There are other types of unsaturated hydrocarbons; the alkenes are only one sub-set of this group of compounds.

Using bromine to test for unsaturation

Suppose you are given two liquids. One is a saturated alkane and the other an unsaturated alkene. How could you tell them apart? The experiment with hydrogen and a catalyst is difficult to set up. There is, however, a simpler way, using bromine. Bromine molecules, like hydrogen, can add on to a double bond and the reaction does not need heat or a catalyst. In addition, there is a change in colour when the reaction takes place.

Usually, a solution of bromine in water is used for the test. Bromine solution is an orange colour. The product of the reaction between bromine and an alkene is colourless. We say that the bromine has been 'decolorised' when it reacts with an alkene. The reaction between the alkene pentene and bromine is shown in figure 4.3.

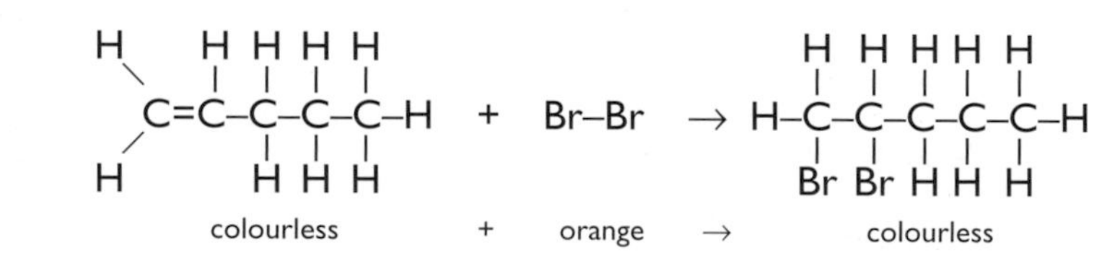

Figure 4.3 Reaction between pentene and bromine – the bromine test

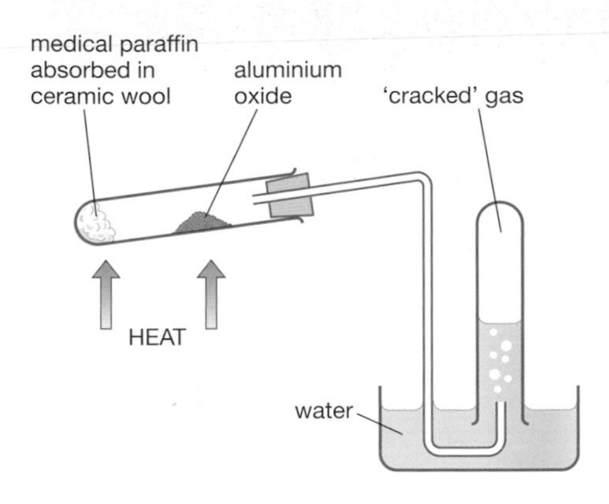

Figure 4.4 Laboratory cracking experiment

Cycloalkanes

Like the alkanes, the cycloalkanes do not possess C=C bonds. Cycloalkanes are therefore saturated hydrocarbons, so they do not decolorise bromine water.

Cracking

Breaking down long-chain alkanes into smaller molecules is called **cracking**. In oil refineries, the fractional distillation of oil generally produces more long-chain alkanes than are needed by industry. Cracking turns them into the smaller molecules that are in greater demand. For example octane, the alkane with eight carbon atoms, is used in making petrol. Octane is produced by cracking long alkane molecules.

Cracking also gives another useful type of product – alkenes. When a long-chain alkane is cracked it produces a mixture of smaller alkanes and alkenes. The alkenes are often used in the plastics industry. Crude oil is therefore a source for both alkanes and alkenes.

Cracking can be carried out in the laboratory, as shown in figure 4.4.

Why does cracking produce a mixture of alkanes and alkenes? Consider, for example, the cracking of the alkane decane, $C_{10}H_{22}$ to give two products. If one of the products is the alkane hexane (C_6H_{14}) then that has accounted for 6 of the 10 carbon atoms and 14 of the 22 hydrogen atoms. This means that the other product must have the molecular formula C_4H_8. This product is butene:

$$\underset{\text{decane}}{C_{10}H_{22}} \rightarrow \underset{\text{hexane}}{C_6H_{14}} + \underset{\text{butene}}{C_4H_8}$$

When an alkane breaks there are not enough hydrogen atoms to produce two alkanes and so one of the products is an unsaturated alkene.

In practice, when an alkane is cracked the carbon chain may break at different points on different molecules. This means that usually a mixture of products is obtained.

Industrially, the use of the catalyst aluminium oxide enables the reaction to take place at a lower temperature, so saving energy and making the process more economical.

Questions

Q1 It is possible that the product with the formula C_4H_8 obtained by cracking decane could be butene or cyclobutane. What simple test could you carry out to show that the product was an alkene?

Q2 When a molecule of nonane (C_9H_{20}) was cracked, two products were obtained. One was propane. Give the name and molecular formula for the other product.

Q3 Which alkane would produce ethane and butene on cracking, assuming that these are the only products of the reaction?

Figure 4.5 Cracking in industry. A catalytic cracking plant

Chapter 6 Study Questions

1 Some hydrocarbons have been suggested for use in ozone-friendly aerosols.

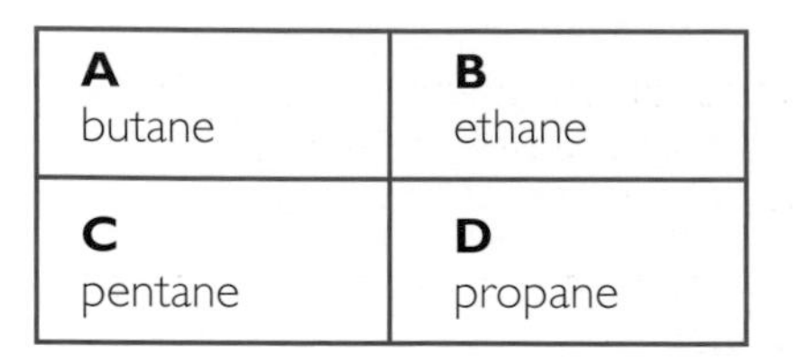

A butane	**B** ethane
C pentane	**D** propane

a) Identify the hydrocarbon with two carbon atoms in each molecule.

b) Identify the hydrocarbon(s) with a boiling point greater than 10°C. You may wish to use the SQA Data Booklet.

SEB GENERAL (PS)

2 Naphtha can be cracked to give a number of useful products. A typical result is shown in the table.

Product	Methane	Ethene	Propene	Petrol	Other products
% by mass	15	25	16	28	15

a) Draw a bar graph to show this result.

b) Name the family of hydrocarbons to which ethene and propene belong.

c) Propene reacts with hydrogen.

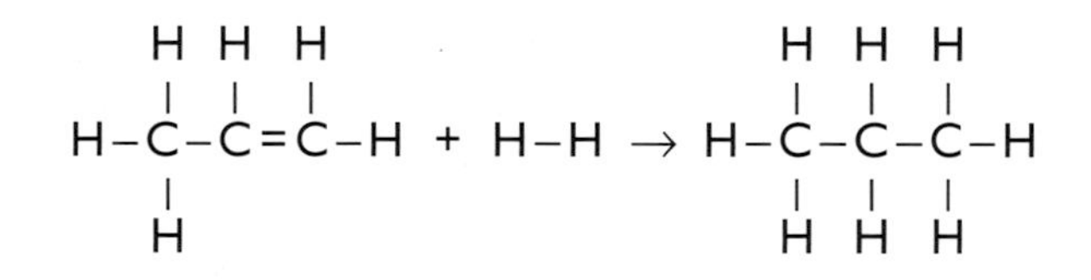

Name the type of reaction taking place.

d) One of the other products is a compound called butadiene. It decolorises bromine solution quickly. What does this tell you about butadiene?

SQA GENERAL (PS)

3 When methane is passed over a platinum catalyst, other alkanes and a cycloalkane are produced.

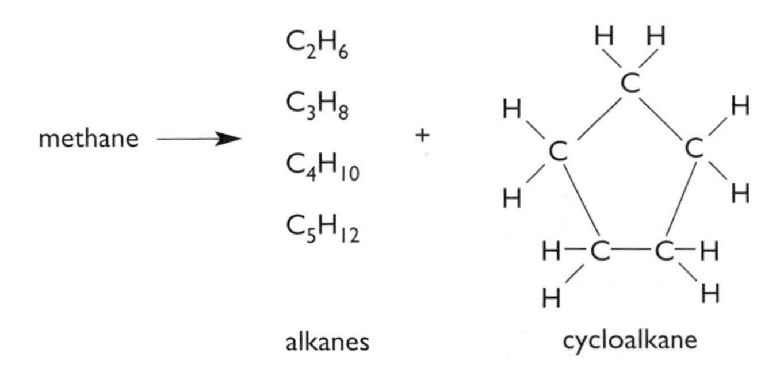

a) State the general formula of the alkanes.

b) Name the cycloalkane produced in the above reaction.

c) Draw the full structural formula of an isomer of this cycloalkane.

SQA CREDIT (PS)

4 There are various sub-sets of hydrocarbons. These include alkanes, alkenes and cycloalkanes. Here are the structural formulae for some hydrocarbons:

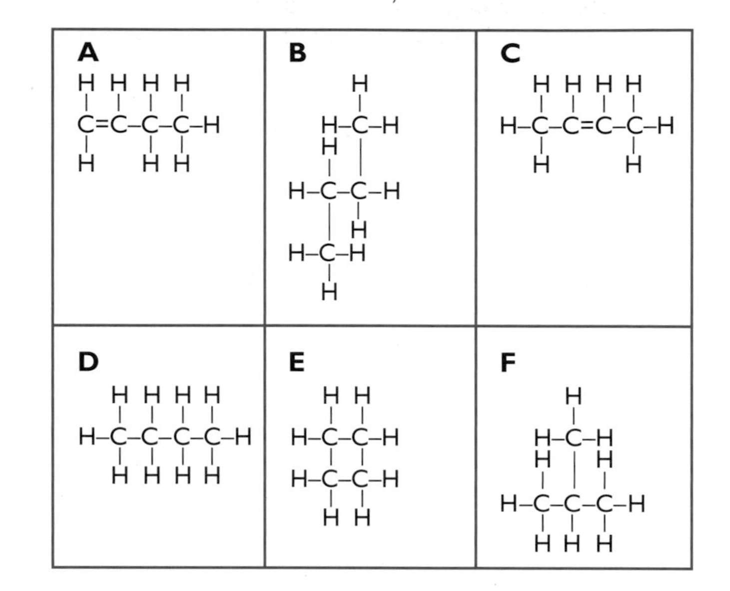

Which hydrocarbon (or hydrocarbons):

a) is an isomer of that shown in box A

b) is identical to that shown in box B

c) would quickly produce

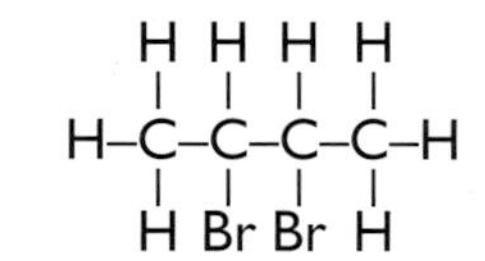

on reaction with bromine?

SEB CREDIT (KU)

5 As part of an investigation into hydrocarbons, Kirsten was given samples of two liquids in bottles labelled as follows:

'heptane – a saturated hydrocarbon'

'heptene – an unsaturated hydrocarbon'

a) Explain what is meant by the word 'saturated' when used in this context.

b) Which bond is present in molecules of heptene but not in those of heptane?

c) If the bottles had *not* been labelled, what chemical test could Kirsten have carried out in order to decide which bottle contained heptane and which one contained heptene? Don't forget to give the result of the test.

SEB GENERAL (KU)

6 The hydrocarbons butene and cyclobutane are isomers. They are members of different homologous series. Because of this they have some reactions in common (because they are both hydrocarbons), but they can also react differently (since they belong to different homologous series).

a) Explain the meaning of the following terms:
   (i) isomer
   (ii) homologous series.
b) Give an example of a reaction which both butene and cyclobutane would undergo.
c) Draw a full structural formula for cyclobutane.

CREDIT (KU)

7 The molecular formulae for the first four alkenes are as follows:

$$C_2H_4 \quad C_3H_6 \quad C_4H_8 \quad C_5H_{10}$$

a) What is the general formula for the alkenes?
b) Use the general formula to work out the molecular formulae for the alkenes which have:
   (i) 16 carbon atoms in each molecule
   (ii) 36 hydrogen atoms in each molecule. CREDIT (KU)

8 The dienes are a homologous series of hydrocarbons which possess two carbon-to-carbon double covalent bonds within each molecule. The first three members of the series have the following formulae:

$$C_4H_6 \quad C_5H_8 \quad C_6H_{10}$$

a) Give the formula of the next member of the series after C_6H_{10}.
b) What is the general formula for this homologous series?

CREDIT (PS)

9

A	B	C
C_5H_{12}	H H | | H–C–C–H | | H H	$CH_3CH_2CH_2CH_3$
D	**E**	**F**
H | H–C–H | H	C_8H_{18}	$CH_3CH_2CH_3$

Identify:

a) octane
b) propane
c) ethane.

GENERAL (KU)

10 The general formula for the alkenes is C_nH_{2n}. The first member of the alkenes is ethene, which has the molecular formula C_2H_4.

Use the general formula to work out molecular formulae for the alkenes which have:

a) 8 carbon atoms in each molecule
b) 10 hydrogen atoms in each molecule.

GENERAL (KU)

11 The following reaction took place during a cracking process:

$$C_5H_{12} \rightarrow C_3H_8 + Y$$

Assume that this represents a balanced equation and that there are no other products.

a) Name compound Y and give its full structural formula.
b) Explain whether Y would decolorise bromine solution or not.
c) Why would it be an advantage to use a catalyst during the cracking process?

CREDIT (PS)

12 During catalytic cracking processes, the surface of the catalyst becomes coated with a substance shown as X in the typical balanced equation below:

$$C_{15}H_{32} \rightarrow C_2H_4 + C_3H_8 + C_9H_{20} + X$$

a) Name substance X.
b) Suggest a way of removing X from the surface of the catalyst (bearing in mind that the catalyst is a relatively unreactive, finely divided powder).

CREDIT (PS)

13 During a cracking process, ethane was found to produce only ethene and a substance Y which, on combustion, produced water but no carbon dioxide.

a) Name substance Y.
b) Give an equation for this cracking process, using full structural formulae for all substances.

CREDIT (PS)

CHAPTER SEVEN

Properties of Substances

SECTION 7.1 Ionic bonding

Figure 1.1 Sodium chloride (common salt) is an ionic compound

This chapter looks at the **properties** of substances – what they are like and how they behave. You will see that most properties depend on the type of bonding which is present in a substance. This first section introduces a new type of chemical bond – the **ionic bond**.

There are two main types of chemical bond. In Chapter 4 you saw that some atoms join by sharing pairs of electrons to form covalent bonds. In most cases this gives each atom the electron arrangement of a noble gas. This is a very stable arrangement. For example, a chlorine atom with an electron arrangement of 2,8,7 will form a covalent bond to obtain an electron arrangement of 2,8,8, the same as that of the noble gas argon.

In ionic bonding, atoms also obtain the electron arrangement of a noble gas, but they do this by losing or by gaining electrons rather than by sharing them. Sodium chloride (common salt) contains ionic bonds. Sodium atoms have an electron arrangement of 2,8,1. The nearest noble gas arrangement to this is 2,8. Sodium atoms obtain this arrangement by losing one outer electron. For a chlorine atom to form an ionic bond with a sodium atom, it needs to *gain* one electron, changing the electron arrangement from 2,8,7 to 2,8,8.

Atoms are electrically neutral. However, electrons are negatively charged. So if an atom gains an electron it will also gain a negative charge. Charged particles such as this are called **ions**. Atoms which gain electrons become negatively charged ions. Atoms which lose electrons become positively charged ions. A positive sodium ion can be written as Na^+ and a negative chloride ion as Cl^-. The formation of sodium chloride can be summed up as follows:

sodium atom	+	chlorine atom	→	sodium ion	+	chloride ion
Na	+	Cl	→	Na^+	+	Cl^-
2,8,1		2,8,7		2,8		2,8,8

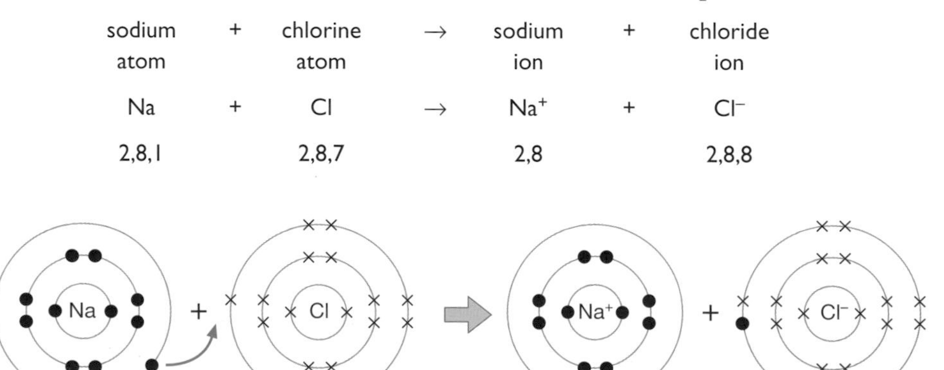

Figure 1.2 Simplified picture showing electron arrangements in the formation of sodium chloride

A magnesium atom has an electron arrangement of 2,8,2. It needs to lose two electrons to obtain the noble gas electron arrangement of 2,8. This produces a doubly charged positive ion, written as Mg^{2+}. An oxygen atom with an electron arrangement of 2,6 will need to gain two electrons to form a doubly charged negative ion, written as O^{2-}.

The charge which an ion will have depends on which group it belongs to in the periodic table. For example, all atoms in Group 1 can obtain a noble

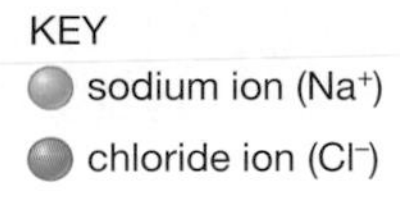

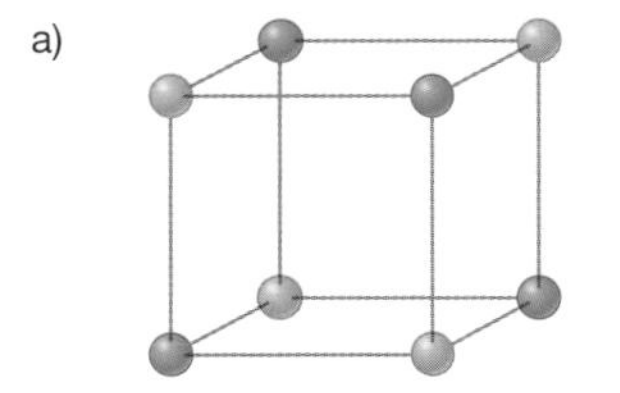

Figure 1.3 a) The basic unit in the ionic lattice of sodium chloride, b) each sodium ion is surrounded by six chloride ions, c) each chloride ion is surrounded by six sodium ions, d) a larger section of the sodium chloride lattice

gas electron arrangement by losing one outer electron. This means that they all form singly charged positive ions such as Na^+ and K^+. Table 1.1 shows the charges on the ions for Groups 1 to 7 in the periodic table. Note that the elements in Group 4 do not usually form ions. This is because they would have to lose or gain four electrons to obtain a noble gas electron arrangement. This is much more difficult than losing or gaining one, two or three electrons.

The formation of ions can be shown by equations, where $\mathbf{e^-}$ is the symbol for an electron. The equation for the formation of a sodium ion is:

$$Na \rightarrow Na^+ + e^-$$

and for a barium ion

$$Ba \rightarrow Ba^{2+} + 2e^-$$

Chlorine is a diatomic element; its formula is Cl_2. The equation for the formation of chloride ions shows that two Cl^- ions are produced from one diatomic molecule:

$$Cl_2 + 2e^- \rightarrow 2Cl^-$$

Group number	1	2	3	4	5	6	7
Charge on ion	1+	2+	3+	—	3–	2–	1–

Table 1.1

Questions

Q1 The magnesium ion is written as Mg^{2+}. Give the formulae for the ions formed by:
a) calcium,
b) sulphur,
c) aluminium,
d) bromine.

Ionic lattices

So far we have looked at individual ions. However, a minute amount of an ionic compound will contain millions of ions. Ionic bonds consist of forces of attraction between positive and negative ions. These forces are so great that the ions cluster together. The positive ions surround the negative ions and the negative ions surround the positive ions. This produces a regular-ordered structure called an **ionic lattice**.

Sodium chloride, for example, is one of the many ionic compounds that form cubic crystals. In the sodium chloride lattice, each chloride ion is surrounded by six sodium ions and each sodium ion is surrounded by six chloride ions.

Questions

Q2 What is the shape of a sodium chloride crystal and the basic unit in its ionic lattice?

Comparing covalent and ionic substances

Compound	Melting point (mp)/°C
barium chloride	963
lithium bromide	547
potassium iodide	686
sodium chloride	801

Table 2.1 Melting points of some ionic compounds

Physical states

All ionic compounds are solids at room temperature – they have high melting and boiling points (see table 2.1). A high melting point means that it takes a lot of heat energy to break the bonds which hold the ionic lattice together.

Substances which contain covalent bonds show a wide range of melting and boiling points, from very low to very high. For example, the elements chlorine, bromine and iodine all contain covalent bonds, but at room temperature chlorine is a gas, bromine is a liquid and iodine is a solid. Table 2.2 gives their melting and boiling points, together with those of some compounds. As a general rule, if a compound is a gas or a liquid at room temperature then it contains covalent bonds.

Figure 2.1 Tincture of iodine is a solution of iodine in ethanol. Several non-aqueous solvents can be used to make nail polish and its remover

Name	mp/°C	bp/°C	State at 25°C
Some covalently bonded elements:			
chlorine	−101	−35	gas
bromine	−7	59	liquid
iodine	114	184	solid
Some covalently bonded compounds:			
sulphur dioxide	−73	−10	gas
water	0	100	liquid
silicon dioxide	1880	2500	solid

Table 2.2

Solubility

Most covalent substances do not dissolve easily in water. They are more likely to dissolve in non-aqueous solvents, that is solvents other than water. For example, candle wax is insoluble in water but will dissolve in petrol. Nail polish will not dissolve in water. However, it does dissolve in the solvents used in nail polish remover.

Ionic compounds show a range of solubilities in water. Some, like sodium chloride, are very soluble whereas others, such as calcium carbonate, are highly insoluble. The table on page 5 in the SQA Data Booklet gives the solubilities of selected compounds.

When an ionic compound dissolves in water the lattice breaks down. Water molecules are able to separate the ions and to prevent them from joining together again.

Questions

Q1 Look at the label on a bottle of nail polish remover. What are the names of the solvents used?

Q2 Refer to page 5 in the SQA Data Booklet and answer the following questions:
a) Name an insoluble metal chloride.
b) What statement can you make about the solubility of metal nitrates?

Conduction of electricity

One important difference between covalent and ionic substances is in their abilities to conduct electricity. Before looking at this in detail, here are some key points concerning conduction.

Conduction of electricity by elements

To find out if an element conducts electricity we can put it into an electric circuit like the one in figure 2.2.

If the bulb lights up there must be a current flowing. We then know that the element is a conductor. Table 2.3 shows some results from this kind of test.

Questions

Q3 Refer to the SQA Data Booklet and give the names of ten elements which are electrical conductors and five which are non-conductors.

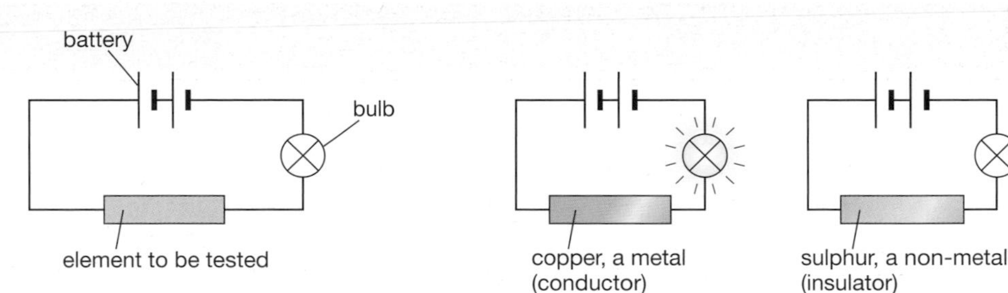

Figure 2.2 Testing an element for electrical conduction

Conductors	Non-conductors
copper	sulphur
iron	phosphorus
aluminium	iodine
silver	
carbon (graphite)	

Table 2.3 Conductors and non-conductors

Notice that all the conductors (except for graphite) are metals. In fact, it has been found that *all* metal elements are conductors of electricity. Most non-metal elements, like those in the right-hand column of table 2.3, are non-conductors of electricity.

Note that carbon is an unusual element in that it can exist in different forms. Graphite is the only form of carbon which conducts.

When an electric current flows through a substance, it is due to the movement of electrically charged particles. When metals and graphite conduct electricity, it is electrons that move through them.

Conduction of electricity by compounds

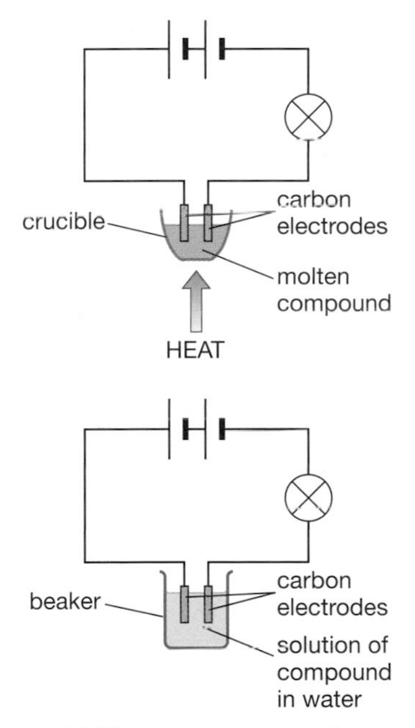

Figure 2.3 Testing a compound for electrical conduction

The circuit shown in figure 2.3 can be used to find out if compounds conduct electricity. With the exception of graphite, covalent substances (elements or compounds) do not conduct. This is because their electrons are not free to move. Solid ionic compounds also do not conduct electricity. This may be surprising since ionic compounds do contain charged particles – ions. However, there has to be a *flow* of charged particles for conduction to occur. Remember that the ions are all held together in the ionic lattice. This means they are not free to move and so cannot conduct electricity. To make an ionic compound conduct, the ions have to be separated and be free to move. This can be done in two ways:

- Aqueous solutions of ionic compounds conduct electricity. As mentioned earlier in this section, when an ionic compound dissolves in water, the lattice breaks down. The ions are then freed.
- Molten ionic compounds also conduct electricity. Therefore a second way to free the ions in an ionic compound is to melt it. The molten compound, which is called a **melt**, will conduct because the ions are free to move.

The results of some conduction experiments are shown in table 2.4.

Compound	Formula	Conduction of electricity		
		Solid	Molten	Dissolved in water
sodium iodide	NaI	no	yes	yes
glucose	$C_6H_{12}O_6$	no	no	no
silicon dioxide	SiO_2	no	no	insoluble

Table 2.4 Results of some conduction experiments

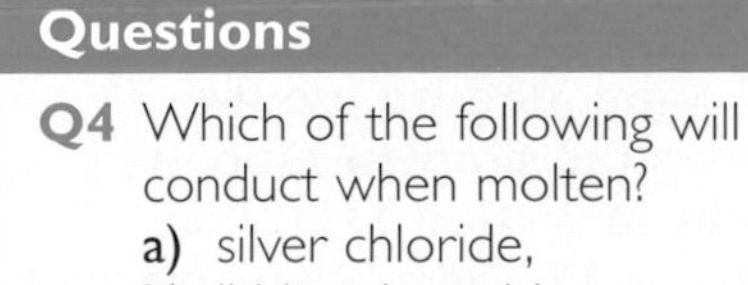
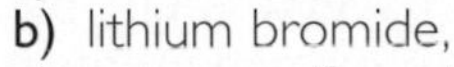

Questions

Q4 Which of the following will conduct when molten?
a) silver chloride,
b) lithium bromide,
c) sucrose – formula $C_{12}H_{22}O_{11}$.

As a general rule, if a compound contains a metal and a non-metal in its formula, for example copper chloride, then it is most likely to be an ionic compound. This means it will conduct electricity when molten and when dissolved in water.

Covalent networks

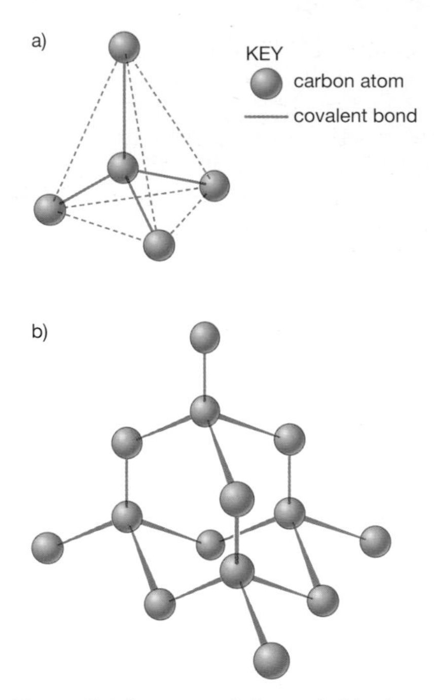

Figure 3.1 Structure of diamond a) basic tetrahedral unit, b) the diamond structure

Covalent network elements

Diamonds are made of only one type of atom – carbon atoms. Diamond is the hardest naturally occurring substance and has a very high melting point. This suggests that there must be very strong forces holding the carbon atoms together.

Figure 3.1 shows part of the structure of diamond. All of the atoms are joined together by covalent bonds. The atoms are held in position rigidly, which gives diamond its hardness and high melting point.

Diamond is an example of a **covalent network element**. The network is based on a tetrahedral arrangement of carbon atoms linked by single covalent bonds. In a covalent network, millions of atoms are linked in a regular structure which tends to have very high melting and boiling points.

Questions

Q1 How many bonds does each carbon atom have inside the diamond structure?

Q2 Elements in the same group in the periodic table tend to have similar properties. Refer to the periodic table on page 1 in the SQA Data Booklet and select another non-metal which you think would also possess a covalent network structure. Draw the basic unit of five atoms to show how they are joined. Use the structure shown in figure 3.1 to help you do this.

Covalent network compounds

Some compounds exist as covalent networks. For example, silicon dioxide is a solid with a very high melting point of 1880°C. Its structure, which is similar to that of diamond, is shown in figure 3.2.

The arrangement of the atoms in silicon dioxide makes the structure very rigid. Substances that are composed of silicon dioxide, such as sand and quartz, therefore have high melting points.

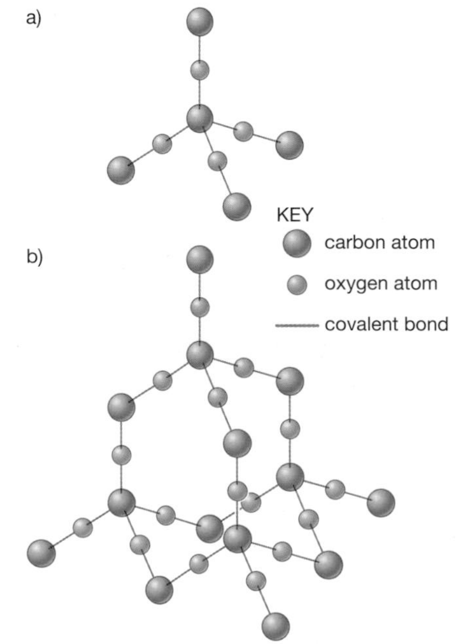

Figure 3.2 Structure of silicon dioxide, a) basic unit, b) the silicon dioxide structure

Questions

Q3 You may have noticed in DIY shops a very dark form of sandpaper called *carborundum paper*. Carborundum is the covalent network compound silicon carbide. Work out the simple formula for this compound and suggest a structure for the atoms.

To give the formula of a covalent network compound we use the **empirical formula**. The empirical formula gives the simplest ratio of atoms in a compound. It is explained more fully in section 11.4. For example, if we count up all the atoms of silicon and oxygen in any piece of silicon dioxide then we find that the ratio of silicon to oxygen atoms is 1:2. We can therefore give an empirical formula for silicon dioxide of SiO_2.

SECTION 7.4

Electrolysis

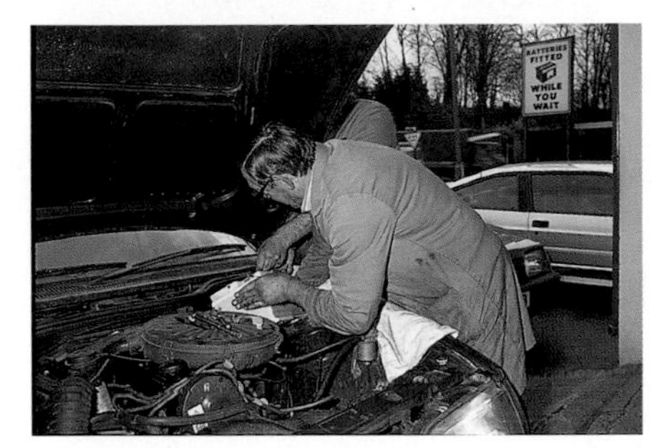

Figure 4.1 Why are car repair shops 'no smoking' areas?

Charging a car battery can be dangerous. An explosive mixture of gases is produced when the electric current passes through the battery acid. In other words, the acid is changed and new products are formed.

Many ionic substances are changed by electrical energy. When an ionic substance is dissolved in water, the lattice breaks up completely allowing the ions to move about in the solution. Figure 4.2 shows an experiment in which electrodes pass an electric current through such a solution. Notice that one electrode is positive and the other is negative. The positive metal ions in the solution are attracted to the negative electrode, and the negative non-metal ions are attracted to the positive electrode. When the ions reach the electrodes they are changed into new products.

Figure 4.3 shows a similar experiment with copper chloride solution. Copper metal is formed at the negative electrode and chlorine gas is formed at the positive electrode.

The same process occurs when electricity is passed through a molten ionic substance. As with a solution, positive ions are able to move towards the negative electrode and negative ions towards the positive electrode.

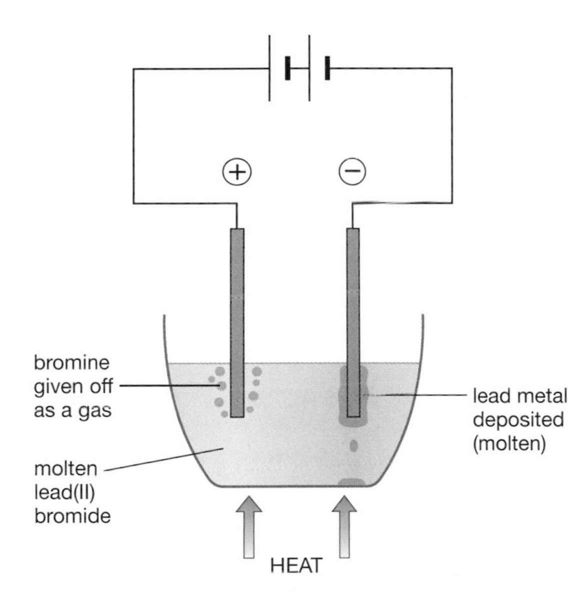

Figure 4.2 Passing a direct current through molten lead(II) bromide

Figure 4.3 Passing a direct current through copper(II) chloride solution

Questions

Q1 In the experiment in figure 4.3, why do you think copper metal forms at the negative electrode?

Compounds which conduct electricity when molten or in solution are called **electrolytes**. The process of 'breaking up' an ionic compound by means of electricity to form new substances is called **electrolysis**. Electrolysis is used for many purposes, for example all the calcium metal used in school laboratories is made from the electrolysis of calcium compounds.

In electrolysis, it is useful to think of the negative electrode as a source of electrons, supplying them to positive ions. The positive electrode, on the other hand, removes electrons from negative ions.

In the electrolysis of copper chloride solution shown in figure 4.3, the copper ions each gain two electrons from the negative electrode. This can be shown in an equation where electrons are represented by the symbol e^-.

$$Cu^{2+} + 2e^- \rightarrow Cu$$

At the positive electrode, chloride ions lose electrons and form atoms of chlorine:

$$Cl^- \rightarrow Cl + e^-$$

The chlorine atoms then join in pairs to form molecules of chlorine gas:

$$2Cl \rightarrow Cl_2$$

These two steps are usually shown as a single process:

$$2Cl^- \rightarrow Cl_2 + 2e^-$$

Direct current and electrolysis

All electrolysis experiments use direct current and not alternating current. If direct current is used then one electrode remains positive and the other negative. If an alternating current is used then the electrical sign of each electrode is constantly changing between positive and negative. This means that both products are formed at each electrode. Using direct current, the products are formed at separate electrodes, making it possible to separate and identify the products.

Questions

Q2 Why is alternating current not normally used in electrolysis experiments?

Q3 The equations for the reactions at the negative and positive electrodes when molten lead(II) bromide is electrolysed are very similar to those when copper(II) chloride solution is electrolysed. Write equations for these electrode reactions.

SECTION 7.5 Formulae for ionic compounds

Simple formulae

You probably know that the formula for sodium chloride (common salt) is NaCl. This type of formula, referred to as a **simple formula**, shows the ratio of the ions present. In other words, the formula NaCl means that in sodium chloride there is one Na^+ ion for every Cl^- ion.

Simple formulae can be worked out by considering the charges on the ions present. In the case of sodium chloride there are two ions.

The single positive charge on each sodium ion is cancelled out by the single negative charge on each chloride ion. In the case of barium chloride, the barium ion has a double positive charge, therefore two chloride ions will be required for every barium ion.

Using valencies

An alternative way of working out formulae is to use the system of valency numbers, introduced in section 4.2. For any ion, the valency is the same as the charge on the ion. For example, Mg^{2+} has a valency of 2.

Table 5.1 gives the relationship between group number in the periodic table and valency, dealt with in Chapter 4.

Group number	1	2	3	4	5	6	7	0
Valency	1	2	3	4	3	2	1	0

Table 5.1

Here is how the formula for magnesium chloride is worked out:

Step 1 symbols — Mg Cl

Step 2 valencies — 2 1

Step 3 cross over valencies — Mg_1 Cl_2

Step 4 cancel out any common factors (not required in this case)

Step 5 omit '1' if present — Mg Cl_2

The simple formula for magnesium chloride is $MgCl_2$.

Ions containing more than one kind of atom

So far we have looked at ions consisting of only one kind of atom, for example the chloride ion. However, there are also ions with two or more different kinds of atom. The sulphate ion, for example, contains sulphur and oxygen. Its formula is SO_4^{2-}. The '2–' shows that it has a double negative charge. Table 5.2 shows some other ions with more than one kind of atom, together with their charges. A fuller list is shown on page 4 in the SQA Data Booklet.

Questions

Q1 Use the valency method to work out the formula for:

a) lithium fluoride,
b) calcium iodide,
c) aluminium oxide,
d) sodium sulphide.

One positive	One negative	Two negative	Three negative
ammonium NH_4^+	hydroxide OH^-	carbonate CO_3^{2-}	phosphate PO_4^{3-}

As before, the valency method can be used to find the formulae of these ions, for example potassium sulphate. The sulphate ion has a double negative charge, therefore its valency is two.

Step 1 symbols — K SO_4
Step 2 valencies — 1 2
Step 3 cross over valencies — K_2 $(SO_4)_1$
Step 4 cancel out any common factor
Step 5 omit '1' if present — K_2 SO_4

The simple formula for potassium sulphate is K_2SO_4.

Questions

Q2 Use the valency method to work out the formula for:
- **a)** sodium carbonate,
- **b)** lithium hydroxide,
- **c)** ammonium chloride. *Note:* the ammonium ion has a single positive charge, so its valency is one.

You have seen that a formula such as $MgCl_2$ means there are two chloride ions for every magnesium ion. To show that there are two hydroxide ions in magnesium hydroxide, we have to use brackets: $Mg(OH)_2$.

Questions

Q3 Use the valency method to work out the formula for:
- **a)** magnesium nitrate,
- **b)** lead(II) hydroxide,
- **c)** ammonium sulphate.

All of these formulae require the use of brackets.

Ionic formulae

The **ionic formula** shows the charges of the ions which are present in an ionic substance. For example, the simple formula for sodium chloride is NaCl. The ionic formula is Na^+Cl^-.

There are two main rules for changing a simple formula into an ionic formula:

1 Put the charges back into the simple formula, for example lithium chloride, LiCl, becomes Li^+Cl^-.

2 When there is more than one ion of a particular type, it has to be put in brackets. For example, the simple formula for barium chloride is $BaCl_2$. The ionic formula is $Ba^{2+}(Cl^-)_2$. Note that the number 2 is placed outside the bracket. Similarly, the simple formula for ammonium sulphate is $(NH_4)_2SO_4$. The ionic formula is $(NH_4^+)_2SO_4^{2-}$.

Questions

Q4 Give the ionic formula for each of the following:
- **a)** potassium iodide,
- **b)** sodium oxide,
- **c)** ammonium phosphate.

The colour of ionic compounds

Figure 6.1 a) Copper (II) sulphate, b) nickel (II) sulphate, c) sodium chromate

Sodium chromate crystals are a delicate yellow colour. What makes them yellow? The crystals contain two types of ions: sodium ions, which have no colour, and chromate ions, which are yellow. The colour, then, is due to the chromate ions. Table 6.1 shows the colour of certain ions. From this you can see that if a compound such as sodium chloride contains only colourless ions then it will have no colour. It will be a white solid and will make colourless solutions.

In the case of copper(II) chloride, the compound contains the blue-coloured copper(II) ion and the colourless chloride ion. Its colour is therefore blue. In fact, most copper(II) compounds are blue. The only exceptions tend to be those in which the other ion has its own colour. For example, in copper(II) chromate, the blue of the copper(II) ion and the yellow of the chromate ion combine to give a compound which is green in solution.

Compound	Colour	Ion responsible for the colour
copper(II) sulphate copper(II) nitrate	blue	copper(II), Cu^{2+}
nickel(II) sulphate nickel(II) chloride	green	nickel(II), Ni^{2+}
sodium chromate potassium chromate	yellow	chromate, CrO_4^{2-}

Table 6.1 Colours of some ionic compounds

Questions

Q1 Iron(II) sulphate is green. Give the name and formula of the ion which is responsible for this colour.

Q2 Ammonium dichromate, sodium dichromate and potassium dichromate are all orange. With the help of information on page 4 in the SQA Data Booklet, give the name and formula of the ion responsible for this colour.

Ion migration experiments

As already discussed, positive ions are attracted to negative electrodes and negative ions are attracted to positive electrodes. Chemists have used this fact to construct **ion migration experiments** in which the movement of ions can be seen as different colours moving towards each electrode. Figure 6.2 shows one of these ion migration experiments.

In figure 6.2, a current is passed through a green solution of copper(II) chromate. Slowly, a blue colour moves through the colourless electrolyte towards the negative electrode. At the same time a yellow colour moves towards the positive electrode.

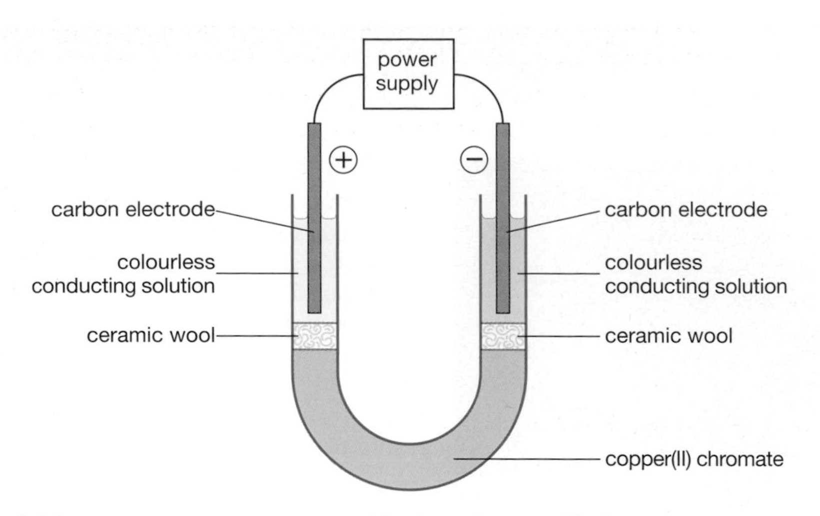

Figure 6.2 Ion migration experiment using U-tube and copper(II) chromate solution

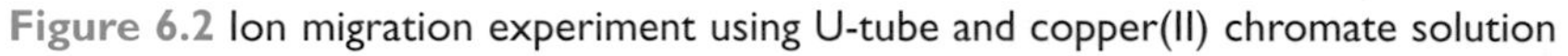

Questions

Q3 a) Explain what is responsible for the blue colour and why it moves towards the negative electrode.

b) Explain what is responsible for the yellow colour and why it moves towards the positive electrode.

c) Do you think that direct current or alternating current would be used for this experiment? Explain your answer.

A second type of ion migration experiment uses filter paper soaked in an electrolyte such as sodium chloride solution. Here, bands of colour can be seen to move slowly across the filter paper.

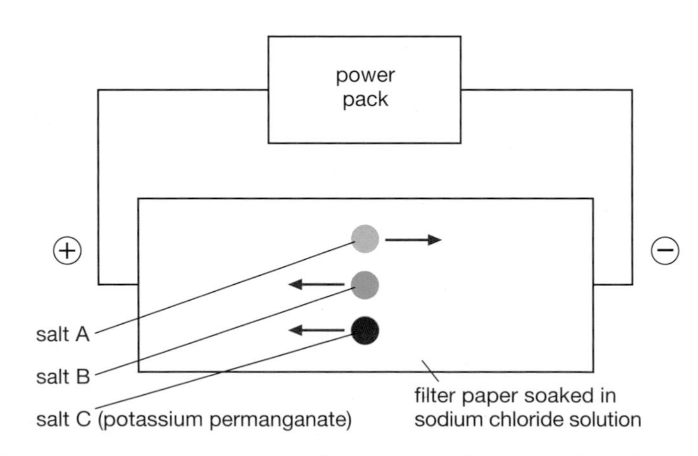

Figure 6.3 Ion migration experiment using filter paper soaked in an electrolyte

Questions

Q4 An ion migration experiment was set up as shown in figure 6.3. From the first salt (A), a green colour moved towards the negative electrode. From the second salt (B), an orange colour moved towards the positive electrode. The third salt (C) was potassium permanganate, from which a purple colour moved towards the positive electrode. Give the charges carried by each of the coloured ions present and suggest a possible name and formula for each. For salt C, refer to page 4 in the SQA Data Booklet.

Chapter 7 Study Questions

1 Copy the following table and complete it by adding 'conducts' or 'does not conduct' in the spaces.

Substance	Solid state	Liquid state
phosphorus	—	—
copper	conducts	—
calcium oxide	—	—

GENERAL (KU)

2 Name, as precisely as you can, the particles which carry an electric current through each of the following:

a) molten potassium iodide
b) mercury
c) sodium chloride solution. GENERAL (KU)

3 Explain why molten sodium chloride conducts electricity but solid sodium chloride does not. GENERAL (KU)

4 Explain why, although neither lithium chloride ($LiCl$) nor tetrachloromethane (CCl_4) conduct electricity in the solid state, only lithium chloride conducts in the liquid state. GENERAL (KU)

5 Explain why, having established that a given substance is a *compound*, a good way of finding out the type of bonding present is to discover whether the substance conducts electricity in the liquid state. GENERAL (KU)

6 Martin set up the following experiment.

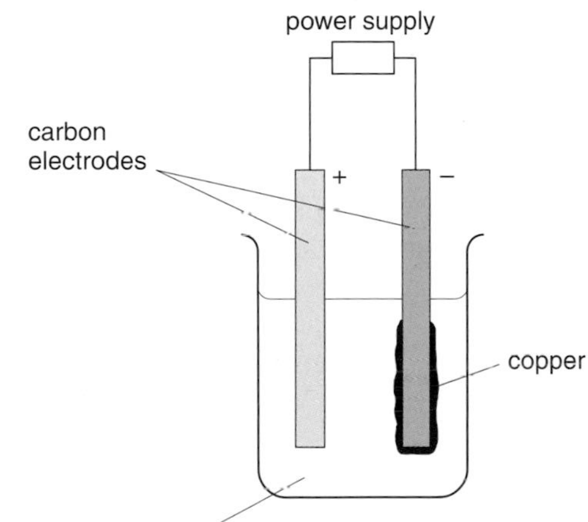

a) What type of experiment did Martin carry out?
b) Why does copper form at the negative electrode?
c) Carbon is unreactive and insoluble in water. Give another reason why it is suitable for use as electrodes.

SQA CREDIT (KU)

7 When molten potassium iodide is electrolysed, potassium forms at the negative electrode and iodine at the positive electrode.

Write ion-electron equations for the formation of:

a) potassium
b) iodine. CREDIT (KU)

8 The properties of substances are related to bonding.

Substance	Melting Point/°C	Boiling Point/°C	Ability to conduct as a solid	Ability to conduct as a liquid
A	2617	4607	yes	yes
B	782	1600	no	yes
C	81	218	no	no
D	−183	−164	no	no
E	6	80	no	no
F	2300	2550	no	no

a) Identify the substance which is a solid at room temperature (20°C) and would melt if placed in a test tube surrounded by boiling water.
b) Identify the substance with ionic bonding.
c) Identify the substance with a covalent network.

SQA CREDIT (KU)

9 Salt solution conducts electricity but glucose solution does not.

a) Draw and label a diagram of the apparatus which you would use to show that salt solution conducts electricity.
b) Why does glucose solution not conduct electricity?

SQA CREDIT (KU)

10 Each box in the grid below shows the formula of a substance.

A CH_4	**B** H_2S	**C** N_2
D O_2	**E** $CaCl_2$	**F** NH_3

Which box (or boxes) shows the formula of a substance which is an ionic compound?

SEB CREDIT (KU)

11 Mendeleev, a Russian chemist, proposed his first Periodic Table of Elements in 1869. From this our present periodic table has been developed. Here are six common elements from the periodic table:

A lithium	B chlorine	C magnesium
D potassium	E nitrogen	F oxygen

a) Atoms of which element (or elements) form ions with the same electron arrangement as argon atoms?
b) Which *two* elements will combine to form an ionic compound with the formula X_3Y_2?

SEB CREDIT (KU)

12 Ions are formed when atoms lose or gain electrons.

A $^{18}_{8}O$	B $^{23}_{11}Na$	C $^{24}_{12}Mg$
D $^{23}_{11}Na^{+}$	E $^{16}_{8}O$	F $^{16}_{8}O^{2-}$

Identify the particle(s) in the grid above with an electron arrangement of 2,8. CREDIT (KU)

13 Water in swimming pools can be purified using a chlorinating cell. Sodium chloride solution is electrolysed in the cell to produce chlorine.

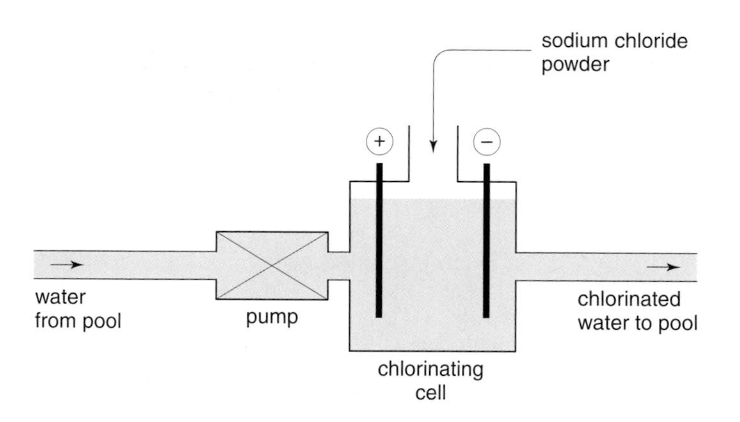

a) (i) What is meant by electrolysis?
(ii) Why can solid sodium chloride not be electrolysed?
b) Write the ion electron equation showing the formation of chlorine.
c) Chlorine reacts with water.
$Cl_2(g) + H_2O(l) \rightarrow 2H^+(aq) + OCl^-(aq) + Cl^-(aq)$
What happens to the pH of the water in the swimming pool when it is chlorinated?

SQA CREDIT (PS)

14 a) Use the SQA Data Booklet to give
(i) melting points
(ii) boiling points
for carbon dioxide and silicon dioxide. GENERAL (PS)
b) Explain why, although both of these compounds contain covalent bonds, carbon dioxide is a gas at room temperature, whereas silicon dioxide is a high melting point solid. CREDIT (KU)

15 Write chemical formulae for the following compounds:

a) aluminium chloride
b) strontium oxide
c) potassium sulphide
d) magnesium iodide
e) sodium nitrate
f) calcium sulphate
g) potassium carbonate
h) lithium phosphate. GENERAL (KU)

16 Write chemical formulae for the following compounds:

a) copper(II) hydroxide
b) silver(I) carbonate
c) chromium(III) sulphate
d) ammonium carbonate
e) aluminium nitrate
f) magnesium hydrogensulphite
g) calcium hydrogen carbonate
h) ammonium dichromate. CREDIT (KU)

17 Write ionic formulae for the following compounds:

a) magnesium hydroxide
b) lead(II) nitrate
c) ammonium phosphate
d) barium hydrogencarbonate
e) calcium phosphate
f) iron(III) sulphate
g) copper(II) permanganate
h) ammonium chromate. CREDIT (KU)

CHAPTER EIGHT

Acids and Alkalis

SECTION 8.1 Making acids

Figure 1.1 All these products contain acids

Every day you will come across, and even drink, several different **acids**. Fizzy drinks such as lemonade contain carbonic acid. Vinegar contains an acid. Even inside your mouth, bacteria change sugar into acids which can cause tooth decay. Of course, there are also the acids that you use in your chemistry experiments in school. In fact, there are hundreds of different acids. However, they all have certain things in common. This section looks at how acids can be made.

Questions

Q1 Make a table to classify the following compounds as either metal or non-metal oxides. *Hint:* use the periodic table on page 167 to help you decide:
a) barium oxide, **b)** nitrogen dioxide, **c)** lithium oxide, **d)** calcium oxide.

Q2 All of the following oxides will dissolve in water. Which one will give an acidic solution when it dissolves?
a) potassium oxide,
b) phosphorus oxide,
c) barium oxide,
d) sodium oxide.

Non-metal oxides and acidity

Many non-metals burn in oxygen to form non-metal oxides. For example, carbon burns to give carbon dioxide:

$$C + O_2 \rightarrow CO_2$$

When non-metal oxides dissolve in water they form acidic solutions. Carbon dioxide dissolves to produce a solution of carbonic acid:

CO_2	+	H_2O	→	H_2CO_3
carbon dioxide	+	water	→	carbonic acid

A fountain experiment

This is an interesting way of showing just how soluble some gases are in water. For example, in the apparatus shown in figure 1.2, a little water is squirted into a flask which contains sulphur dioxide gas. Quite a lot of the gas dissolves in this water and as a result the pressure in the flask drops. Atmospheric pressure then pushes water from the beaker below up into the flask. Usually this happens with such force that the in-coming water looks like a fountain. The experiment works because sulphur dioxide is very soluble in water, so the large volume of sulphur dioxide in the flask is able to dissolve in the tiny amount of water which is squirted in at the start.

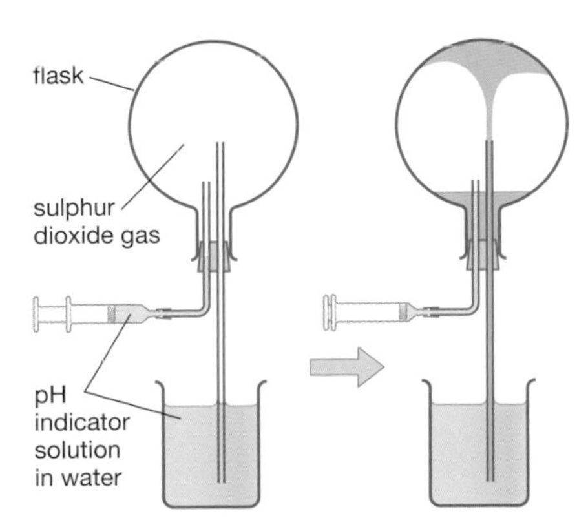

Figure 1.2 Fountain experiment

Acid rain

Recently, people have become worried about acid rain. Rain water with unusually high acidity has been falling for some years in many parts of the world. Table 1.1 shows some of the problems caused by acid rain. More information on this is given in section 8.4.

How does acid get into the rain? The main substance responsible is the gas sulphur dioxide. It is produced mostly by burning coal. All coal contains a little sulphur as an impurity. When it burns the sulphur combines with oxygen to produce sulphur dioxide:

Acid rain:
◆ kills fish by poisoning lakes and rivers ◆ makes fields too acidic for crops to grow in them ◆ damages trees ◆ wears away buildings and statues made of limestone and marble attacks the steel in car bodies, bridges, etc.

Table 1.1 Damage caused by acid rain

Questions

Q3 In most areas, rain water is more acidic than tap water. Plan and carry out an experiment using two iron nails to find out if rain water causes more rusting than tap water. How would you make your experiment fair?

$$\text{sulphur} + \text{oxygen} \rightarrow \text{sulphur dioxide}$$
$$S + O_2 \rightarrow SO_2$$

Sulphur dioxide is a poisonous gas with a sharp, choking smell. You have just learnt that sulphur dioxide is a very soluble gas. It dissolves in rain water to produce an acid called sulphurous acid. After a while, this acid changes into sulphuric acid. The effects of this can be dramatic. On certain days in Mexico City, for example, the rain contains so much sulphuric acid that it burns holes in clothing!

Figure 1.3 Trees damaged by acid rain

Reducing acid rain

Most of the coal we burn in the UK is used for making electricity in coalfired power stations. New ways are being introduced to remove sulphur dioxide from the gases which come out of these power stations. Using different methods of producing electricity could also mean less sulphur dioxide being produced. For example, hydro-electric power stations do not burn fuels and therefore do not produce sulphur dioxide. Also, wind power can be used to make electricity (figure 1.4).

Cars also help to make acid rain. The nitrogen oxides that they produce in exhaust fumes dissolve in rain water to produce acids such as nitric acid (see Chapter 5).

We can also try to save energy by insulating our houses, switching off lights, etc. The less energy we use, the less sulphur dioxide will be produced by the power stations.

Figure 1.4 Wind farms can generate electricity without producing acid rain

Figure 1.5 How can we cut down on the amount of sulphur dioxide produced by power stations?

The pH scale, acids and alkalis

It is impossible to tell simply by looking at an acid whether it is strong enough to be very corrosive or if it is a weak acid such as citric acid. Chemists measure the difference between weak and strong acids by using a scale of numbers called the **pH scale** which runs from about 1 to 14.

How to measure pH

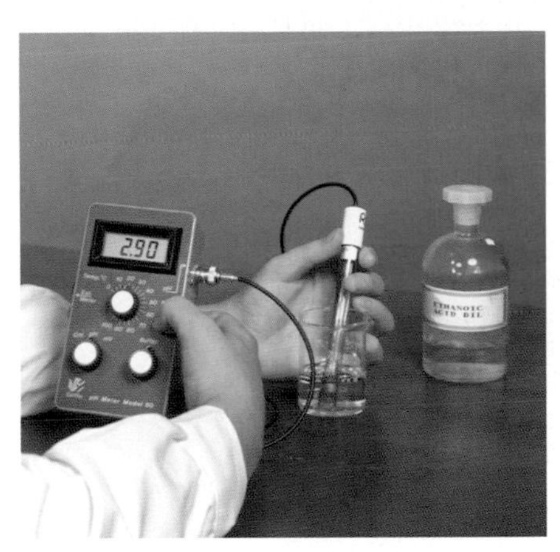

Figure 2.1 Measuring pH with a pH meter

There are two main ways of measuring pH. You will probably have used pH paper (see figure 2.1). If this paper is dipped in the solution to be tested then it will usually change colour. To find the pH of the solution you match the colour against a chart. For example, a deep red colour means it is pH 1 (see figure 2.2).

The other main method uses a pH meter. A pH meter measures pH more accurately than pH paper. You simply dip the probe of the meter into the solution to be tested and the meter gives a pH number, usually to one decimal place.

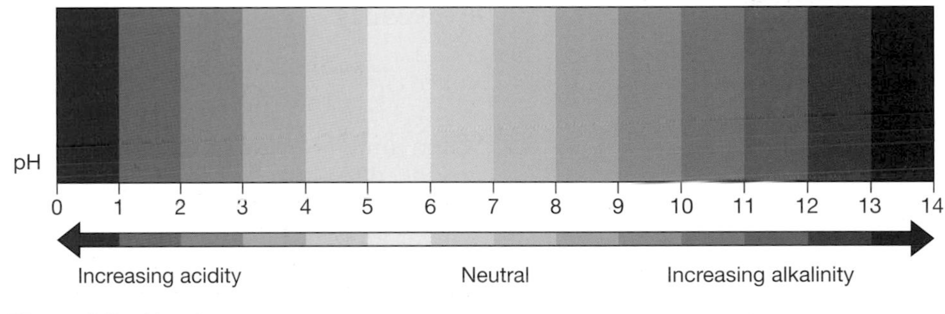

Figure 2.2 pH scale

What the pH numbers mean

Any substance with a pH of less than 7 is an acid. The lower the number the *stronger* the acid. Car battery acid has a pH of less than 1, while vinegar has a pH of around 3.

Alkalis are a group of compounds which dissolve in water to give solutions with pH values greater than 7. A strong alkali has a pH of around 14. As with acids, there are strong and weak alkalis. A weak alkali might have a pH of around 9.

The hydrogen ion and acid solutions

All acid solutions are ionic: they all contain the hydrogen ion. Table 2.1 shows the ions which are present in the three dilute acids most commonly used in school laboratories.

Acid solution	Formula	Ions present
dilute hydrochloric acid	HCl	H^+ and Cl^-
dilute nitric acid	HNO_3	H^+ and NO_3^-
dilute sulphuric acid	H_2SO_4	H^+ and SO_4^{2-}

Table 2.1 Ions present in some common acids

Questions

Q1 Ask your teachers if you can have some pH paper to test some household substances. You could try soft drinks, fruit juices, soap, toothpaste, washing-up liquid, bicarbonate of soda, etc. If you want to test a powder, add water to it first to make a solution.

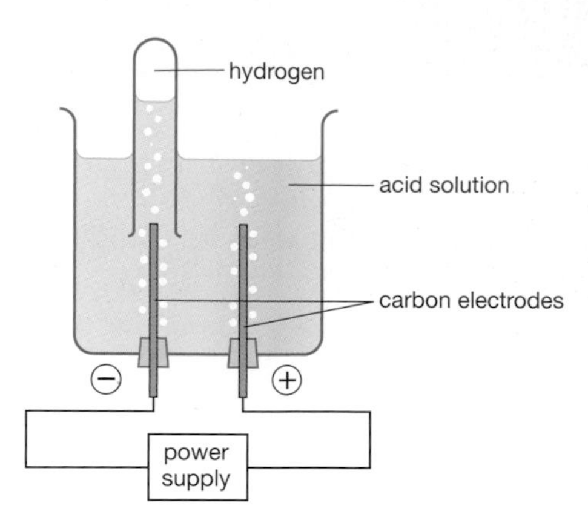

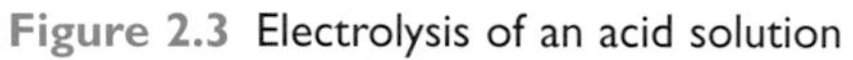

Figure 2.3 Electrolysis of an acid solution

The electrolysis of dilute acids

In Chapter 7 you saw that ionic solutions can be electrolysed. When an acid solution is electrolysed, the positively charged hydrogen ions are attracted to the negative electrode. Here they gain electrons and form molecules of hydrogen gas:

$$2H^+(aq) + 2e^- \rightarrow H_2(g)$$

How can you prove that the gas collected over the negative electrode is hydrogen? If a burning splint is placed at the mouth of the test tube, the hydrogen will burn with a 'pop'. This is the standard test for hydrogen.

All acid solutions produce hydrogen at the negative electrode when they are electrolysed.

Questions

Q2 Give the names of the negative ions present in each of the following acids:
a) hydrochloric acid,
b) nitric acid,
c) sulphuric acid.

Use table 2.1 and the table on page 4 in the SQA Data Booklet showing ions to help you answer this question.

Hydroxide ions and alkali solutions

Many metals burn in oxygen to form metal oxides. For example:

$$\text{magnesium} + \text{oxygen} \rightarrow \text{magnesium oxide}$$
$$2Mg + O_2 \rightarrow 2MgO$$

Some metal oxides will react with water to produce substances called **metal hydroxides**:

$$\text{sodium oxide} + \text{water} \rightarrow \text{sodium hydroxide}$$
$$Na_2O + H_2O \rightarrow 2NaOH$$

When a metal oxide or hydroxide dissolves in water, an alkaline solution is produced. All alkaline solutions contain hydroxide ions. Table 2.2 shows the ions present in some alkaline solutions.

Not all metal oxides and hydroxides will dissolve in water – page 5 in the SQA Data Booklet gives information on the solubilities of these and other compounds.

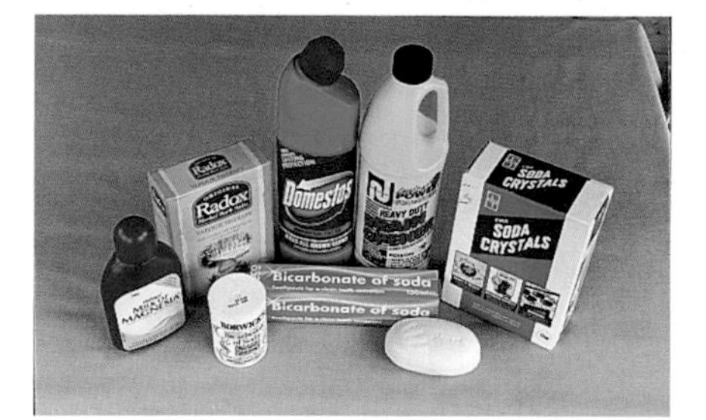

Figure 2.4 All of these contain alkalis

Alkaline solution	Formula	Ions present
sodium hydroxide	NaOH	Na^+ and OH^-
potassium hydroxide	KOH	K^+ and OH^-
calcium hydroxide	$Ca(OH)_2$	Ca^{2+} and OH^-

Table 2.2 Ions present in some common alkaline solutions

Questions

Q3 When potassium burns it combines with oxygen to form potassium oxide. This metal oxide dissolves in water, reacting with it to produce potassium hydroxide solution. Write equations for these two reactions.

Summary

Acidic solutions
all contain hydrogen ions H^+

Alkaline solutions
all contain hydroxide ions OH^-

pH and the dilution of acids and alkalis

Most fish and chip shops take deliveries of quite concentrated vinegar. This is an acidic solution that is too acidic to use with food. The vinegar is diluted by adding water to it. This makes it less acidic. It also changes the pH value from about 2.5 to 3.5. In other words, for acids, the pH value rises on dilution.

Put simply, when you add water to an acidic solution it becomes a little more like water. That is, its pH moves towards pH 7 – the pH of pure water or any neutral solution.

When alkaline solutions are diluted, the pH also moves towards 7. In other words, for alkalis, the pH *falls* on dilution.

Formula masses and moles

Have you ever wondered how the quantities of the chemicals you use in your experiments are worked out? For example, why are you asked to add 4 g of calcium carbonate to a certain acid and not 40 g or even 400 g? In fact, most of the quantities are calculated using quite simple arithmetic. This section looks at two important tools for this kind of calculation – formula masses and moles.

Questions

Q1 Calculate the formula masses for the following:
a) sodium bromide (NaBr),
b) lithium oxide (Li_2O),
c) nitric acid (HNO_3).
The top table on page 4 in the SQA Data Booklet will help you.

Formula masses

The formula mass of a compound is obtained by adding together the relative atomic masses of the elements present in the formula. For example, to calculate the formula mass of sodium carbonate:

formula Na_2CO_3
formula mass $= (2 \times 23) + 12 + (3 \times 16)$
$= 106$

Notice that the relative atomic mass for sodium (23) is multiplied by 2 and the relative atomic mass for oxygen is multiplied by 3.

Moles

You probably noticed that no units are given for formula masses. However, in order to carry out experiments you need to know which units to use. Chemists use a special term called the **mole** to change formula masses to grams.

A mole of a substance is the formula mass given in grams. A mole is often called the **gram formula mass of a substance**. For example, a mole of sodium carbonate (formula mass 106) weighs 106 g. Table 3.1 shows the masses of one mole of some other substances.

Questions

Q2 Give the mass of one mole of each of the following:
a) calcium oxide, CaO,
b) magnesium nitrate, $Mg(NO_3)_2$,
c) water, H_2O,
d) helium, He.

Substance	Formula	Formula mass	Mass of 1 mole
lithium oxide	Li_2O	$(2 \times 7) + 16 = 30$	30 g
carbon dioxide	CO_2	$12 + (2 \times 16) = 44$	44 g

Table 3.1

The formula for magnesium hydroxide is $Mg(OH)_2$. When you carry out calculations using a formula which contains brackets, you must multiply all the atoms inside the brackets by the number which is outside. In the case of $Mg(OH)_2$, the masses for both oxygen and hydrogen must be multiplied by 2. The mass of one mole of magnesium hydroxide is therefore $24.5 + [(16 + 1) \times 2] = 58.5$ g.

So far we have looked at the mass of one single mole of a substance. If a different number of moles is present the mass of the substance is found as follows:

mass = number of moles present × the mass of one mole

Example:
How many grams are present in 1.5 moles of sodium carbonate? (mass of one mole of sodium carbonate, Na_2CO_3, is 106 g)

mass = 1.5 × 106 g = 159 g

Questions

Q3 a) What is the mass of the following? (i) 5 moles of potassium oxide (K_2O), (ii) 0.5 moles of sulphur dioxide (SO_2).
b) How many moles are present in each of the following? (i) 9.8 g of sulphuric acid (H_2SO_4), (ii) 8 g of sodium hydroxide (NaOH), (iii) 16.8 g of potassium hydroxide (KOH).

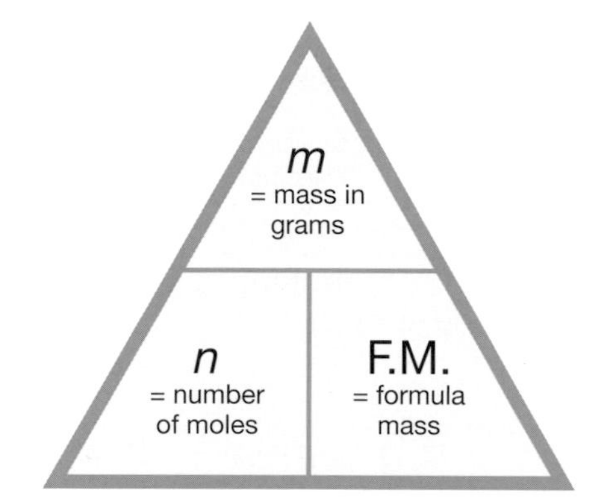

Figure 3.1 This triangle gives:

$m = n \times F.M.$ $n = \frac{m}{F.M.}$ $F.M. = \frac{m}{n}$

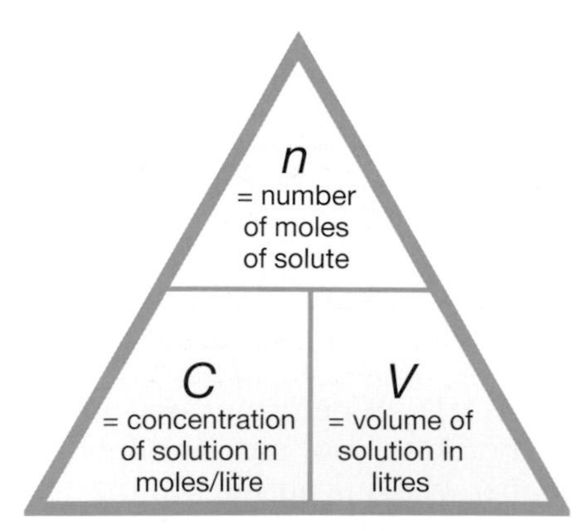

Figure 3.2 This triangle gives:

$n = C \times V$ $C = \frac{n}{V}$ $V = \frac{n}{c}$

Questions

Q4 222g of calcium chloride were dissolved in water and the volume of the solution was made up to 4 litres. Calculate the concentration of the solution in moles/litre (mol/l).

The number of moles present in a given mass of a substance is found as follows:

$$\text{number of moles} = \frac{\text{mass of substance (measured in g)}}{\text{mass of one mole of substance}}$$

Example:
Find the number of moles of hydrogen bromide present in 56 g of hydrogen bromide. (mass of 1 mole of hydrogen bromide, HBr, is 81 g)

$$\text{number of moles} = \frac{56\text{ g}}{81\text{ g}} = 0.691$$

The formula mass of a substance can be found using the following:

$$\text{formula mass of a substance} = \frac{\text{mass of substance in grams}}{\text{number of moles of substance}}$$

Example:
2 moles of a liquid weigh 36 g. Calculate the formula mass and suggest what the liquid might be.

$$\text{formula mass} = \frac{36}{2}$$
$$= 18 - \text{The liquid could possibly be water } (H_2O).$$

These relationships can be put into a triangle of knowledge (figure 3.1).

Calculations involving moles and solutions

The concentration of a solution is measured in moles/litre. This means that in a solution there is a connection between the number of moles which are dissolved, the volume of the solution in litres and the concentration of the solution. The triangle in figure 3.2 shows this relationship.

In any solution, the number of moles is found as follows:

$$n = C \times V$$

Suppose you had to make up 2 litres of sodium hydroxide solution with a concentration of 0.2 moles/litre. How many grams of sodium hydroxide would you need? First, the number of moles needed can be found as follows

$$n = 0.2 \times 2 = 0.4 \text{ moles}$$

Now that you know the number of moles needed, the mass of sodium hydroxide that has to be used can be found using the triangle in figure 3.1.

Note that for sodium hydroxide, NaOH, the formula mass is (23 + 16 + 1) = 40.

$$m = n \times F.M.$$

$$\text{mass} = \text{number of moles} \times \text{formula mass of} = 0.4 \times 40 = 16\text{ g}$$

Therefore 16g of sodium hydroxide would be required to make up the solution.

The triangle in figure 3.2 can also be used to find the concentration of solutions where the mass of solute and the volume in litres are known. For example, what is the concentration of 0.5 litres of a potassium nitrate solution which contains 20.2 g of potassium nitrate?

First find the number of moles using the triangle in figure 3.1.

formula = KNO_3

formula mass = 39 + 14 + (3 × 16) = 101

$$\text{number of moles in 20.2 g} = \frac{m}{F.M.} = \frac{20.2}{101} = 0.2$$

$$\text{concentration} = \frac{n}{V} = \frac{0.2}{0.5} = 0.4 \text{ moles/litre (abbreviated to 0.4 mol/l)}$$

SECTION 8.4 More about solutions

The triangles which we used in section 8.3 can also be used to calculate the volume of solution you can prepare. For example, suppose you had 42.5 g of sodium nitrate and you were told to make up a solution with a concentration of 0.1 moles/litre. What volume of solution could you prepare?

First you need the formula mass of sodium nitrate and the number of moles present in 42.5 g.

formula of sodium nitrate = $NaNO_3$
formula mass = 23 + 14 + (16 × 3) = 85

number of moles in 42.5g of sodium nitrate = $\frac{m}{F.M.}$

this gives $\frac{42.5}{85} = 0.5$

Next, from the other triangle you can see that the volume of solution is found as follows:

$$V = \frac{n}{C}$$

$$V = \frac{0.5}{0.1} = 5 \text{ litres}$$

Questions

Q1 What volume of solution could be produced if 10 g of sodium hydroxide are used to make up a solution with a concentration of 0.5 moles/litre?

Acids, alkalis and water

Pure water is a neutral liquid. It is made up almost entirely from covalent molecules. Water contains a very small concentration of ions. There is about one ion for every 250 million water molecules. The ions come from the water molecules themselves; a molecule of water can split up to form a hydrogen ion and a hydroxide ion:

$$H_2O \rightarrow H^+ + OH^-$$

H^+ and OH^- ions are found in acid and alkali solutions. However, water is a neutral liquid with a pH of 7. This is because water contains exactly the same numbers of H^+ and OH^- ions.

Acid solutions contain more hydrogen ions than water. Or more precisely, acid solutions have *a greater concentration* of H^+ ions than water. Alkaline solutions have a greater concentration of OH^- ions than water. This is summarised in table 4.1.

water and neutral solutions	equal concentrations of H^+ and OH^-
acid solutions	greater concentration of H^+ than water
alkaline solutions	greater concentration of OH^- than water

Table 4.1

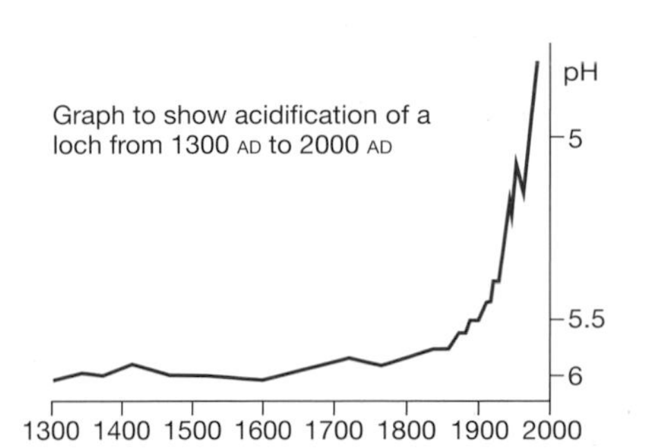

Figure 4.1 Acidification of a loch

Dilution

When an acid solution is diluted, the concentration of H^+ ions decreases and the pH moves towards 7, the pH of pure water. Similarly, the concentration of OH^- ions decreases when an alkali is diluted, and again the pH moves towards 7.

Figure 4.2 Many of the features of the Sphinx have been eroded by acid rain

Acid rain

Acid rain, lakes and fish

Many lochs and lakes in Scotland, Norway, Poland, Canada and elsewhere are now fishless. The pH in some cases is as low as 4. However, the acid does not kill the fish directly. In fact, they are poisoned by aluminium. This is because the sulphate ion in acid rain is able to release aluminium ions from the soil surrounding the lakes. When a fish is poisoned by aluminium, its gills become covered with mucus and eventually it cannot get enough oxygen to breathe.

Trees

Acid rain can harm trees in two main ways. First, acidic particles can damage leaves and pine needles making them turn yellow and drop off. This means the tree is unable to grow properly. Second, the acid can affect the soil in the way described above. It is thought that nutrients which the trees need, such as magnesium and calcium ions, are washed out of the soil, and the trees then absorb poisonous aluminium ions instead.

As a result of all these effects, trees become generally weaker and more likely to be attacked by diseases or damaged by frosts. Surveys carried out in 1986 suggest that 29 per cent of British trees have been damaged in this way. A 1992 survey suggested that around two-thirds of trees in Germany showed signs of damage from acid rain.

Questions

Q2 **a)** List three ways in which acid rain damages living things.
b) Acid rain falling in a city was analysed and found to be made up of sulphuric acid (75 per cent) and nitric acid (25 per cent). Present this information in a suitable way.

Buildings

Acid rain will react with iron and with stone, especially limestone and marble. In Britain, many old stone buildings were sand-blasted in the 1980s to remove soot and grime. However, the stone in some of these buildings is now covered with spots and marks where acid pollution has attacked them. The dirt and soot had been protecting them from the effects of acid rain. The Scott monument in Edinburgh was covered with plastic sheeting for over a year while it was decided whether or not to clean it. In the end, it was thought better to let the monument stay dirty rather than expose it to the acid rain.

In Egypt, the 4600-year-old statue of the Sphinx is made mostly of limestone. For many years it was buried up to its neck in the sand, but now it is exposed to pollution, including acid rain, and the whole structure is crumbling and flaking. Old photographs show how many of the features of the Sphinx have been eroded in the past fifty years (see figure 4.2).

In India, the Taj Mahal has beautiful domes made of marble. The palace is only 48 km from an oil refinery, which is allowed to release one tonne of sulphur dioxide into the air every day. In addition, the Taj Mahal is close to very busy roads with thousands of cars a day either passing by or stuck in traffic jams. The white marble is now turning yellow and many of the iron structures have rusted away.

Many countries are now taking steps to reduce acid rain, but the problem is huge. The EU has ruled that Britain should cut its emissions of sulphur dioxide by 60 per cent by the year 2003. In Yorkshire, over £400 million has been spent on building the world's largest plant for removing sulphur dioxide from gases emitted by the Drax power station.

Figure 4.3 This tree has been damaged by acid rain

Chapter 8 Study Questions

1 The chart shows the pH of some common substances.

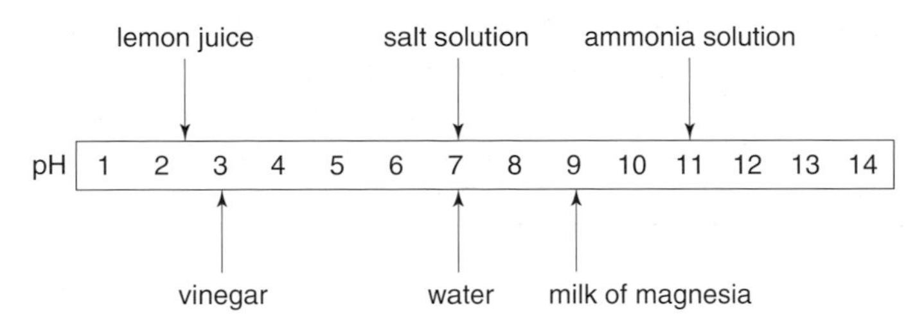

A ammonia solution	**B** lemon juice	**C** milk of magnesia
D salt solution	**E** vinegar	**F** water

a) Identify the **two** substances which are acids.
b) Identify the **two** substances which will show a decrease in pH when they are diluted with water.

SQA GENERAL (PS)

2

A barium hydroxide	**B** calcium oxide	**C** nitrogen dioxide
D potassium nitrate	**E** sodium chloride	**F** sulphur dioxide

The grid shows compounds which dissolve in water.

a) Identify the **two** compounds which produce alkaline solutions.
b) Identify the **two** covalent compounds.

SQA GENERAL (PS)

3

A	The pH of the solution will rise.
B	The solution will become more concentrated.
C	The pH of the solution will fall towards 7.
D	Adding water will have no effect on the solution.

Class 4C made some statements about the effect of adding water to an alkaline solution.

Identify the correct statement.

SQA GENERAL (KU)

4 Toners for colouring black and white photographs are made by mixing two solutions. The solutions should be made up using warm water.

Solution 1
Sodium sulphide solution concentration 100g/l

Solution 2
Sodium hydroxide solution concentration 80g/l

a) Why is **warm** water used to make up the solutions? (PS)

b) Calculate the concentration of sodium hydroxide in mol/l.
Show your working clearly. (KU)

c) Different colours are obtained by mixing the two solutions:

Colour	Toner Composition	
	Volume of sodium sulphide solution/cm^3	Volume of sodium hydroxide solution/cm^3
cold brown	10	40
brown	20	30
warm brown	30	20

What mass of sodium sulphide will be required to make 100 cm^3 of toner for a cold brown colour?
Show your working clearly.

SQA CREDIT (PS)

5 A school technician has been asked to make up a solution of sodium hydroxide in water with an exact concentration of 1 mole per litre. (Formula mass of NaOH = 40)

Identify the procedure(s) in the grid below which would result in a solution with the correct concentration.

A Dissolve 40 g of sodium hydroxide in 100 cm^3 of water.	**B** Dissolve 40 g of sodium hydroxide in water and make up to 1 litre of solution by the addition of more water
C Dissolve 1 mole of sodium hydroxide in 1 litre of water.	**D** Dissolve 0.1 mole of sodium hydroxide in water and make up to 100 cm^3 using more water.

CREDIT (KU)

6

A H_2O	**B** H^+	**C** H_2
D OH^-	**E** O_2	

a) Identify the entities in the grid above which are found in equal concentrations in a neutral solution.
b) Identify the entity whose concentration is reduced when water is added to an alkaline solution.

CREDIT (KU)

7

A sodium hydroxide	B sulphuric acid	C potassium chloride
D sodium chloride	**E** nitric acid	**F** potassium hydroxide

a) Identify the **two** compounds in the grid above that produce hydrogen ions when they dissolve in water.
b) Identify the **two** compounds that could produce a solution with a pH value of 11 when dissolved in water. GENERAL (KU)

8 If acidic water gets into a central heating system, it can attack the steel radiators, producing hydrogen gas.

a) How could a plumber test the gas that sometimes collects in radiators to see if hydrogen is present?
b) What would be the result of the test if hydrogen was present? GENERAL (KU)

9 Calculate the formula mass of each of the following substances:

a) oxygen, O_2
b) methane, CH_4
c) lead(II) sulphate, $PbSO_4$
d) calcium nitrate, $Ca(NO_3)_2$. CREDIT (KU)

10 Calculate the mass of one mole of each of the following substances:

a) bromine, Br_2
b) zinc carbonate, $ZnCO_3$
c) magnesium hydrogencarbonate, $Mg(HCO_3)_2$
d) ammonium sulphate, $(NH_4)_2SO_4$. CREDIT (KU)

11 Calculate the mass of each of the following:

a) 4 moles of ethane, C_2H_6
b) 2.5 moles of ammonium carbonate, $(NH_4)_2CO_3$
c) 3 moles of iron(III) sulphate, $Fe_2(SO_4)_3$
d) 100 moles of ammonium phosphate, $(NH_4)_3PO_4$ CREDIT (KU)

12 Calculate the number of moles of each of the following substances which will have a mass of 100 g:

a) iodine, I_2
b) propene, C_3H_6
c) copper(II) nitrate, $Cu(NO_3)_2$
d) magnesium nitride, Mg_3N_2 CREDIT (PS)

13 Calculate the number of moles of solute which are present in each of the following solutions:

a) 2 litres of 1 mol/l sodium hydroxide
b) 0.25 litres of 3 mol/l ammonia
c) 250 cm^3 of 2 mol/l lithium nitrate
d) 500 cm^3 of 0.1 mol/l potassium iodide. CREDIT (KU)

14 Calculate the mass, in grams, of solute present in each of the following solutions:

a) 1 litre of 1 mol/l Na_2CO_3
b) 2.5 litres of 4 mol/l $NaOH$
c) 500 cm^3 of 5 mol/l H_2SO_4
d) 100 cm^3 of 0.01 mol/l $Ca(OH)_2$. CREDIT (PS)

15 Calculate the concentration of each of the following as solutions in mol/l:

a) 3 mol of hydrochloric acid in 2 litres of solution
b) 0.4 mol of sodium nitrate in 500 cm^3 of solution CREDIT (KU)
c) 222 g of calcium chloride in 4 litres of solution
d) 20.2 g of potassium nitrate in 250 cm^3 of solution. CREDIT (PS)

16 Calculate the volume of solution that could be produced in each of the following cases:

a) 6 mol of sulphuric acid are used to prepare a 2 mol/l solution
b) 1 mol of hydrochloric acid is used to prepare a 0.1 mol/l solution CREDIT (KU)
c) 80 g of sodium hydroxide are used to prepare a 4 mol/l solution
d) 1.008 kg of potassium hydroxide are used to prepare a 0.2 mol/l solution. CREDIT (PS)

CHAPTER NINE

Reactions of Acids

SECTION 9.1 Neutralisation

Figure 1.1 Why is lime being dropped into this lake?

The lake shown in figure 1.1 is acidic. Fish, plants and other animals in the lake are dying because of the acidic water. Lime is a substance which **neutralises** acids. It is hoped that if enough lime is added, then the water will become less acidic and the plants and animals will survive.

Acids can be neutralised by different types of substances such as alkalis and insoluble carbonates. We call such substances **neutralisers**. The lime in figure 1.1 is a neutraliser of the acids in the lake. Farmers add lime to their fields to neutralise acidity in the soil.

Indigestion is caused by having too much acid in the stomach. Indigestion tablets include neutralisers such as calcium carbonate. These neutralise the stomach acid and so relieve the indigestion (see figure 1.2).

Neutralising acids with alkalis

Sodium hydroxide is an alkali. It neutralises hydrochloric acid solution to produce sodium chloride (common salt) and water. The word equation for this neutralisation is:

$$\text{hydrochloric acid} + \text{sodium hydroxide} \rightarrow \text{sodium chloride} + \text{water}$$

In Chapter 8 you learnt that acidic solutions have excess hydrogen ions (H^+). During neutralisation, these ions react to produce water molecules. As a general rule, when an alkali neutralises an acid, the products are a salt and water.

$$\text{acid} + \text{alkali} \rightarrow \text{salt} + \text{water}$$

Salts are ionic compounds which can be produced by neutralisation reactions. They usually consist of a negative ion, which comes from the acid, and a positive ion, which comes from the neutraliser.

The chemical name for a salt is derived from the particular acid and neutraliser from which it is made. For example, hydrochloric acid produces *chloride* salts. The alkali potassium hydroxide produces *potassium* salts. Therefore the neutralisation reaction of these two substances gives water plus the salt *potassium chloride*. Tables 1.1 and 1.2 give the names of salts produced by some common acids and alkalis.

Figure 1.2 Indigestion tablets contain calcium carbonate as a neutraliser. Why do you think they taste chalky?

Questions

Q1 Write word equations for the reaction between:
- **a)** potassium hydroxide and nitric acid,
- **b)** calcium hydroxide and hydrochloric acid.

Q2 Which acid and which alkali could be used to produce sodium sulphate?

Acid	Salt
hydrochloric acid	... chloride
nitric acid	... nitrate
sulphuric acid	... sulphate

Table 1.1 Naming salts from acids

Alkali	Salt
sodium hydroxide	sodium ...
potassium hydroxide	potassium ...
ammonia solution	ammonium ...
calcium hydroxide	calcium ...

Table 1.2 Naming salts from alkalis

Ions and neutralisation

Acidic solutions contain an excess of hydrogen ions (H^+). Alkalis contain an excess of hydroxide ions (OH^-). During neutralisation these ions react to form water molecules:

$$H^+ + OH^- \rightarrow H_2O$$

The other ions present in the acid and the alkali *do not react*. This can be shown by using an **ionic equation** for the reaction. In this type of equation, the ions in each dissolved ionic compound are shown separately. For example, calcium chloride has the ionic formula $Ca^{2+}(Cl^-)_2$. When it is in solution it is written in an ionic equation as:

$$Ca^{2+}(aq) + 2Cl^-(aq)$$

In some cases, the state symbols (aq) are left out.

The neutralisation of hydrochloric acid by sodium hydroxide can be shown as follows:

$$H^+ + Cl^- + Na^+ + OH^- \rightarrow Na^+ + Cl^- + H_2O$$

You can see that the Na^+ and Cl^- ions are not changed. They appear in exactly the same form on both sides of the arrow in the equation. We call ions such as these, which do not take part in a chemical reaction, **spectator ions**.

If we remove the spectator ions from the above equation then the fundamental ionic equation for the reaction is seen to be:

$$H^+ + OH^- \rightarrow H_2O$$

Questions

Q3 For the reaction between sodium hydroxide solution (NaOH) and dilute nitric acid (HNO_3) write:

a) a word equation,

b) a balanced equation using simple formulae,

c) an ionic equation. Identify the spectator ions by putting a line through them.

Bases and alkalis

A base is a substance which neutralises an acid producing water. It does this by joining with hydrogen ions from the acid to produce water as one of the products. Examples of bases include metal carbonates, metal oxides, metal hydroxides and ammonia.

Alkalis are a type of base. An alkali is a base which dissolves in water. The carbonates, oxides and hydroxides of all Group 1 metals are dissolved easily in water, giving alkaline solutions. Sodium hydroxide, for example, dissolves to give sodium hydroxide solution, which is strongly alkaline. Ammonia gas is also soluble in water giving an alkaline solution.

More about neutralising acids

Figure 2.1 Acid rain attacks unprotected iron and steel; wax polish helps to protect steel

Acid/metal reactions

Acid rain attacks iron. If a bridge, a car or a building contains iron which is not protected by paint, then any acid rain falling on it will start a chemical reaction.

The reaction that occurs is a neutralisation reaction. For example, iron can react with the sulphuric acid in rain to produce the salt iron(II) sulphate. Also in the reaction, hydrogen ions (H^+) from the acid are converted into hydrogen gas (H_2). The reaction can be summarised as:

$$\text{iron} + \text{sulphuric acid} \rightarrow \text{iron(II) sulphate} + \text{hydrogen}$$

$$Fe + H_2SO_4 \rightarrow FeSO_4 + H_2$$

Zinc reacts with hydrochloric acid as follows:

$$\text{zinc} + \text{hydrochloric acid} \rightarrow \text{zinc chloride} + \text{hydrogen}$$

$$Zn + 2HCl \rightarrow ZnCl_2 + H_2$$

As a general rule, metals react with acids to give a salt and hydrogen, although there are exceptions to this.

metal + acid → salt + hydrogen

Questions

Q1 In a reaction between a metal and an acid, bubbles of gas were given off. How could you show that the gas was hydrogen?

Q2 Write a word equation for the reaction between aluminium and hydrochloric acid.

Acid/metal oxide reactions

Gardeners and farmers often have soil which is too acidic for their crops to grow properly. They use lime to neutralise the acidity in the soil. Lime is a metal oxide. Its chemical name is calcium oxide. Like all metal oxides it will react with acids to form a salt plus water.

metal oxide + acid → salt + water

For example, zinc oxide reacts with sulphuric acid as follows:

$$\text{zinc oxide} + \text{sulphuric acid} \rightarrow \text{zinc sulphate} + \text{water}$$

$$ZnO + H_2SO_4 \rightarrow ZnSO_4 + H_2O$$

Figure 2.2 Lime is used to neutralise acidity in the soil

Questions

Q3 Copy and complete the following reactions:

a) aluminium oxide + nitric acid → . . . + . . .

b) magnesium oxide + hydrochloric acid → . . . + . . .

Figure 2.3 An acid/carbonate reaction

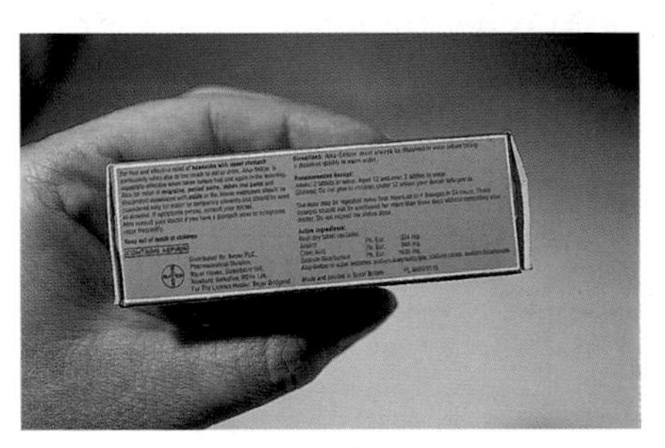

Figure 2.4 These 'health salts' contain a mixture of acids and carbonates

During the neutralisation, the H^+ ions from the acid are removed and the acid's pH moves towards 7.

Acid/carbonate reactions

Carbonate rocks are often used for building, for example limestone, which is mostly calcium carbonate. When acid rain falls on this type of rock, the carbonate reacts with the acid and the building gradually wears away.

Geologists use hydrochloric acid to help identify carbonate rocks. After a few drops of the dilute acid are added to the rock, a carbonate rock will produce bubbles of carbon dioxide.

For example, calcium carbonate reacts with hydrochloric acid as follows:

calcium carbonate	+	**hydrochloric acid**	→	**calcium chloride**	+	**water**	+	**carbon dioxide**
$CaCO_3$	+	$2HCl$	→	$CaCl_2$	+	H_2O	+	CO_2

In general, the reaction between a metal carbonate and an acid takes the form:

metal carbonate + acid → salt + water + carbon dioxide

In all acid/carbonate reactions, hydrogen ions from the acid react with carbonate ions to produce water and carbon dioxide. This can be shown by writing an ionic equation for an acid/carbonate reaction and removing the spectator ions. For example, the ionic equation for the reaction between sodium carbonate and hydrochloric acid is as follows:

$$2Na^+ + CO_3^{2-} + 2H^+ + 2Cl^- \rightarrow 2Na^+ + 2Cl^- + H_2O + CO_2$$

Removing the Na^+ and Cl^- spectator ions leaves:

$$CO_3^{2-} + 2H^+ \rightarrow H_2O + CO_2$$

Note: If the carbonate is soluble in water, the state symbol after CO_3^{2-} would be (aq).

Questions

Q4 For the reaction between potassium carbonate solution and sulphuric acid
a) write a word equation,
b) write an ionic equation.
Identify the spectator ions. Rewrite the equation leaving out the spectator ions.

Salts

In each of the neutralisation reactions given above, one of the products has been a salt. We can now define a salt as a substance in which the hydrogen ion of an acid has now been replaced by a metal ion (or the ammonium ion). For example, the salt formed when hydrochloric acid reacts with sodium hydroxide is sodium chloride. The sodium ion takes the place of the hydrogen ion in the acid.

Acid	Salt
HCl	NaCl
hydrochloric acid	sodium chloride

The ammonium ion (NH_4^+) is found in neutralisers such as ammonium carbonate. Hydrochloric acid is neutralised by ammonium carbonate to produce water, carbon dioxide and the salt ammonium chloride (NH_4Cl).

Preparation of salts

Preparation of insoluble salts by precipitation

An insoluble salt is one where less than 1g of the salt can be dissolved in 100 g of water at 20°C. We can make insoluble salts by using a type of reaction called a **precipitation reaction**.

A precipitation reaction is the reaction of two solutions to form an insoluble product called a **precipitate**. Silver iodide, for example, is insoluble and can be made by the following method:

A soluble silver salt is dissolved in water, and a soluble iodide is also dissolved in water. When these two solutions are mixed, a precipitate of insoluble silver iodide forms (see figure 3.1). In order to obtain a pure, dry sample of this salt, it is filtered off, washed with water and allowed to dry in air. The reaction taking place in this experiment is:

silver nitrate + sodium iodide → silver iodide + sodium nitrate

The sodium nitrate which is formed in the reaction is a soluble salt – it stays in solution.

If we remove the spectator ions then the reaction can be written as follows:

$$Ag^+(aq) + I^-(aq) \rightarrow Ag^+I^-(s)$$

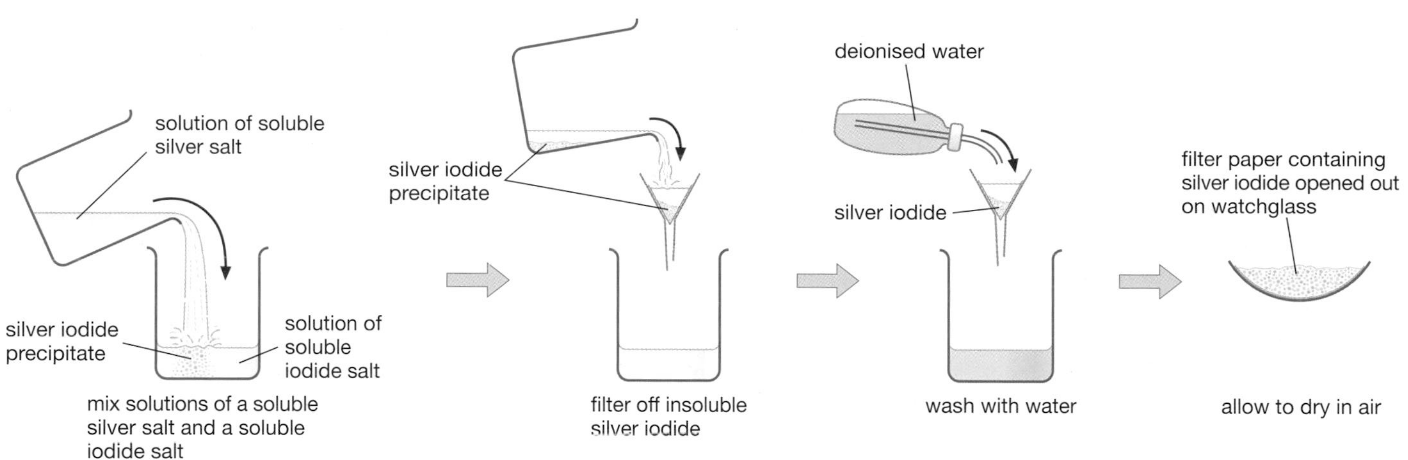

Figure 3.1 Preparing the insoluble salt, silver iodide

Questions

Q1 Name the insoluble product formed when aqueous solutions of the following are mixed:
a) nickel chloride and potassium carbonate,
b) sodium chloride and silver nitrate.

Making soluble salts by neutralisation

Using a soluble base as the neutraliser

You saw in section 9.1 that the following type of reaction occurs when an acid is neutralised by a soluble base:

acid + alkali → salt + water

In this type of neutralisation, an indicator is used to show when the acid has been neutralised. Figure 3.2 shows a neutralisation in which pH paper is used as the indicator.

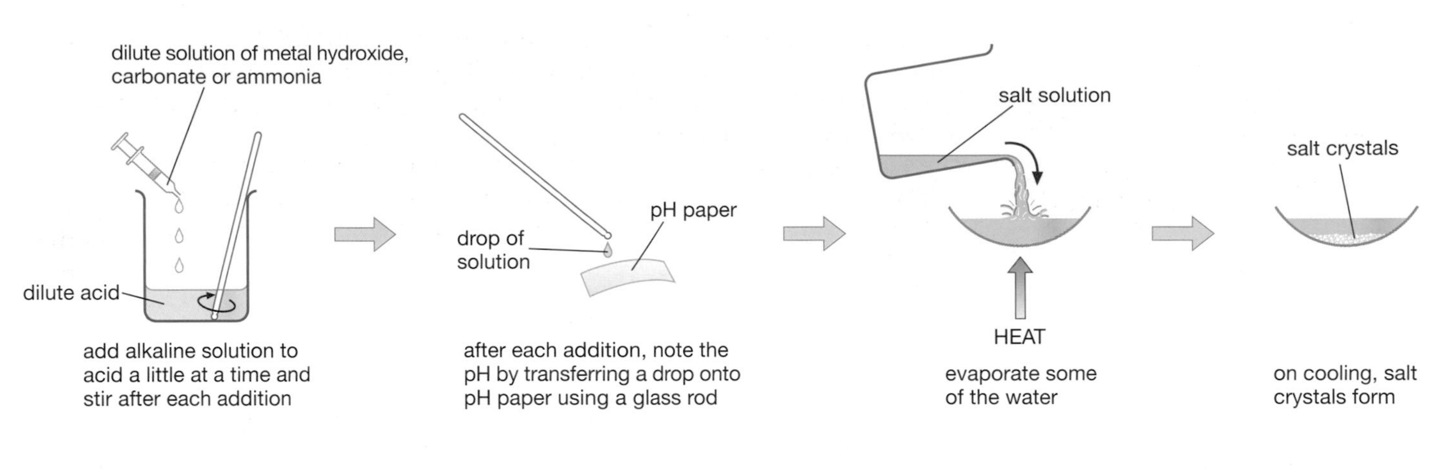

Figure 3.2 Preparing soluble salts using an acid and a soluble base

Questions

Q2 Using the method shown in figure 3.2, it is possible to make sodium chloride crystals by the reaction of dilute hydrochloric acid with **a)** sodium hydroxide solution, or **b)** sodium carbonate solution. Write balanced chemical equations for both these reactions.

Using metals and insoluble bases

It is quite easy to make soluble salts from acids using metals. For example, magnesium chloride can be produced by adding magnesium metal powder to dilute hydrochloric acid. You simply have to add the powder to the acid until no more reacts. Excess magnesium is filtered off and the salt is obtained by crystallisation from the solution which is produced (see figure 3.3).

This method can also be used with insoluble bases. The solid base is added to the acid until there is no further reaction. This is followed by filtration of the remaining solution and crystallisation.

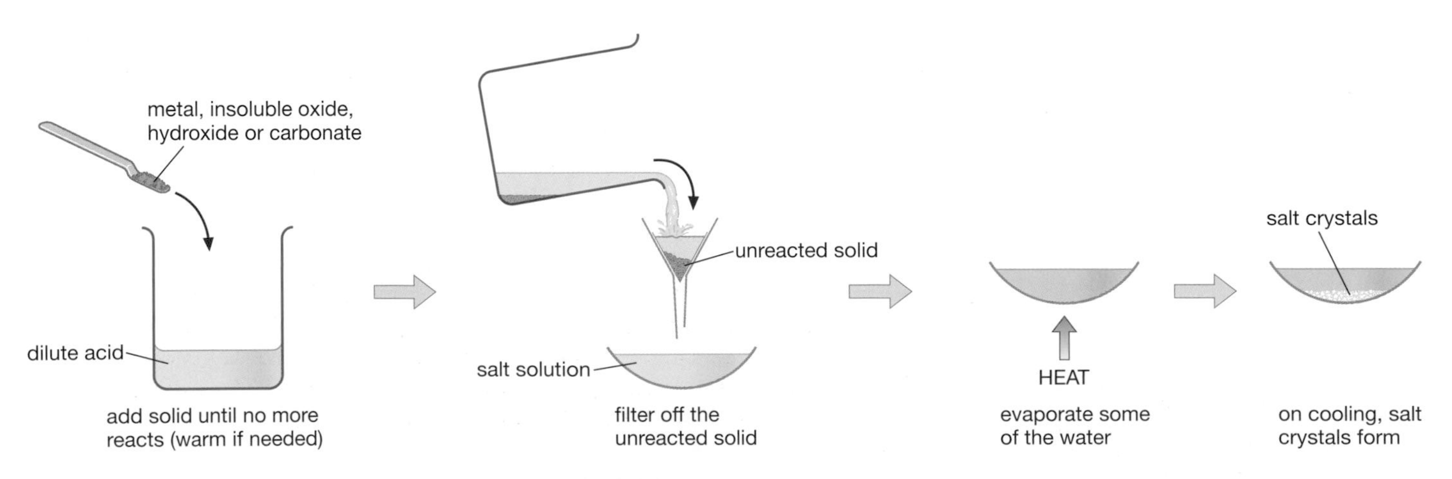

Figure 3.3 Preparing soluble salts using a metal or an insoluble base

Questions

Q3 Magnesium sulphate crystals can be made using the method shown in figure 3.3, by adding any of the following to dilute sulphuric acid: (i) magnesium, (ii) magnesium oxide, (iii) magnesium hydroxide, (iv) magnesium carbonate. **a)** Identify the reaction(s) in which a gas is given off. **b)** Write a chemical equation for each reaction.

Titrations

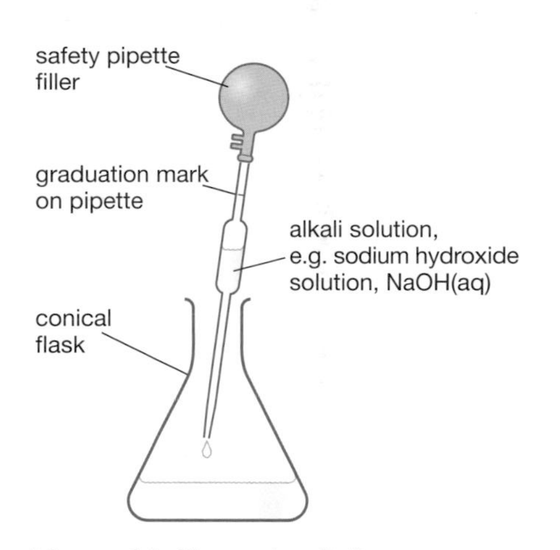

Figure 4.1 Pipette the alkali

How can you work out the concentration of a solution if you don't know how much solute has been dissolved? The answer is to carry out a **titration**. This is a technique in which the volumes of reacting solutions are measured accurately.

Usually, but not always, a titration involves a neutralisation reaction between an acid and an alkali where the concentration of one solution is known and the concentration of the other can be found using the results of the titration.

How to carry out a titration

Two special pieces of equipment are usually required. A **pipette** is used to measure out exact volumes of the alkali (see figure 4.1). Liquid is sucked into the pipette using the safety pipette filler. A **burette** is used to measure the volume of acid which is required to neutralise the alkali (see figure 4.2).

Indicators are solutions which change colour just at the point where the reaction between the acid and alkali is complete (see figure 4.3). For example, phenolphthalein changes from pink in the alkali solution to colourless when the acid is *just* in excess. It can be used in titrations involving the acids and alkalis listed in table 4.1.

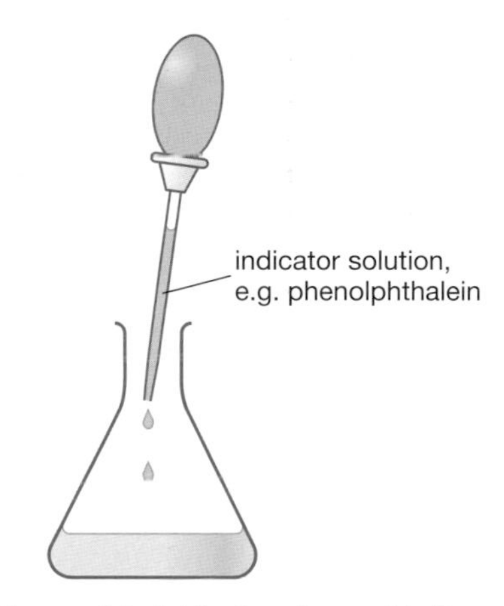

Figure 4.2 Add a few drops of indicator

Acid	Alkali
sulphuric acid hydrochloric acid nitric acid	sodium hydroxide potassium hydroxide

Table 4.1 Acids and alkalis used with phenolphthalein

To perform a titration, a known volume of alkali solution is pipetted into a conical flask and a few drops of a suitable indicator are added. Acid is then run in from the burette until the indicator just shows the desired colour change. This is called the **end-point** of the titration. It occurs when the reaction between the acid and the alkali is complete. The process of titration is shown in figures 4.1 to 4.3.

The volume of the acid used is read from the burette scale. Usually, a trial titration is first carried out quite quickly. The titration is then repeated two or three times, with the acid being added drop-by-drop as the end-point is approached. The volumes of acid required each time are averaged out, with any very high or very low results being ignored.

Using the average value for the volume of acid required to neutralise the alkali, it is then possible to calculate the concentration of the unknown solution.

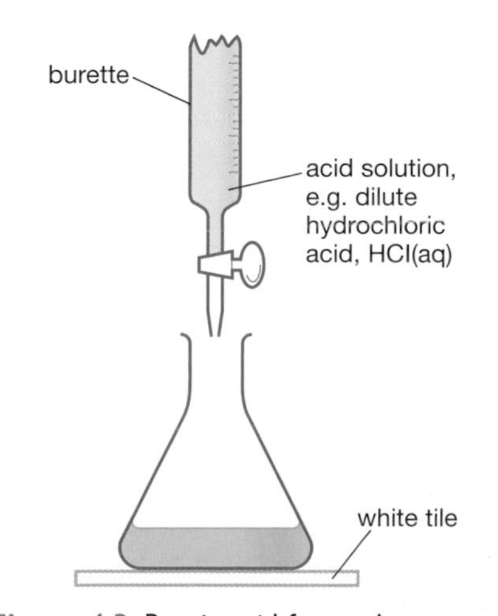

Figure 4.3 Run in acid from a burette

Questions

Q1 Marie obtained the following burette readings in a series of titrations using the same acid and alkali solutions: 23.4 cm^3, 22.6 cm^3, 22.8 cm^3, 21.3 cm^3, 22.7 cm^3.

a) Which results should be averaged?

b) Calculate the average burette value which should be used in the calculation.

Calculations based on titrations

Consider the reaction between any acid solution and any alkali. Remember that an alkali is a soluble base. We can use the terms *a* for the number of moles of acid and *b* for the number of moles of base involved in the neutralisation. The balanced equation for the reaction would look like this:

$$a \text{ acid} + b \text{ base} \rightarrow \text{products}$$

This means that *a* moles of acid plus *b* moles of base react to give products.

Let C_A be the concentration of the acid in moles per litre
C_B be the concentration of the base in moles per litre
V_A be the volume of the acid solution in litres
V_B be the volume of the base solution in litres.

Using $n = C \times V$, where *n* is the number of moles, then

$$\text{number of moles of acid} = C_A \times V_A$$
$$\text{number of moles of base} = C_B \times V_B$$

The ratio of the number of moles of acid to number of moles of base required for netrisation is then given as:

$$\frac{\text{number of moles of acid}}{\text{number of moles of base}} = \frac{C_A \times V_A}{C_B \times V_B} = \frac{a}{b}$$

The equation

$$\frac{C_A \times V_A}{C_B \times V_B} = \frac{a}{b}$$

can be used to calculate *any* of the six quantities in it. If any five of the quantities are known then the sixth can be calculated by rearranging the equation.

In a titration, the balanced equation for the reaction gives the values for *a* and *b*. The volume of the base solution (alkali solution), V_B, is usually the capacity of the pipette. The volume of the acid solution, V_A, is the average burette reading. Either the concentration of the acid, C_A, or that of the base, C_B, will be known, so the equation can be rearranged to solve one of them.

In practice, there is no need to convert the volumes, V_A and V_B, into litres. The equation involves a *ratio* of volumes and so the precise units do not matter as long as they are both the same, for example cm^3.

Example

In a titration, it was found that 20 cm^3 of potassium hydroxide solution was neutralised by 15 cm^3 of dilute sulphuric acid with a concentration of 0.1 moles per litre. Calculate the concentration of the potassium hydroxide solution in moles per litre.

balanced equation: $H_2SO_4 + 2KOH \rightarrow K_2SO_4 + 2H_2O$

rearranging:

$$\frac{C_A \times V_A}{C_B \times V_B} = \frac{a}{b} \qquad C_B = \frac{C_A \times V_A \times b}{V_B \times a}$$

$$= \frac{0.1 \times 15 \times 2}{20 \times 1} = 0.15 \text{ mol/l}$$

Thus the concentration of the potassium hydroxide solution is 0.15 moles per litre.

Questions

Q2 a) Write a balanced equation for the neutralisation of sodium hydroxide solution by dilute hydrochloric acid.
b) In a titration, 10 cm^3 of sodium hydroxide solution with a concentration of 0.2 mol/l was neutralised by 25 cm^3 of dilute hydrochloric acid. Calculate the concentration of the acid solution in mol/l.

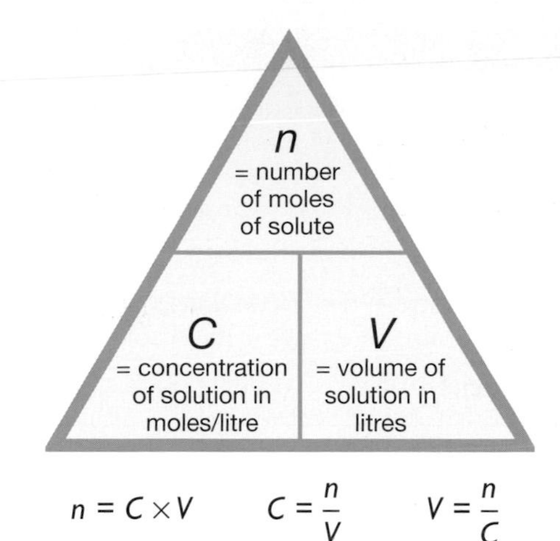

Figure 4.4

Calculation – alternative method

In section 8.3 on page 81 you read about the relationship between the concentration of a solute (C), the number of moles of solute (n) and the volume of the solution (V). These are shown below, along with *a triangle of knowledge in figure 4.4.*

These relationships can be used in calculations involving the results of titrations.

Example

In a titration, it was found that 20 cm^3 of sodium hydroxide was neutralised by 15 cm^3 of dilute sulphuric acid of a concentration 0.1 moles per litre. Calculate the concentration of the sodium hydroxide solution in moles per litre. (Note: the volume has to be given in litres, therefore 15 cm^3 becomes 0.015 litres.)

balanced equation: $H_2SO_4 + 2NaOH \rightarrow Na_2SO_4 + 2H_2O$
1 mol 2 mol

Number of moles of acid reacting: $n_{(H_2SO_4)} = C \times V$
$= 0.1 \times 0.015$
$= 0.0015$ moles acid

From the equation: 1 mol H_2SO_4 reacts with 2 mol NaOH

Therefore: 0.0015 mol H_2SO_4 reacts with 0.0030 mol NaOH

Concentration of NaOH, $C_{(NaOH)} = \frac{n}{V_{(NaOH)}}$
$= \frac{0.0030}{0.02}$
$= 0.15$ mol/l

Chapter 9 Study Questions

1 The label on a bottle of mineral water listed ions present in the water.

Ion	Concentration in mg/l
calcium	70
magnesium	40
sodium	120
chloride	220
sulphate	20

a) Draw a **bar graph** of *Ion* against *Concentration*.
b) A sample of the mineral water was evaporated to dryness. The remaining solid was collected. Name a compound which could have been present in the solid.
c) Barium nitrate solution was added to a sample of the mineral water.
A white precipitate was obtained.
Name this precipitate.
You may wish to use page 5 of the SQA Data Booklet.

SEB GENERAL (PS)

2 John spilled sulphuric acid from a car battery on his garage floor. First he poured water over the acid. Then he sprinkled washing soda (sodium carbonate) on it before mopping it up.

a) Why did John add water to the acid?
b) (i) Name the type of reaction which takes place when washing soda is added to sulphuric acid.
(ii) Name the gas produced in the reaction.
c) Calculate the mass of one mole of sodium carbonate, Na_2CO_3.
Show your working clearly.

SQA GENERAL (KU)

3 The following is a report given by the Wellcare Company to a fish farmer.

> **WELLCARE REPORT**
>
> Water test
>
> *This showed that the water was pH 5.*
>
> Action needed
>
> *The water should be treated to increase the pH.*

a) Describe how you would use Universal indicator or pH paper to measure the pH of the water.
b) Suggest a cause for the low pH.
c) Give the chemical name of a substance which could be used to increase the pH of the water.

SEB GENERAL (PS)

4 There are many different magnesium compounds.

A	B	C
magnesium bromide	magnesium carbonate	magnesium chloride
D	**E**	**F**
magnesium nitrate	magnesium oxide	magnesium sulphate

a) Identify the **two** compounds which are insoluble. (You may wish to use page 00 of the SQA Data Booklet to help you.)
b) Identify the compound(s) which will neutralise an acid.

SQA GENERAL (KU)

5 a) Calculate the mass of sodium hydroxide which would be required to make 1 litre of exactly 1 mol/l solution.

Show your working clearly.

> INSTRUCTION CARD
>
> Preparation of approximately 1 mol/l sodium hydroxide solution
>
> CARE: Sodium hydroxide is highly corrosive.
>
> 
>
> 1. Weigh out 41 g of sodium hydroxide pellets.
> 2. Dissolve in about 400 cm^3 of water.
> 3. Add a few drops of barium chloride solution to remove the carbonate ions.
> 4. Filter.
> 5. Make the volume of solution up to exactly 1 litre.

b) Solid sodium hydroxide often contains sodium carbonate as an impurity. This is caused by some of the sodium hydroxide reacting with carbon dioxide from the air.
The equation for Step 3 is:

$$BaCl_2(aq) + Na_2CO_3(aq) \rightarrow BaCO_3(s) + 2NaCl(aq)$$

(i) Name an ion which will be an impurity in the final sodium hydroxide solution.
(ii) Rewrite the equation as an ionic equation omitting spectator ions.
(iii) 25.0 cm^3 of the solution was neutralised by 25.5 cm^3 of 1 mol/l hydrochloric acid.
Calculate the exact concentration of the sodium hydroxide solution.
Show your working clearly.
(Some knowledge from Chapter 8 is required to answer this question.)

SQA CREDIT (KU)

6 Common salt is extracted from salt mines by pumping in water to dissolve it.

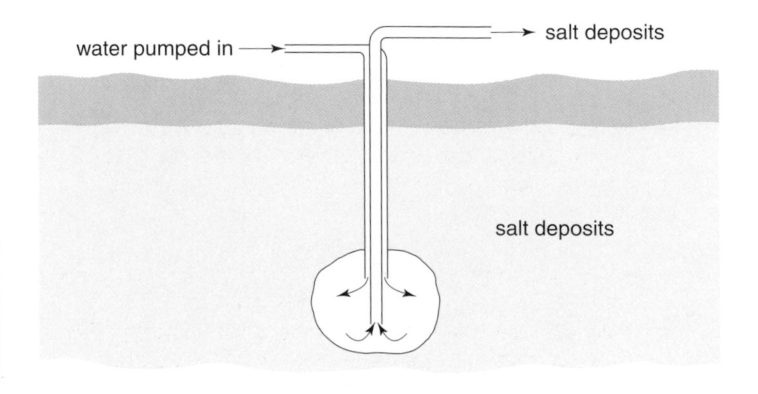

The salt solution which is obtained contains magnesium sulphate as an impurity.
The magnesium ions are removed by adding sodium hydroxide solution.

$$Mg^{2+}(aq) + SO_4^{2-}(aq) + 2Na^+(aq) + 2OH^-(aq) \rightarrow Mg^{2+}(OH^-)_2(s) + 2Na^+(aq) + SO_4^{2-}(aq)$$

a) (i) Name the type of reaction taking place between the magnesium sulphate solution and the sodium hydroxide solution.
(ii) Name the spectator ions in the reaction. (KU)

b) The products of the reaction can be separated by filtration.
Draw and label the apparatus you would use in the laboratory to carry out this process.
Indicate on the diagram where each product is collected.

SQA CREDIT (PS)

7 An indigestion remedy is found to contain the following neutralisers. The masses given show the amount of each neutraliser in a single tablet.

calcium carbonate	200 mg
magnesium silicate	60 mg
magnesium carbonate	60 mg
sodium bicarbonate	80 mg

a) Present this information as a labelled bar graph.
b) The bicarbonate ion has the formula HCO_3^-. Refer to the table of ions on page 4 of the SQA Data Booklet and give an alternative name for it.

GENERAL (PS)

8 Taking indigestion remedies to cure acid indigestion is an example of neutralisation. Give another everyday example of neutralisation which involves the use of lime.

GENERAL (KU)

9 Copy and complete the following word equations:

a) calcium hydroxide + hydrochloric acid → ...
b) sodium hydroxide + nitric acid → ...
c) potassium hydroxide + sulphuric acid → ...

GENERAL (KU)

10

A oxygen	B hydrogen	C nitrogen
D carbon dioxide	**E** sulphur dioxide	**F** chlorine

a) Using the grid above, identify the gas produced when hydrochloric acid reacts with calcium carbonate.
b) Identify the gas produced when zinc reacts with sulphuric acid.

GENERAL (KU)

11 The diagram top right shows a bottle of 'milk of magnesia', which consists of water and magnesium hydroxide.

a) Milk of magnesia is an acid indigestion remedy. Use the solubility data on page 5 of the SQA Data Booklet to explain why the white powdery solid at the bottom of the bottle has a greater neutralising ability than the colourless liquid above it.

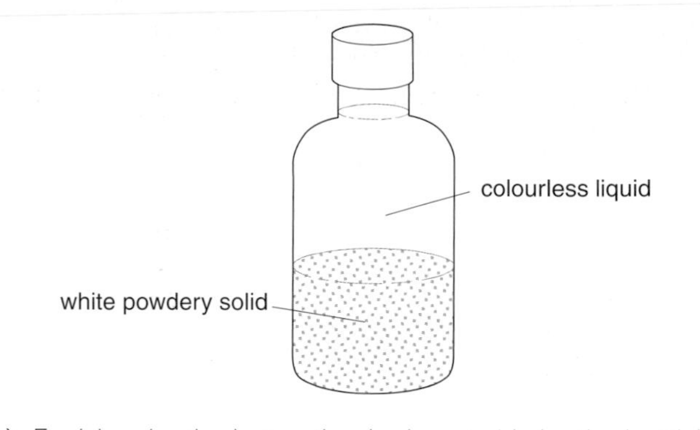

b) Explain why the instruction is given to 'shake the bottle'.

GENERAL (PS)

12 a) Copper sulphate solution and potassium carbonate solution react to produce insoluble copper carbonate. What name is given to this type of reaction?
b) Name the insoluble product formed when silver nitrate solution is mixed with potassium chloride solution. (Refer to the table of solubilities on page 5 of the SQA Data Booklet.)

GENERAL (KU)

13 Explain the meaning of each of the following terms:

a) a base
b) an alkali.

CREDIT (KU)

14 a) Write an ionic equation for the reaction between hydrochloric acid and potassium hydroxide solution.
b) Identify the spectator ions in this reaction.
c) Which ions join to produce water in this reaction?

CREDIT (KU)

15 Iain titrated citric acid solution against sodium hydroxide solution. He found that 15 cm^3 of 0.1 mol/l citric acid were needed to neutralise 45 cm^3 of 0.1 mol/l sodium hydroxide.

a) How many moles of citric acid reacted? (KU)
b) How many moles of sodium hydroxide reacted? (KU)
c) How many moles of sodium hydroxide react with 1 mole of citric acid?

CREDIT (PS)

16 Sodium hydroxide and nitric acid react as follows:

$$NaOH + HNO_3 \rightarrow NaNO_3 + H_2O$$

In a titration, 30 cm^3 of nitric acid neutralised 20 cm^3 of 0.2 mol/l sodium hydroxide. Calculate the concentration of the nitric acid.

CREDIT (PS)

17 25 cm^3 of 0.2 mol/l sodium hydroxide solution was neutralised by 16 cm^3 of hydrochloric acid. Calculate the concentration of the hydrochloric acid.

18 10 cm^3 of potassium hydroxide solution was neutralised by 12 cm^3 of 0.1 mol/l nitric acid. Calculate the concentration of the potassium hydroxide solution.

19 20 cm^3 of 0.5 mol/l lithium hydroxide solution was neutralised by 30 cm^3 of sulphuric acid. Calculate the concentration of the sulphuric acid.

CHAPTER TEN

Making Electricity

SECTION 10.1 Batteries and cells

Figure 1.1 All these appliances use batteries

What are batteries used for? You might use them in personal stereos, computer games, cameras and bicycle lights. In each case the batteries are being used to produce electricity.

How do batteries work? Inside a battery there is a mixture of chemicals. You learnt in Chapter 1 that energy is released when chemical reactions take place. When the chemicals in a battery react, they produce electrical energy. Figure 1.2 shows the inside of a battery.

When a battery produces electricity, electrons flow from the battery, through the wires, to the device it is connected to. In most batteries, the electrons come from a layer of zinc metal. The electrons are produced when a zinc atom reacts and forms a positive zinc ion and two electrons. This can be shown in a special kind of equation called an **ion-electron equation**.

$$Zn \rightarrow Zn^{2+} + 2e^-$$

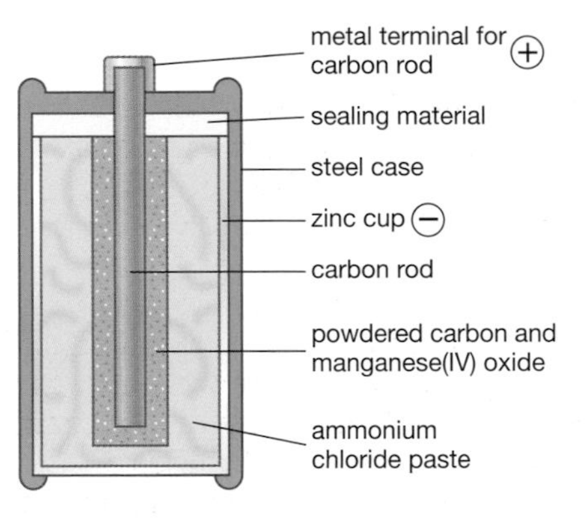

Figure 1.2 Dry cell battery

The battery in figure 1.2 is a **dry cell battery**. This is the type most commonly used for radios, torches, etc. The zinc cup forms the negative terminal of the battery. The carbon rod is the positive terminal. In-between the two terminals there is a paste of ammonium chloride. This completes the circuit by allowing ions to move through it. A substance which does this is called an **electrolyte**.

One reason why a battery 'goes flat' is that the zinc and other chemicals become used up, and so no more electrons can be released. In **rechargeable batteries**, the chemicals can be restored by giving the battery a supply of electrons. For example, a car battery contains lead metal. When the battery is being used the lead metal atoms turn into lead ions. During recharging, the ions are turned back into lead atoms.

Modern batteries are quite complicated, compact devices. To appreciate how they work it is helpful to look at simpler arrangements for producing electricity. Figure 1.3 shows two metals, zinc and copper, connected together. Electrons flow from the zinc to the copper. The sodium chloride solution is an electrolyte and completes the circuit. An arrangement such as this is called a **cell**. In most cases, electricity will be produced when any two different metals are connected in a cell.

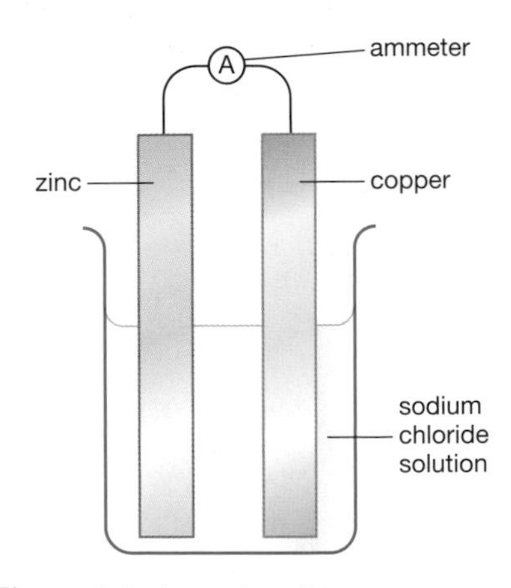

Figure 1.3 A simple cell

Figure 1.4 shows a 'lemon clock'. The two strips of metals are pushed into the fruit, creating a chemical cell that powers the clock.

Different pairs of metals produce different voltages. Figure 1.5 shows how the voltage between two metals can be measured. Using voltages, metals can be placed in a kind of 'league table of metals' called the **electrochemical series**. This is discussed further in section 10.2. Table 1.1 shows some voltages produced in this way.

Cells can also be set up by connecting two **half-cells** together. A half-cell consists of a metal in contact with a solution of its ions, such as a strip of copper metal in a beaker of copper(II) sulphate solution. Electricity is produced when two half-cells containing different metals are connected together. The metals are joined by wires and the two solutions are connected using an **ion bridge**. A length of filter paper soaked in sodium

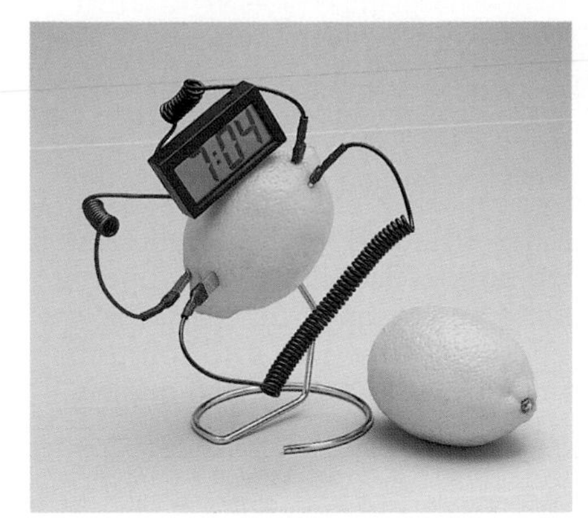

Figure 1.4 A 'lemon' clock

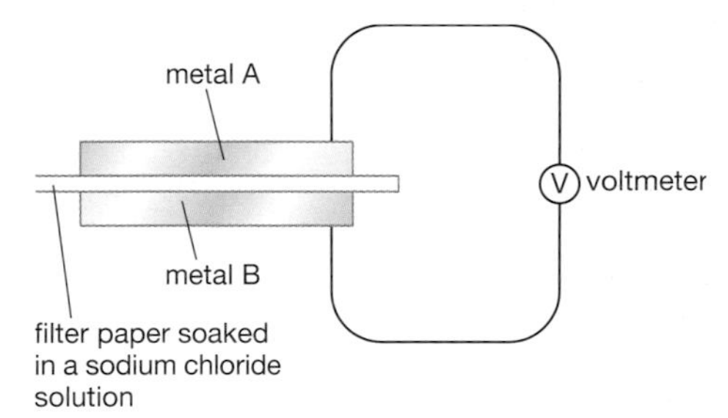

Figure 1.5 Measuring the voltage between two metals

Metal A	Voltage/V
magnesium	+2.1
zinc	+1.0
iron	+0.7

Table 1.1 Some voltages produced, using copper as metal B

Questions

Q1 When an old battery 'leaks' the ammonium chloride paste seeps out. Explain why this happens.

Q2 Make a list of the advantages and disadvantages of using batteries instead of mains electricity.

chloride solution is often used for the ion bridge. Figure 1.6 shows a cell built in this way. The ion bridge completes the circuit by connecting the two half-cells together.

The ion bridge works by letting ions move across it. If the ion bridge is removed, the circuit is broken and no electricity is produced. Any ionic solution can be used for the ion bridge, as long as it does not react with the other solutions used in the half-cells.

In the cell in figure 1.6, zinc atoms lose electrons and form zinc ions. On the other side of the cell, copper(II) ions gain electrons and turn into atoms of copper metal. This can be shown using ion-electron equations as follows:

$$Zn \rightarrow Zn^{2+} + 2e^-$$

$$Cu^{2+} + 2e^- \rightarrow Cu$$

The cell in figure 1.6 could easily be used to light a small bulb. However, after some time the zinc metal would be used up as it turned into zinc ions. The number of copper(II) ions would also decrease and eventually the cell would stop making electricity.

This takes us back to modern batteries. You can, perhaps, see now that the battery shown in figure 1.2 is really a compact, modified version of two half-cells joined together. The zinc cup is the metal which *loses* electrons, and the ammonium chloride paste acts in a similar way to the ion bridge. The carbon rod and manganese(IV) oxide act as the other half-cell, and *accept electrons* lost by the zinc.

The costs of using batteries

The advantages of using batteries are obvious. They let us run electrical appliances without having to plug them into the mains. Batteries in heart pace-makers save lives. However, there are many costs involved in using them. First, batteries are a very expensive way to obtain electricity. It costs less than 3p to run a cassette player from the mains for 24 hours. Using batteries, it could cost around £6 – about 200 times as much!

Second, when batteries are being made, they use up valuable chemicals, especially metals, some of which will eventually run out. Of course, there are problems involved in producing mains electricity. In Britain, mains electricity comes mostly from nuclear, coal and gas power stations. All of these can damage the environment. However, making batteries requires large amounts of energy, and once made they have to be transported, usually by trucks – this uses up fossil fuels.

Finally, after they have gone 'flat', batteries have to be disposed of. Many batteries contain compounds which are very poisonous. What happens to these poisons after a battery has been thrown out?

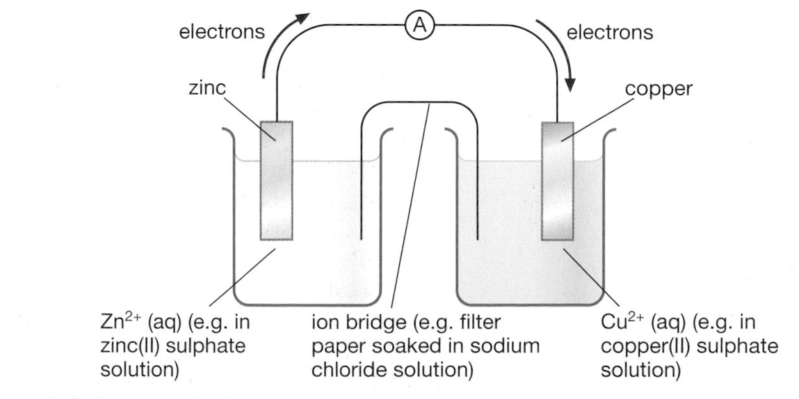

Figure 1.6 A chemical cell made by joining copper and zinc half-cells

SECTION 10.2 Displacement reactions

Questions

Q1 Explain the following in the zinc and copper displacement reaction described above:

a) Why does the zinc become smaller?

b) What is the brown solid substance on the zinc?

c) Why does the blue colour disappear?

Here is a chemical puzzle. What happens if you place a piece of zinc weighing 1 g in a solution of copper(II) sulphate? After some time, the zinc becomes smaller and is covered with a brown solid. Also, the blue copper(II) sulphate solution loses its colour. Why is this? The reason is that the zinc atoms have lost electrons and turned into zinc ions, which go into solution. The copper(II) ions in the solution take up electrons and turn into copper metal. We can show what happens using ion-electron equations:

Zn	$\rightarrow$	Zn^{2+}	+	$2e^-$
zinc atoms	$\rightarrow$	zinc ions	+	2 electrons
Cu^{2+}	+	$2e^-$	$\rightarrow$	Cu
copper(II) ions	+	2 electrons	$\rightarrow$	copper atoms

The overall reaction is

$$Zn + Cu^{2+} \rightarrow Zn^{2+} + Cu$$

This is called a **displacement reaction.**

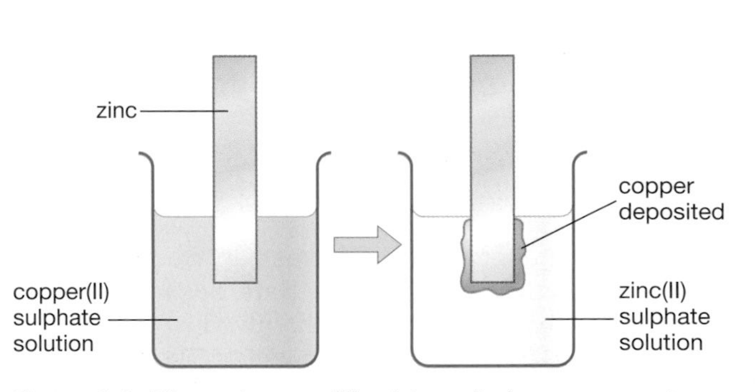

Figure 2.1 Zinc and copper(II) sulphate displacement reaction

Figure 2.2 Displacement reactions of (a) copper and zinc sulphate solution, (b) magnesium and zinc sulphate solution

In the above reaction, the zinc *displaces* the copper. If the experiment is changed and copper metal is added to a solution of zinc ions, you find that there is no reaction. Copper cannot displace zinc. However, if a piece of magnesium is placed in a solution of zinc ions, the zinc will be displaced by the magnesium. This is shown in figure 2.2.

How can you predict if one metal will displace another? This can be done using the electrochemical series mentioned in section 10.1. You remember that metals produce a voltage when they are connected in pairs with an electrolyte between them. In the electrochemical series, the order of the metals is obtained by measuring the voltage produced by each metal when connected in a special circuit.

The highest voltages are obtained when there is a large gap between the positions of the metals in the series. In other words, large voltages are produced when metals near the top of the electrochemical series are connected to those near the bottom of the series.

A simplified version of the electrochemical series is shown in table 2.1. Looking at this, you can see that a pair of metals such as magnesium and silver will produce a bigger voltage than zinc and iron. In fact, magnesium and silver will usually give a voltage of about 3.1 volts. The voltage from zinc and iron is normally about 0.3 volts.

As a general rule, a metal will displace a metal *lower than itself* in the electrochemical series. For example, iron is above silver in the series and so will displace silver ions from solution. Lead is lower than tin and therefore will not displace tin ions from solution.

Name	Symbol
lithium	Li
potassium	K
calcium	Ca
sodium	Na
magnesium	Mg
aluminium	Al
zinc	Zn
iron	Fe
nickel	Ni
tin	Sn
lead	Pb
copper	Cu
silver	Ag
mercury	Hg
gold	Au

Table 2.1 Electrochemical series of metals

A displacement reaction, therefore, is one in which a metal is formed from a solution of its ions when a metal higher than itself in the electrochemical series is added to it.

Questions

Q2 For each of the following say whether or not a displacement reaction will occur:

a) zinc metal in a solution of tin ions,

b) tin metal in silver nitrate solution,

c) magnesium metal in sodium chloride solution.

More on displacement

Hydrogen can be placed in the electrochemical series by considering the reactions of metals with dilute acids. You have probably carried out experiments such as those shown in figure 2.3. Metals down to lead in the electrochemical series react with dilute acids to produce hydrogen gas. This means that they displace hydrogen ions from acids. This can be shown by an ion-electron equation:

$$2H^+ + 2e^- \rightarrow H_2$$

Given the results shown in figure 2.3, hydrogen can be placed below lead but above copper in the electrochemical series (see table 2.2). A fuller version of the electrochemical series is shown on page 7 in the Data Section.

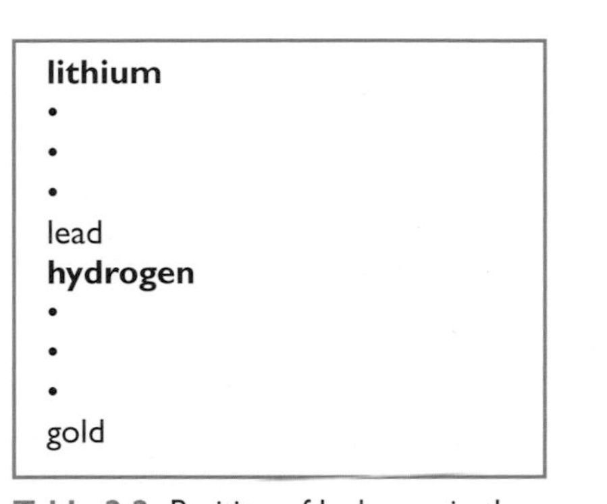

lithium
•
•
•
lead
hydrogen
•
•
•
gold

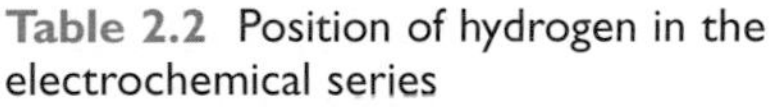

Table 2.2 Position of hydrogen in the electrochemical series

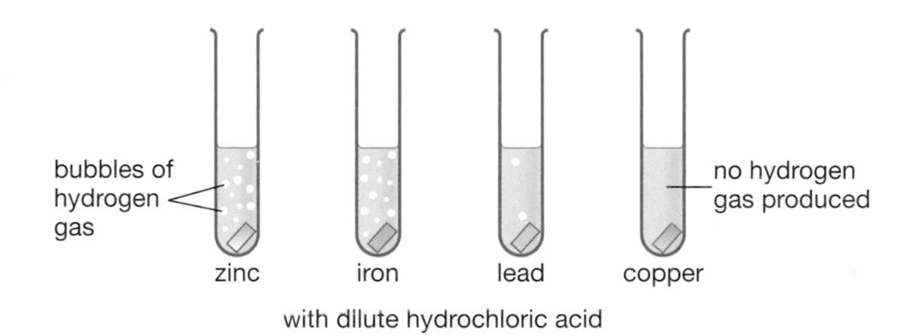

Figure 2.3 Reactions of metals with dilute hydrochloric acid

Cells involving non-metals

The fuller electrochemical series includes reactions for non-metals, such as bromine molecules. Non-metals can be used in half-cells if a carbon rod is used to make electrical contact with the solution.

Example

In the cell shown in figure 2.4, the two half-cells are a solution of iodide ions and a solution of iron(III) ions.

Electrons flow from the iodide ions through the meter to the iron(III) ions. As this happens, the iodide ions turn into iodine molecules:

$$2I^- \rightarrow I_2 + 2e^-$$

The iron(III) ions gain electrons and turn into iron(II) ions:

$$\text{(yellow) } Fe^{3+} + e^- \rightarrow Fe^{2+} \text{ (green)}$$

Questions

Q3 Iodide ions are colourless. Iodine molecules in solution are reddish-brown. Describe the changes in appearance which would occur in the experiment in figure 2.4.

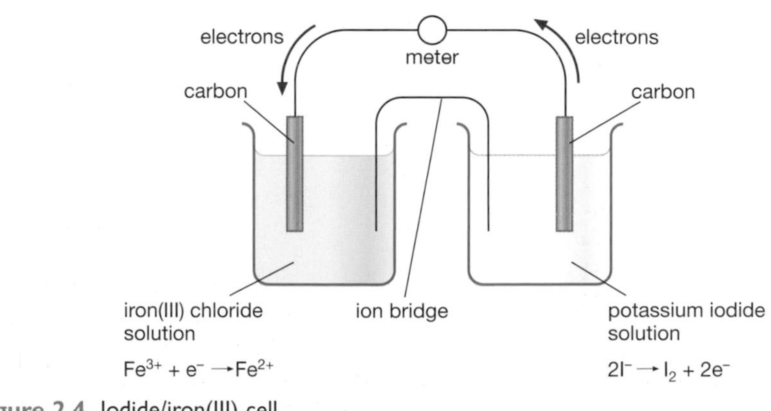

Figure 2.4 Iodide/iron(III) cell

Redox reactions

So far in this chapter, we have looked at two main types of reaction – those in which electrons are lost and those in which they are gained by another substance. **Oxidation** is a term used to describe reactions in which electrons are *lost*. This occurs, for example, when a metal such as magnesium joins with oxygen to form an oxide:

$$2Mg + O_2 \rightarrow 2(Mg^{2+}O^{2-})$$

We can use an ion-electron equation to show what happens to the magnesium. Each magnesium atom loses two electrons:

$$Mg \rightarrow Mg^{2+} + 2e^-$$

We say that the magnesium atoms have been **oxidised**.

Oxidation occurs whenever a metal forms a compound. For example, calcium loses electrons and forms calcium ions when it reacts with sulphur to give calcium sulphide.

Reduction reactions are those in which electrons are *gained*. Reduction takes place when a metal is obtained from one of its compounds. For example, when copper metal is produced from copper(II) oxide, the copper(II) ions gain electrons and turn into copper atoms:

$$Cu^{2+} + 2e^- \rightarrow Cu$$

The word 'OILRIG' can help you remember about electron loss and gain during oxidation and reduction, as shown in figure 3.1. You can also use the simple rule that in an oxidation reaction, electrons are shown on the *right-hand side* of the ion-electron equation. In a reduction reaction, the electrons are shown on the *left-hand side*.

Oxidation
Is
Loss
Reduction
Is
Gain

Figure 3.1 Oil rig – a useful way to remember that electrons are lost during oxidation and gained during reduction

Questions

Q1 State whether the following are oxidation or reduction reactions:

a) $Na \rightarrow Na^+ + e^-$

b) $I_2 + 2e^- \rightarrow 2I^-$

c) $S + 2e^- \rightarrow S^{2-}$.

Redox

Wherever a reaction involves both oxidation and reduction, it is called a **redox reaction**. In a redox reaction, electrons lost by one substance during oxidation are gained by another substance during reduction.

The formation of a compound by a metal is a redox reaction. For example, when sodium joins with chlorine, sodium chloride is formed. The oxidation and reduction parts of the overall reaction are:

$2Na \rightarrow 2Na^+ + 2e^-$ **oxidation**

$Cl_2 + 2e^- \rightarrow 2Cl^-$ **reduction**

All displacement reactions are redox reactions, for example the reaction between zinc metal and a solution of copper(II) ions. The zinc atoms lose electrons to the copper(II) ions. In other words, the zinc metal atoms are oxidised and at the same time the copper(II) ions are reduced:

overall reaction: $\mathbf{Zn + Cu^{2+} \rightarrow Zn^{2+} + Cu}$

oxidation part: $\mathbf{Zn \rightarrow Zn^{2+} + 2e^-}$

reduction part: $\mathbf{Cu^{2+} + 2e^- \rightarrow Cu}$

The reaction between a metal and a dilute acid can also be considered as a redox reaction. For example, in the reaction between magnesium and hydrochloric acid, the magnesium is oxidised and forms magnesium ions. The hydrogen ions present in the acid gain electrons and form hydrogen gas:

overall reaction: $\mathbf{Mg + 2H^+ \rightarrow Mg^{2+} + H_2}$

oxidation part: $\mathbf{Mg \rightarrow Mg^{2+} + 2e^-}$

reduction part: $\mathbf{2H^+ + 2e^- \rightarrow H_2}$

As a general rule, if electrons are transferred from one substance to another then a redox reaction has occurred.

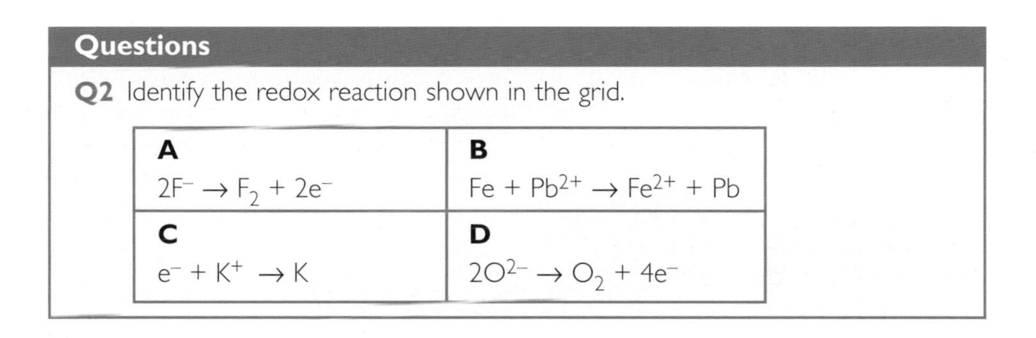

Questions

Q2 Identify the redox reaction shown in the grid.

A	**B**
$2F^- \rightarrow F_2 + 2e^-$	$Fe + Pb^{2+} \rightarrow Fe^{2+} + Pb$
C	**D**
$e^- + K^+ \rightarrow K$	$2O^{2-} \rightarrow O_2 + 4e^-$

Chapter 10 Study Questions

1 Pairs of metals can be used to produce a voltage.

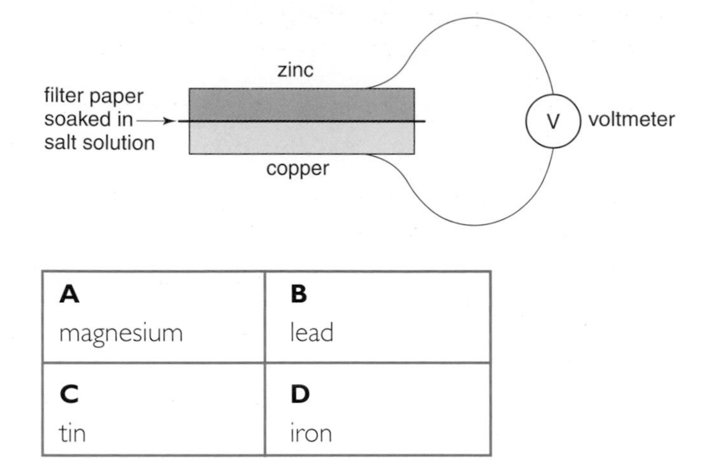

A magnesium	**B** lead
C tin	**D** iron

Identify the metal which would produce the **smallest** voltage if used in place of zinc.

You may wish to use page 7 of the SQA Data Booklet to help you.

SQA GENERAL (PS)

2 Batteries can be used to power everyday items.

A battery is a number of cells joined together.

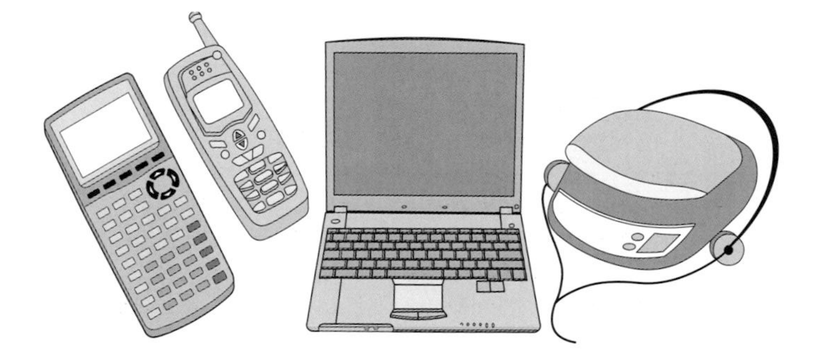

a) Suggest an advantage in using a battery rather than mains electricity. (KU)

b) A simple cell can be made from everyday objects.

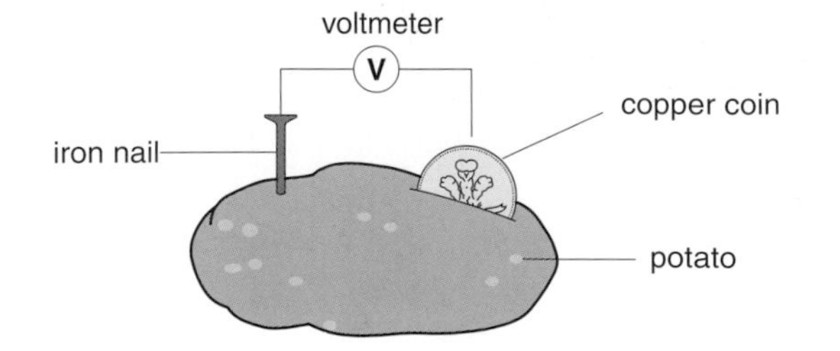

(i) What is the direction of electron flow?

(ii) What would happen to the voltage if the iron nail was replaced with aluminium foil?

You may wish to use page 7 of the SQA Data Booklet.

SQA GENERAL (PS)

3 You have been asked to investigate whether the concentration of the electrolyte affects cell voltage.

Which **two** cells could be compared to give this information?

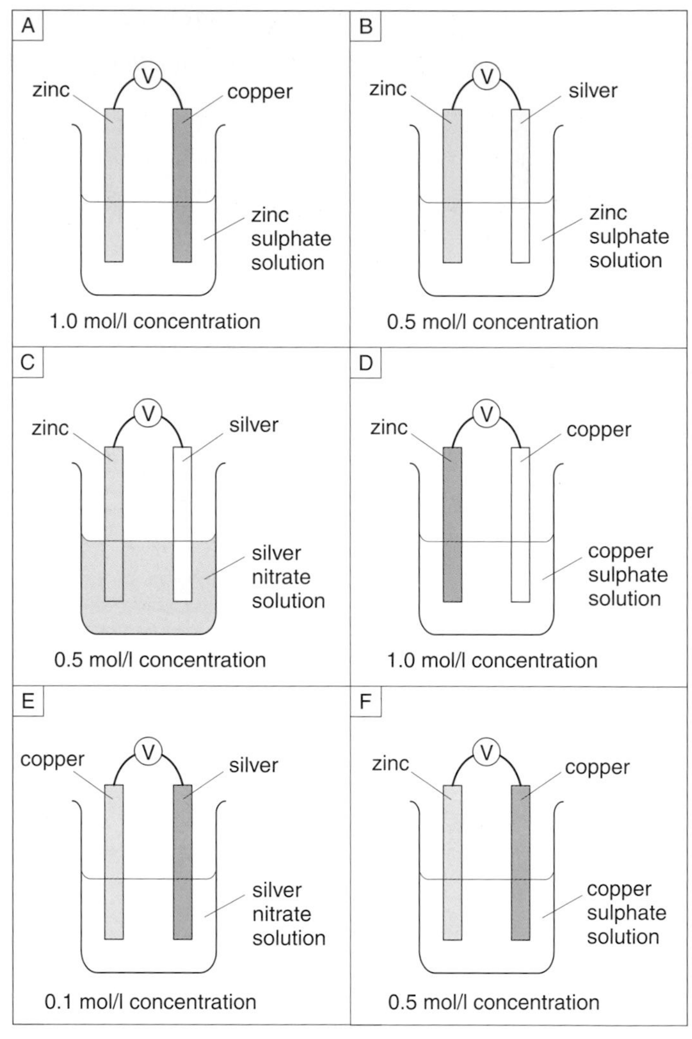

SEB CREDIT (PS)

4 Sodium sulphite solution reacts with bromine solution. The sulphite ions are oxidised. The ion-electron equation for the oxidation reaction is:

$$SO_3^{2-}(aq) + H_2O(l) \rightarrow SO_4^{2-}(aq) + 2H^+(aq) + 2e^-$$

This reaction takes place in the cell as shown top left of the next page.

a) Label the two solutions at (i) and (ii) so that the flow of electrons is in the direction shown. (PS)

b) What is the purpose of the filter paper (X) soaked in electrolyte between the beakers?

c) In the reaction, the bromine solution is reduced. Write the ion-electron equation for this reaction. (KU)

d) If samples of the two solutions were mixed, what would you see happening?

SQA CREDIT (PS)

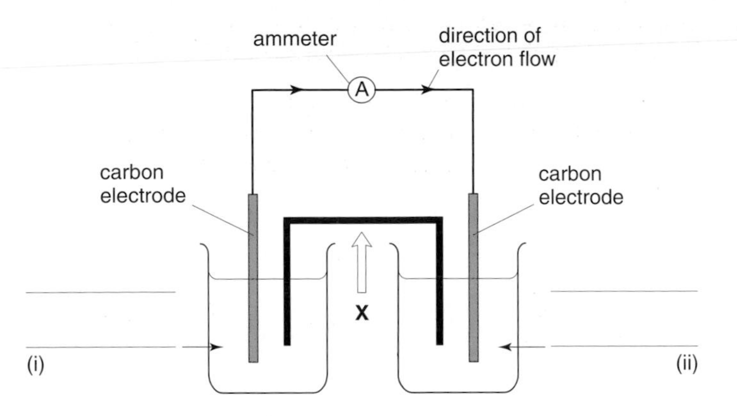

5 The batteries that are used to power torches, personal stereos, etc., are energy changers. What energy change takes place in a battery when it is being used?

GENERAL (KU)

6 The following metals are found in the electrochemical series:

zinc
iron
tin
lead
copper

a) Which *one* of the above metals would produce the smallest voltage when joined with zinc in a chemical cell?
b) Which *two* metals would give the greatest voltage if they were used in a chemical cell?

GENERAL (KU)

7 Hydrogen is placed in the electrochemical series between lead and copper. What are metals reacted with in order to establish this position?

CREDIT (KU)

8 State what is meant by the following terms:

a) reduction
b) oxidation
c) redox reaction.

CREDIT (KU)

9 In which of the following examples will displacement take place?

a) magnesium + zinc sulphate solution
b) iron + tin chloride solution
c) copper + lead nitrate solution
d) gold + silver nitrate solution
e) mercury + gold chloride solution
f) iron + sodium chloride solution.

GENERAL (KU)

10 When zinc reacts with silver(I) nitrate solution, the reactants are zinc atoms and silver(I) ions.

a) Complete and balance the equation:
$Zn(s) + Ag^{+}(aq) \rightarrow ...$
b) Give the formula of the spectator ions in this reaction.
c) State which reactant undergoes oxidation.
d) Write an ion-electron equation for this oxidation reaction.
e) Write an ion-electron equation for the reduction reaction.

CREDIT (KU)

11 When zinc metal is added to copper(II) sulphate solution a displacement reaction takes place.

a) Describe *two* changes in appearance that you would expect to occur.
b) (i) Write an ion-electron equation for the oxidation reaction taking place.
(ii) Write an ion-electron equation for the reduction reaction taking place.
(iii) Write the overall redox equation.
(iv) Name the spectator ions in this reaction.

CREDIT (KU)

12 A metal Q will displace a metal R from a solution containing ions of metal R if Q is above R in the electrochemical series.
Some results of displacement experiments using metals A, B and C are given in the table below:

Reactants	Result
C + ions of B	no reaction
B + ions of A	A displaced
A + ions of C	C displaced
A + ions of B	no reaction

a) What conclusions can be drawn from each of these four experiments, taken separately?
b) What is the order of these metals in the electrochemical series?

GENERAL (PS)

13 Apply the terms oxidation, reduction or redox to each of the following:

a) silver changing to silver(I) nitrate
b) iron(III) oxide changing to iron
c) sodium changing to sodium chloride
d) $Mg(s) + Fe^{2+}(aq) \rightarrow Mg^{2+}(aq) + Fe(s)$
e) $2H^{+}(aq) + 2e^{-} \rightarrow H_2(g)$
f) $2I^{-}(aq) \rightarrow I_2(s) + 2e^{-}$
g) $4e^{-} + O_2 + 2H_2O \rightarrow 4OH^{-}$

CREDIT (KU)

CHAPTER ELEVEN
Metals

SECTION 11.1 Properties of metals

We are surrounded by metals. Some are easy to see, such as car bodies, steel bridges, etc. Others are present in only tiny amounts. For example, tungsten metal is used in the tips of ball-point pens and in the wire filaments inside light bulbs. Computers contain small amounts of gold in their electrical connections.

We can use metals for so many different jobs because of their **properties**. Some of the most important properties of metals are:

- **Strength** – metals that are strong are used for bicycle frames, railway lines, ships' hulls, etc. Special steels are used for armour-plating in tanks.
- **Malleability** – metals are malleable. This means that they can be hammered and rolled into flat sheets or different shapes. Think about the way gold can be shaped into intricate chains and rings.
- **Conduction of electricity** – all metals are conductors of electricity. Copper wires are commonly used in household cables and flexes in electrical appliances. Aluminium cables carry the electricity in the National Grid.
- **Conduction of heat** – we use metals for making cooking pots and pans because they are able to conduct the heat from the cooker to the food inside. If you look on the back of a fridge you will see a metal grill. Its job is to conduct heat away from the fridge, keeping it cool.
- **Density** – this is the mass of a substance in a given volume. A high density material is much heavier than the same volume of a low density material. Aluminium is a metal with a low density. It is used to build aircraft bodies. Lead, with its high density, is used as weights for some fishing nets and lines.

Figure 1.1 These guitar strings are made from a metal containing nickel and copper

Questions

Q1 Look at figure 1.1. The guitar strings are made of a metal containing copper and nickel. What properties do they need to have?

Figure 1.2 These coins are made from alloys

Alloys

The properties of metals can be changed and improved by making **alloys**. Most alloys are mixtures of two or more metals, but in a few cases non-metals are also added. The usual way to make an alloy is to melt together the elements that make it up. Pure gold is very soft. Eighteen carat gold, which is used in some jewellery, is a harder metal made of an alloy of gold and copper. Modern coins are alloys. A £1 coin is an alloy of copper, nickel and zinc.

Coin	Made from...
'copper'; 1p, 2p	copper-coated steel
'silver'; 5p, 10p, 20p, 50p	alloy of copper and nickel
'gold'; £1	alloy of copper, nickel and zinc

Table 1.1 What our coins are made of

Alloy	Made from...
brass	copper and zinc
bronze	copper and tin
solder	lead and tin

Table 1.2 Some other alloys

Steels are alloys in which the main metal element is iron. They are usually much stronger than pure iron. Mild steel, an alloy of iron and carbon, is widely used in the construction industry. However, this alloy rusts easily. Stainless steel is much more rust-resistant because there is a small amount of chromium metal present in the alloy.

Figure 1.3 Recycling aluminium cans

Recycling metals

Recycling aluminium saves energy and money. To make a can from recycled aluminium can take only 5 per cent of the energy required to make one from new aluminium. This is because of the large amounts of electricity needed to produce new aluminium. Saving electricity also means that less fuel is burned in power stations. Large quantities of the greenhouse gas carbon dioxide are produced when aluminium is manufactured – recycling means less carbon dioxide is added to the atmosphere. Steel cans can also be recycled.

Figure 1.4 You may have seen this logo on aluminium cans. What does it mean?

Questions

Q2 Give at least four reasons why recycling metals is good for the environment.

Another reason for recycling metals is that the world is running out of certain metals. There is only a finite amount of each metal in the world.

The reactivity of metals

Figure 2.1 Why has the golden dome on this mosque stayed shiny for so long?

A reactivity series of metals

Some metals react very quickly and easily with the oxygen in the air. For example, a new piece of sodium metal will react with oxygen to form a layer of sodium oxide in about 30 seconds. However, copper takes many *years* to react with oxygen. You may also have noticed that gold does not react with the oxygen in the air at all. The golden dome of the mosque in figure 2.1, for example, has not reacted with the oxygen in the air even after hundreds of years.

From the observations above, we can put the three metals sodium, copper and gold in order of **reactivity**. We can say that sodium is more reactive than copper, which is more reactive than gold.

Chemists have produced a **reactivity series** – a kind of 'league table of metals' which puts them in order of reactivity. Table 2.1 shows a reactivity series.

In the reactivity series, the metals nearest the top are the most reactive. The reactivity of the metals decreases as you go down the series. Notice that the order of the metals in the reactivity series is very similar to that in the electrochemical series (see section 10.2).

Name	Symbol
potassium	K
sodium	Na
lithium	Li
calcium	Ca
magnesium	Mg
aluminium	Al
zinc	Zn
iron	Fe
tin	Sn
lead	Pb
copper	Cu
mercury	Hg
silver	Ag
gold	Au

Table 2.1 A reactivity series of metals

Reaction with oxygen

All of the metals above silver in the reactivity series combine with oxygen when heated. This produces a metal oxide. The higher a metal is in the series, the more violent the reaction is between the metal and oxygen. In general, this reaction takes the form:

$$\text{metal} + \text{oxygen} \rightarrow \text{metal oxide}$$

For example,

$$\text{magnesium} + \text{oxygen} \rightarrow \text{magnesium oxide}$$
$$2Mg + O_2 \rightarrow 2MgO$$

Equation showing ions and state symbols:

$$2Mg(s) + O_2(g) \rightarrow 2Mg^{2+}O^{2-}(s)$$

The apparatus shown in figure 2.2 can be used to study the reactivities of metals from magnesium down to copper in the reactivity series.

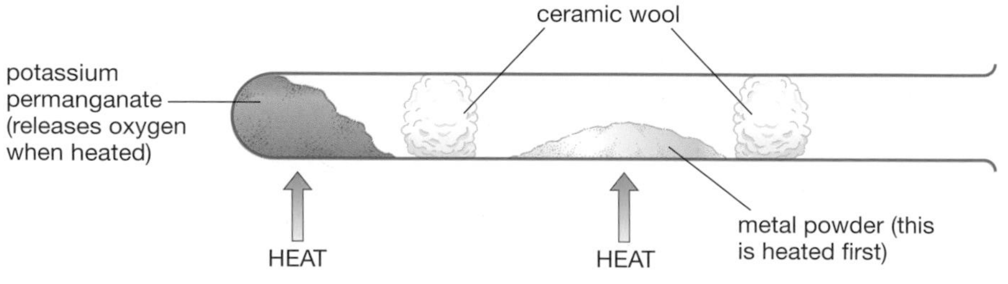

Figure 2.2 Metals glow brightly as they react with oxygen

The three metals at the top of the reactivity series – potassium, sodium and lithium – are so reactive they are stored under oil to prevent them from reacting with oxygen and water in the air.

Questions

Q1 a) Write a balanced equation for the reaction between potassium and oxygen.
b) Write an equation showing the ions present in the reaction. Use the equations given above as a guide.

Reaction with water

The metals above aluminium in the reactivity series react with water to produce hydrogen gas and the corresponding metal hydroxide. The general equation for this is:

$$\text{metal} + \text{water} \rightarrow \text{metal hydroxide} + \text{hydrogen}$$

For example,

$$\begin{array}{ccccccc} \text{sodium} & + & \text{water} & \rightarrow & \text{sodium hydroxide} & + & \text{hydrogen} \\ 2Na & + & 2H_2O & \rightarrow & 2NaOH & + & H_2 \end{array}$$

Equation showing ions and state symbols:

$$2Na(s) + 2H_2O(l) \rightarrow 2Na^+(aq) + 2OH^-(aq) + H_2(g)$$

Reaction with dilute acids

All metals above copper in the reactivity series react with dilute acids such as hydrochloric acid and sulphuric acid to produce a salt and hydrogen:

$$\text{metal} + \text{acid} \rightarrow \text{salt} + \text{hydrogen}$$

For example,

$$\begin{array}{ccccccc} \text{zinc} & + & \text{hydrochloric acid} & \rightarrow & \text{zinc chloride} & + & \text{hydrogen} \\ Zn & + & 2HCl & \rightarrow & ZnCl_2 & + & H_2 \end{array}$$

Equation showing ions and state symbols:

$$Zn(s) + 2H^+(aq) + 2Cl^-(aq) \rightarrow Zn^{2+}(aq) + 2Cl^-(aq) + H_2(g)$$

When a metal reacts with an acid it produces bubbles of hydrogen gas. In most cases, the faster the bubbles are produced, the more reactive the metal. One exception is aluminium. It reacts slowly with acids for about the first 20 minutes, after which it reacts quickly. The reason for this is that it is protected by a thin layer of aluminium oxide, which must first be removed by the acid.

Table 2.1 summarises the reactions of metals.

Metal	Oxygen	Reaction with Water	Diluted acid
potassium sodium lithium calcium magnesium	metal + oxygen ↓ metal oxide	metal + water ↓ metal hydroxide + hydrogen	metal + acid ↓ salt + hydrogen
aluminium zinc iron tin lead		no reaction	
copper			no reaction
mercury silver gold	no reaction		

Table 2.1 Reactions of metals

Questions

Q2 For the reaction between calcium and water, give:
a) a balanced equation,
b) an equation showing ions and state symbols.

Questions

Q3 For the reaction between magnesium and dilute hydrochloric acid, give:
a) a balanced equation,
b) an equation showing ions and state symbols.

Q4 A, B and C are three metals. Metal A reacts with dilute hydrochloric acid but not with water. Metal B does not react with water or dilute acid. Metal C reacts with water and dilute acid.
a) Place the metals in order of reactivity, with the most reactive first.
b) Using the reactivity series in table 2.1, give all the possible metals that A, B and C could be.

SECTION 11.3 Extracting metals

Ore	Metal compound present
bauxite	aluminium oxide
haematite	iron oxide
galena	lead sulphide

Table 3.1 Some common ores

How do we get gold? Most of it comes from gold mines where the gold is found as a metal – it is not combined with other elements. Only unreactive metals such as gold and silver are found uncombined. Most metals are obtained from **ores**, which are compounds of metals that occur naturally. Some common ores are shown in table 3.1.

Metals are obtained from their ores by **extraction.** This involves separating the metals from the other elements with which they have combined.

Looking back through history, we can see that the first metals to be discovered were the least reactive ones: gold, silver, copper, etc. This is because they were the easiest to extract. As a rule, the less reactive a metal, the easier it is to extract from its ore.

One of the most common methods of extraction is that used to obtain a metal from its oxide. Oxides of metals below copper in the reactivity series decompose when heated to give the metal and oxygen:

$$\textbf{metal oxide} \rightarrow \textbf{metal} + \textbf{oxygen}$$

For example,

$$\textbf{silver(I) oxide} \rightarrow \textbf{silver} + \textbf{oxygen}$$
$$2Ag_2O \rightarrow 4Ag + O_2$$

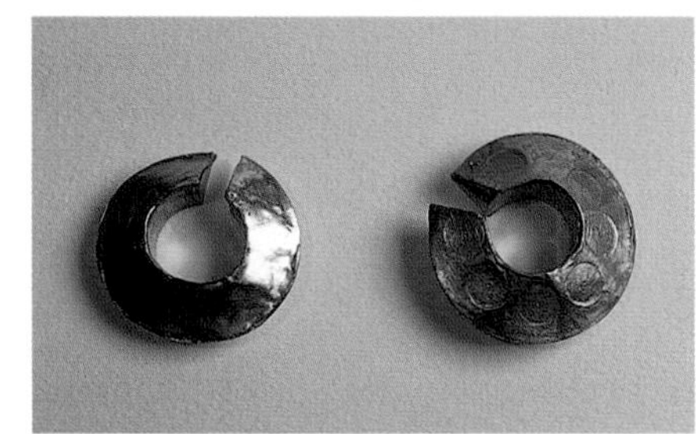

Figure 3.1 Gold hair rings found in Northumberland that date from the Bronze Age

Equation showing ions and state symbols:

$$2(Ag^+)_2O^{2-}(s) \rightarrow 4Ag(s) + O_2(g)$$

Questions

Q1 For the effect of heat on mercury(II) oxide,
a) write a balanced equation,
b) write an equation showing ions and state symbols.

The method described above is useful for the extraction of gold, silver and mercury. However, for oxides of more reactive metals, heating alone produces no reaction. Instead, metals below aluminium in the reactivity series can be extracted from their oxides by heating with carbon. The carbon and oxygen join to give carbon dioxide gas:

$$\textbf{metal oxide + carbon} \rightarrow \textbf{metal + carbon dioxide}$$

For example,

$$\textbf{zinc oxide + carbon} \rightarrow \textbf{zinc + carbon dioxide}$$
$$2ZnO + C \rightarrow 2Zn + CO_2$$

Figure 3.2 Galena (lead sulphide) crystals

Equation showing ions and state symbols:

$$2Zn^{2+}O^{2-}(s) + C(s) \rightarrow 2Zn(s) + CO_2(g)$$

Table 3.2 summaries the reactions of metal oxides.

Since the industrial revolution in the eighteenth century, there has been a great demand for iron metal. This was originally used to make bridges, railway lines, factories, ships, etc. Now most iron is converted into steel

Metal	Effect of heating metal oxide: Alone	Effect of heating metal oxide: With carbon or carbon monoxide
potassium sodium lithium calcium magnesium aluminium	no reaction	no reaction
zinc iron tin lead copper		metal oxide + carbon or carbon monoxide → metal + carbon dioxide
mercury silver gold	metal oxide → metal + oxygen	

Table 3.2 Reactions of metal oxides

before being used. Steel is a stronger metal than iron. In 1999 over 550 million tonnes of iron were produced worldwide.

Iron is extracted from iron ore, which is usually iron(III) oxide. The process is carried out in a **blast furnace** (see figure 3.3). This is a huge structure, up to 70 metres high. Coke, which is mainly carbon, is used in the extraction. Iron ore and coke are fed in at the top of the blast furnace. Some of the iron(III) oxide and the carbon in the coke react to give iron and carbon dioxide, but the important reactions are as follows:

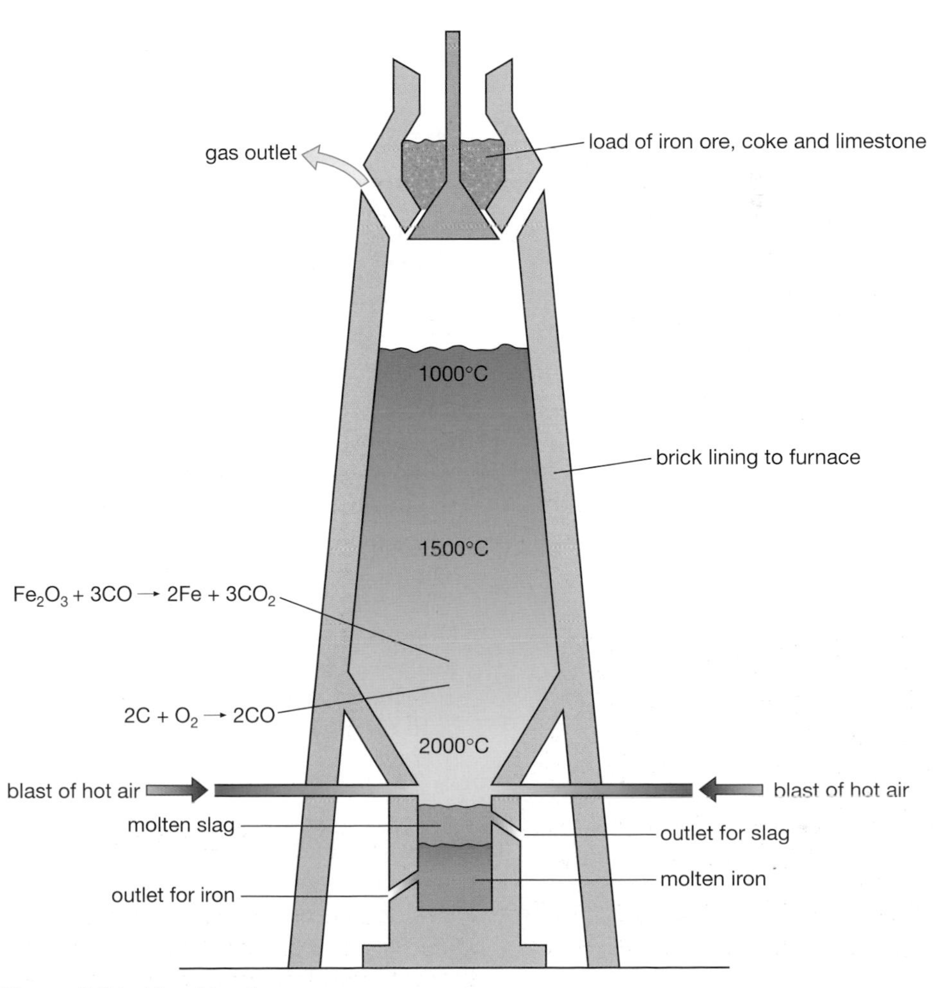

Figure 3.3 Inside a blast furnace

- At the bottom of the furnace, blasts of air are blown in. Carbon in the coke reacts with the oxygen in the air to undergo incomplete combustion. This produces carbon monoxide gas.
- Further up the furnace, carbon monoxide and iron(III) oxide react to produce iron metal and carbon dioxide – the carbon monoxide has removed the oxygen from the iron(III) oxide.

How is carbon monoxide made in a blast furnace?

The oxygen in the blast of hot air reacts with the carbon in the coke to make carbon dioxide (CO_2).

$$C + O_2 \rightarrow CO_2$$

The carbon dioxide then reacts with carbon in the coke to make carbon monoxide.

$$CO_2 + C \rightarrow 2CO$$

Taking these two equations together gives the following overall equation for the reaction.

$$2C + O_2 \rightarrow 2CO$$

The blast furnace is hot enough to melt the newly formed iron that falls to the bottom. On top floats a liquid slag which is formed by reactions between impurities and limestone, which is added with the iron ore and coke.

A blast furnace can produce up to 10 000 tonnes of iron a day.

Questions

Q2 Blast furnaces are used to make iron for the steel industry. Find out at least two places where steel is manufactured in Britain.

Metal extraction involves reduction

Metal oxides are said to be *reduced* to the metal during extraction. This is because the metal ion in the oxide *accepts electrons* when the metal is formed. For example, in the extraction of zinc from zinc oxide, the reduction equation is:

$$Zn^{2+} + 2e^- \rightarrow Zn$$

Reactivity and ease of metal extraction

Reactive metals hold on more strongly to oxygen than less reactive metals. Therefore it is easier to remove oxygen from compounds where it is joined to unreactive metals. For example, oxygen can be removed from mercury oxide by heating. Iron oxide does not change in this way with heat alone. However, when it is heated with carbon, the carbon removes the oxygen from the iron oxide, leaving iron metal behind.

The reactive metals from aluminium upwards in the reactive series hold on to oxygen more strongly than carbon does. Heating with carbon therefore does not work as a method for extracting these metals.

Questions

Q3 Using the reduction equation for zinc to help you, write similar equations for the reduction of:

a) iron(III) ions to form iron metal,

b) silver(I) ions to form silver metal.

Questions

Q1 Calculate the percentage by mass of iron in the ore magnetite, formula Fe_3O_4.

Q2 Calculate the percentage by mass of aluminium in the compound alumina. This can be assumed to be pure aluminium oxide, Al_2O_3.

Percentage composition

There are two important forms of iron ore. One contains a higher percentage of iron than the other. This means that, in a blast furnace, the ores will give different amounts of iron metal.

Chemists often need to know the percentage of each element in a compound. This is called the **percentage composition** of the compound.

For example, haematite consists of a compound of the elements iron and oxygen. The name of the compound is iron(III) oxide. The percentage of each can be calculated as follows:

$$\textbf{percentage mass of a given element in a compound} = \frac{\textbf{mass of element in formula}}{\textbf{formula mass}} \times 100$$

formula mass of haematite (Fe_2O_3) = (2 × 56) + (3 × 16) = 160

$$\textbf{percentage of iron} = \frac{\textbf{mass of iron in formula}}{\textbf{formula mass}} \times 100$$

$$= \frac{(2 \times 56)}{160} \times 100$$

$$= 70\%$$

Note: because there are two iron atoms in the formula, the mass of the iron is found by multiplying its relative atomic mass by two.

Empirical formulae

How are chemists able to work out the formulae of new compounds? One way is to use calculations involving the compound's percentage composition. The calculations are based on the formulae mass of the compound, First, the percentage by mass for each element is converted to the number of moles of atoms of each element present. This is done by dividing the percentage of each element by the element's relative atomic mass. Finally, the formula is obtained by calculating the ratio of moles of each element present. The simplest whole number ratio of atoms gives the empirical fomula.

For example, an oxide of tin was analysed and found to consist of 78.8 per cent tin and 21.2 per cent oxygen. The empirical formula is found as follows:

element:	**Sn**	**O**
mass/g	78.8	21.2
divide by relative atomic mass to find number of moles of atoms	$\frac{78.8}{119} = 0.662$	$\frac{21.2}{16} = 1.325$
divide by smaller number to find simplest whole number ratio	$\frac{0.662}{0.662} = 1$	$\frac{1.325}{0.662} = 2$
empirical formula	SnO_2	

Note: the ratio is found by dividing the number of moles of each atom by the smaller number – in this case 0.662.

Questions

Q3 An oxide of copper was analysed and found to contain 88.8 per cent copper and 11.2 per cent oxygen. Calculate the empirical formula for this oxide.

Q4 0.1225 g of magnesium were found to give 0.2025 g of magnesium oxide on complete combustion. The mass of oxygen joining with the magnesium is therefore (0.2025 – 0.1225) = 0.8000 g. Use this information to work out the empirical formula for magnesium oxide.

Q5 4.78 g of an oxide of lead on reduction gave 4.14 g of lead metal. Use this information to work out the empirical formula for the oxide.

Empirical formulae can also be found using the masses of the elements involved, not just the percentages.

For example, the formula of magnesium oxide can be worked out by heating a known mass of magnesium in a crucible until it is completely converted to magnesium oxide, which is then weighed (see figure 4.1). Subtraction gives the mass of oxygen which has combined with the magnesium.

In an experiment 2.80 g of iron combined with 1.20 g of oxygen to make an oxide of iron.

The empirical formula is found as follows:

	Fe	O
	2.80 g	1.20 g
divide by relative atomic mass to find number of moles of atoms	$\frac{2.80}{56} = 0.05$	$\frac{1.20}{16} = 0.075$
divide by the smaller number to find the simplest whole number ratio	$\frac{0.05}{0.05} = 1$	$\frac{0.075}{0.05} = 1.5$
multiplying by 2 to change 1.5 into a whole number	$1 \times 2 = 2$	$1.5 \times 2 = 3$

This gives the empirical formula of Fe_2O_3

Finding the molecular formula

The empirical formula is not always the same as the molecular formula for a covalent compound. However, given the empirical formula and the formula mass it is possible to work out the molecular formula.

For example, the hydrocarbon benzene has the empirical formula CH and a formula mass of 78. The molecular formula of benzene must be C_nH_n, where n is a whole number.

formula mass of $C_nH_n = (12 \times n) + (1 \times n) = 78$

$$13n = 78$$

$$n = 6$$

Thus the molecular formula of benzene is C_6H_6.

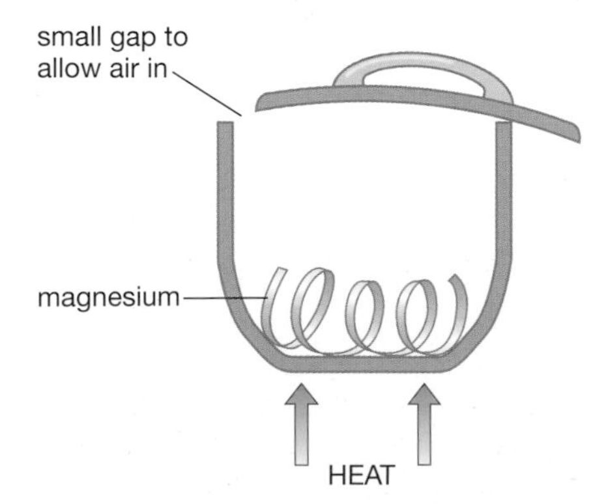

Figure 4.1 Making magnesium by combustion

Calculations based on balanced equations

Balanced equations can be used to make predictions about the masses of substances either reacting or being produced. In order to do this, every formula is taken to represent one mole of the substance concerned.

For example, what mass of hydrogen would be produced when 6 g of magnesium react completely with dilute hydrochloric acid?

balanced equation: $Mg + 2HCl \rightarrow MgCl_2 + H_2$

relate moles of required substances: 1 mole ⟷ 1 mole

replace moles by formula mass in grams: 24 g ⟷ 2 g

use simple proportion: 6 g ⟷ $\frac{2 \times 6}{24}$ = 0.5 g

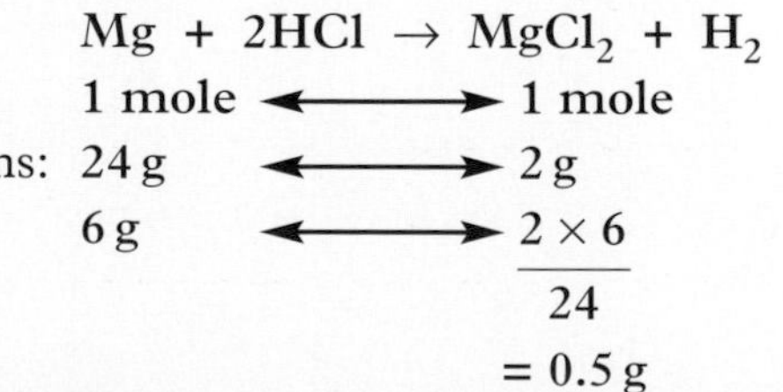

Questions

Q6 The gas cyanogen has the empirical formula CN and a formula mass of 52. What is its molecular formula?

Questions

Q7 Calculate the mass of tin that would be produced if 7.55 g of tin(IV) oxide were reduced by hydrogen. The equation for the reaction taking place is:

$$SnO_2 + 2H_2 \rightarrow Sn + 2H_2O$$

1 Aluminium is a metal with a large number of useful properties.

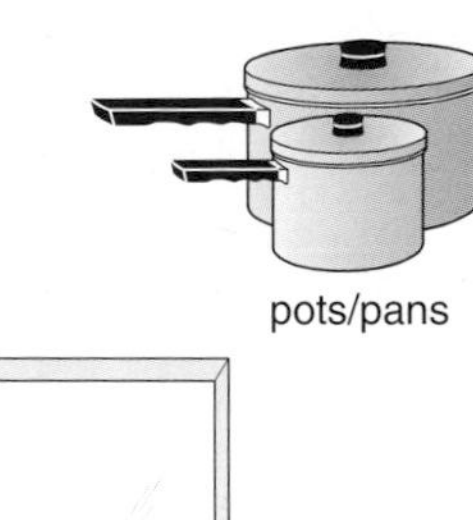

It is a good heat conductor. This makes it useful for making saucepans.

Because it is resistant to corrosion it is used to make window and door frames.

It has a low density, which is one reason for its use in overhead power cables.

One of the reasons it is used for cooking foil is that it is malleable.

a) Present the above information in a table with suitable headings.

b) Duralumin is an alloy of aluminium. It is stronger than aluminium itself.
What is an alloy?

c) The low density of aluminium makes it useful for power cables.
Suggest **another** reason why aluminium is used.

SQA GENERAL (KU)

2 Hydrogen is produced when steam reacts with red hot magnesium ribbon.

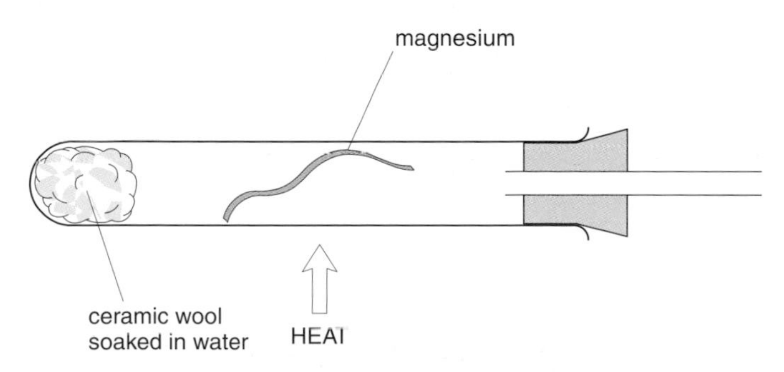

a) Write the equation for the reaction using symbols and formulae.
(There is no need to balance the equation.)

b) Complete the diagram to show how a test tube of hydrogen can be collected.

c)

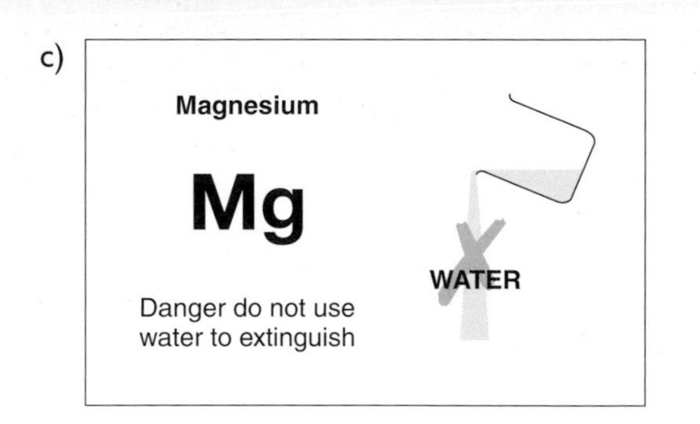

Why should water not be used to extinguish burning magnesium? (PS)

3 When hydrogen gas is passed over heated lead(IV) oxide, silvery beads of lead are produced. The only other product of the reaction is water in the form of steam.

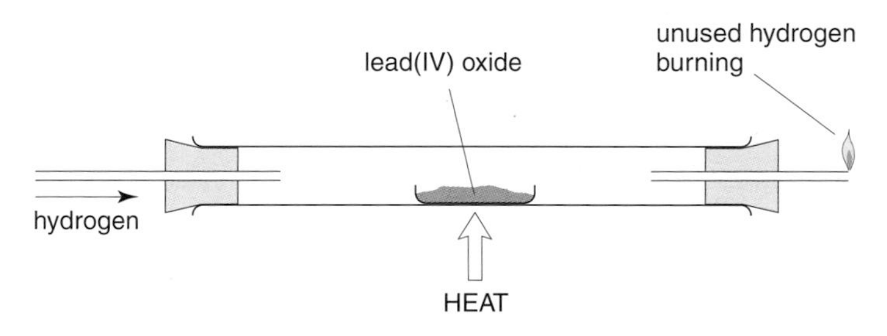

a) Write a balanced equation for the reaction taking place.

b) Name the *type* of reaction taking place.

c) When the experiment was repeated using magnesium oxide instead of lead(IV) oxide, no metal was obtained. Explain this result. CREDIT (KU)

4 The grid below contains formulae for some metal oxides:

A K_2O	**B** CaO	**C** SnO_2
D Ag_2O	**E** Fe_2O_3	**F** Li_2O

Identify the metal oxide which would be the most likely to produce a metal by heating alone. CREDIT (KU)

5 Pieces of zinc and copper are put into separate test tubes each containing dilute hydrochloric acid. Describe what you would expect to see happening in each case. GENERAL (KU)

6 Give (i) word equations and (ii) equations using chemical formulae (not necessarily balanced) for the reactions between:

a) potassium and oxygen

b) lithium and water

c) magnesium and dilute hydrochloric acid. GENERAL (KU)

7 Many different reactions take place in a blast furnace. Identify the reaction in which a reactant **and** a product are elements.

A	$C + O_2 \rightarrow CO_2$
B	$CO_2 + C \rightarrow 2CO$
C	$CaCO_3 \rightarrow CaO + CO_2$
D	$2Fe_2O_3 + 3C \rightarrow 4Fe + 3CO_2$
E	$Fe_2O_3 + 3CO \rightarrow 2Fe + 3CO_2$

SQA GENERAL (PS)

8 a) Platinum metal is extracted from its ores using heat alone. What does this indicate about the reactivity of platinum?

b) Platinum ores also contain copper sulphide. When the ores are heated, the copper sulphide reacts to give sulphur dioxide gas.

$$2CuS + 3O_2 \rightarrow 2CuO + 2SO_2$$

Calculate the mass of sulphur dioxide produced when 96 g of copper sulphide is heated.

c) Some petrol engined cars use catalytic converters containing platinum. What is the purpose of a catalytic converter?

SQA CREDIT (KU)

9 Calculate the percentage by mass of:

a) sodium in sodium chloride ($NaCl$)
b) lead in lead(II) bromide ($PbBr_2$)
c) zinc in zinc carbonate ($ZnCO_3$)
d) calcium in calcium phosphate ($Ca_3(PO_4)_2$). CREDIT (KU)

10 A coin made of pure copper does not last as long as one in which the copper has been alloyed with another metal such as zinc or nickel.

a) State what is meant by the term 'an alloy'.
b) Suggest one possible change which is brought about by alloying that causes the coins to last longer.

GENERAL (KU)

11 3.2 g of sulphur were found to combine with 3.2 g of oxygen. Calculate the empirical formula of the oxide produced. CREDIT (KU)

12 Iron ore is a mixture of compounds. One of these is called siderite and has the composition:

iron	48.3%
oxygen	41.4%
carbon	10.3%

Calculate the empirical formula of siderite.
Show your working clearly.

SQA CREDIT (KU)

13 A hydrocarbon was analysed and found to contain 92.3 per cent by mass of carbon and 7.7 per cent by mass of hydrogen. Calculate the empirical formula of the hydrocarbon. CREDIT (KU)

14 0.39 g of a hydrocarbon was analysed and found to contain 0.36 g of carbon and 0.03 g of hydrogen.

a) Calculate the empirical formula of the hydrocarbon.
b) Using a mass spectrometer, it was found that the formula mass of the compound was 52. Determine the molecular formula of the hydrocarbon. CREDIT (PS)

15 A rocket fuel was analysed and found to be a compound consisting of nitrogen (85.7 per cent) and hydrogen (12.5 per cent).

a) Calculate the empirical formula of the fuel from the percentage composition. (KU)
b) Further tests revealed that the compound had a formula mass of 32. What is the molecular formula of the fuel? CREDIT (PS)

16 2.00 g of an oxide of iron were reduced to the metal by heating in a steam of carbon monoxide. 1.488 g of iron remained when reduction was complete.

a) Work out the empirical formula of the oxide. (PS)
b) Write a balanced equation for the reaction between this oxide and carbon monoxide. CREDIT (KU)

17 Nitrogen is used to fill the air-bags which protect people in car crashes.

It is produced when sodium azide (NaN_3) decomposes rapidly.

$$2NaN_3(s) \rightarrow 2Na(s) + 3N_2(g)$$

a) Why is nitrogen a suitable gas for this purpose? (PS)
b) A driver's air bag contains 60 g of sodium azide. Calculate the mass of nitrogen gas which will be produced.

(You may wish to refer to page 4 of the SQA Data Booklet.)

Show your working clearly.

SQA CREDIT (KU)

18 Consider the following balanced equation:

$$Fe_2O_3 + 3H_2 \rightarrow 2Fe + 3H_2O$$

a) How many moles of iron would be produced from 2 moles of iron(III) oxide?
b) What mass of hydrogen would be required to convert 2 moles of iron(III) oxide to iron? CREDIT (PS)

19 What mass of carbon dioxide is formed when 64 g of methane are burned completely in air?

$$CH_4 + 2O_2 \rightarrow CO_2 + 2H_2O$$

CREDIT (KU)

CHAPTER TWELVE

Corrosion

SECTION 12.1 What is corrosion?

Figure 1.1 Why do cars rust?

If you have an old bicycle then you will probably have seen how corrosion can destroy metals. For example, the steel in the cycle chain can slowly turn to rust, making it weaker and weaker until it eventually snaps.

During corrosion, the surface of a metal changes from being an element into a compound. Copper corrodes to form a green copper compound; this can be seen in the copper domes of old buildings which have turned green over the years.

Most metals corrode. However, *different metals corrode at different rates*. In general, the more reactive a metal is, the faster it corrodes. When a piece of sodium is cut open it has a bright, shiny surface, but this becomes dull after a few seconds as the metal corrodes (figure 1.2). Under normal conditions, lead corrodes very slowly and gold does not corrode at all.

The rusting of iron

Iron is the most widely used metal. Remember that steel is made mostly of iron. **Rusting** is the term used for the corrosion of iron (or steel). Only iron rusts; other metals are said to 'corrode'. **Rust** is the name of the compound which is formed when iron corrodes.

The rusting of iron and steel is a major problem. It costs millions of pounds each year to replace rusted cars, bridges, railings, etc. It is important, therefore, to know what causes rusting. The experiment in figure 1.3 gives some clues to this.

Four identical iron nails are placed in test tubes as shown. After one week they are examined to see if they have rusted. The results of the experiment are shown in figure 1.3. Note that in tube 1, calcium chloride is added to remove any moisture from the air. In tube 2 the water is boiled to drive out any dissolved air from the water. The layer of oil stops any air dissolving back into the water.

Figure 1.2 A freshly cut piece of sodium with a shiny surface. A few minutes later the surface has corroded.

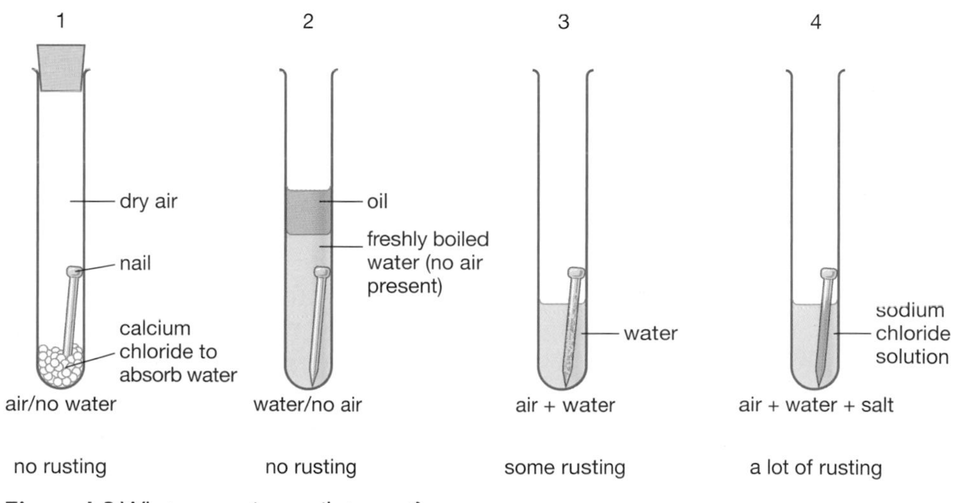

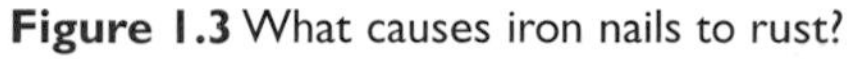

Figure 1.3 What causes iron nails to rust?

Questions

Q1 Look at the experiment in figure 1.4. Why do you think the second test tube with no iron filings was included?

From this experiment it can be concluded that both air *and* water are needed for rusting to occur. Also, salt speeds up rusting. In fact, the presence of most soluble ionic substances will make rusting occur faster.

Rusting, therefore, requires air, but which of the gases present in the air does it need? All of them, or just one? The experiment in figure 1.4 shows that *oxygen* is needed for rusting to take place. As the moist iron filings rust, the water level in the test tube rises. It rises by about one-fifth of the height of the tube – remember that oxygen makes up about one fifth of the air.

From these and other experiments, it can be concluded that for iron to rust, water and oxygen must be present.

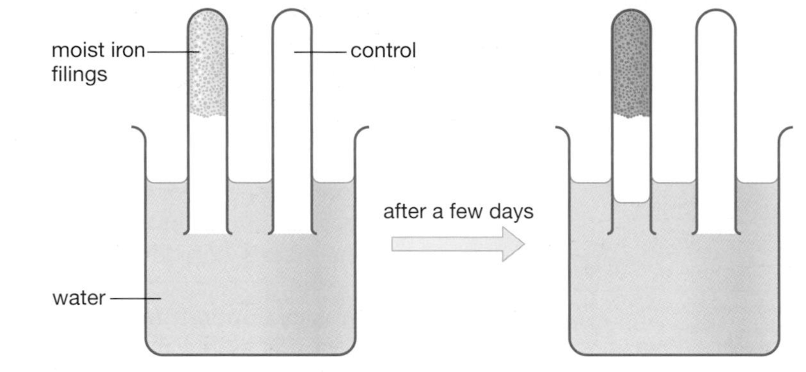

Figure 1.4 What is required to make the moist iron filings rust?

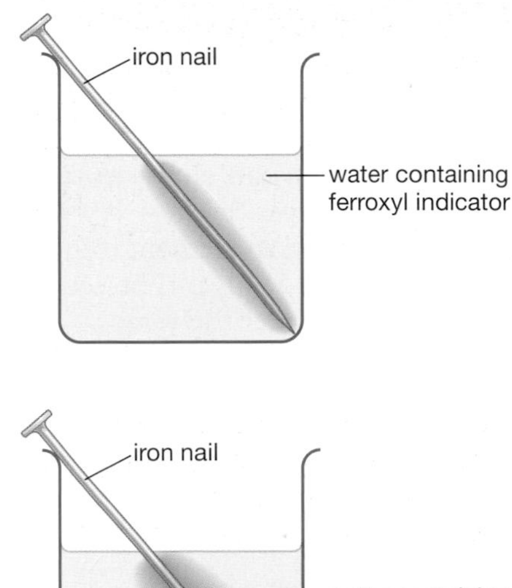

Figure 1.5 In which solution has more rusting taken place?

What happens when iron rusts?

Rusting is a complicated process. However, the first step consists simply of iron atoms losing electrons and forming iron(II) ions:

$$Fe \rightarrow Fe^{2+} + 2e^-$$

Before the iron atoms can lose electrons, there has to be a substance present to accept them, such as oxygen. An electrolyte must also be present. In Chapter 7 you learnt that an electrolyte is an ionic compound which conducts electricity when dissolved in water. Electrolytes are often present in rain water, therefore an outdoor iron gate will rust quite quickly. Sodium chloride (common salt) is an electrolyte, and so rusting occurs faster when salt is present. For example, car bodies and bicycle frames rust more quickly if salt is spread on the roads in winter.

To find out if rusting is taking place, we can use **ferroxyl indicator**. This is a pale yellow-coloured solution which turns blue when it reacts with Fe^{2+} ions (the ions which are formed when iron metal starts to rust). In general, the more blue colour there is, the more rusting has taken place (figure 1.5).

Questions

Q2 Imagine that you have carried out the experiment shown in figure 1.5. What conclusion can you make concerning the results of the experiment?

SECTION 12.2 Preventing corrosion

Physical protection

For a metal to corrode it needs to be exposed to air and water. One way to prevent corrosion is therefore to put a barrier around the metal to keep out air and water. This is why the Forth Railway Bridge girders are painted (figure 2.1). The paint forms a physical barrier which keeps air and water away from the steel underneath. Oil or grease also form a barrier which will cover metal surfaces.

Figure 2.1 The Forth Railway Bridge

Physical barriers to corrosion:
paint oil or grease plastic coating tin-plating electroplating galvanising

Table 2.1 Physical barriers preventing corrosion

Table 2.1 lists some physical barriers which are used to prevent corrosion of metals.

Small metal parts such as bolts and axles are often dipped in molten plastic before being packaged. The plastic solidifies to form a thin, protective layer over the metal. In the case of tin-plating, the metal to be protected is dipped into molten tin. Tin-plating can also be carried out using electrical methods, as described below.

A car body is made of steel. It can be protected against corrosion by dipping it into molten zinc (figure 2.2). This is called **galvanising**. This covers the steel with a tough layer of zinc. The zinc protects the steel from air and water. Steel objects which are protected in this way are called galvanised steel.

Figure 2.2 The dustbins and the cars are both protected from corrosion by zinc coating

Some motorway crash barriers are made from galvanised steel. The steel in the Forth Road Bridge was protected against corrosion by being sprayed with molten zinc (figure 2.3); 500 tonnes of zinc were used.

Galvanising also protects steel by a method called **sacrificial protection** – this is discussed later in this chapter.

Figure 2.3 The Forth Road Bridge

Electroplating

You may have some gold-plated jewellery or perhaps some silver-plated spoons at home. Gold, silver, chromium and other metals can be deposited as shiny coatings on the surface of other metals. The name for this process is **electroplating** (figure 2.4).

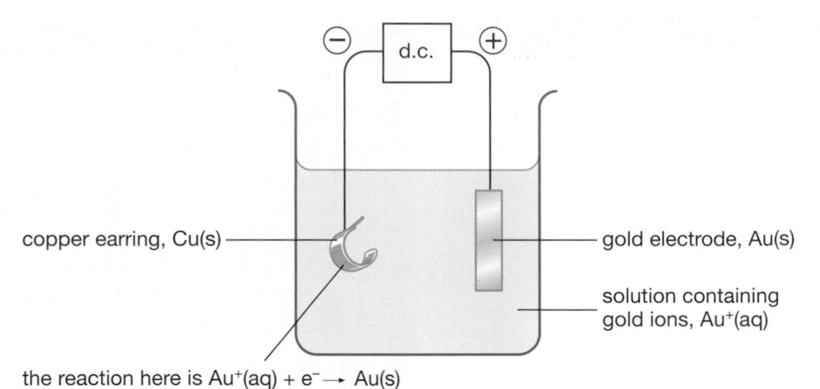

Figure 2.4 Gold plating a copper earring

The object to be plated (for example, a copper earring) is used as the negative electrode. It is in contact with a solution which contains ions of the metal with which it is to be plated. For example, the copper earring could be put in a solution containing gold ions. Metal ions are positively charged, therefore they are attracted to the negative electrode. When they reach the electrode they gain electrons and form a thin layer of metal. Some examples of reactions taking place at the negative electrode are:

silver-plating	$Ag^{+} + e^{-} \rightarrow Ag$
copper-plating	$Cu^{2+} + 2e^{-} \rightarrow Cu$
chromium-plating	$Cr^{3+} + 3e^{-} \rightarrow Cr$

Chromium-plating produces a very hard surface. Chromium-plated parts are used in the engines of military helicopters such as the Lynx.

Tin-plating is used to make the material for 'tin' cans. It is carried out by passing a steel strip through a solution containing tin ions. While it is in the solution, the steel is the negative electrode and a thin layer of tin is deposited on the steel.

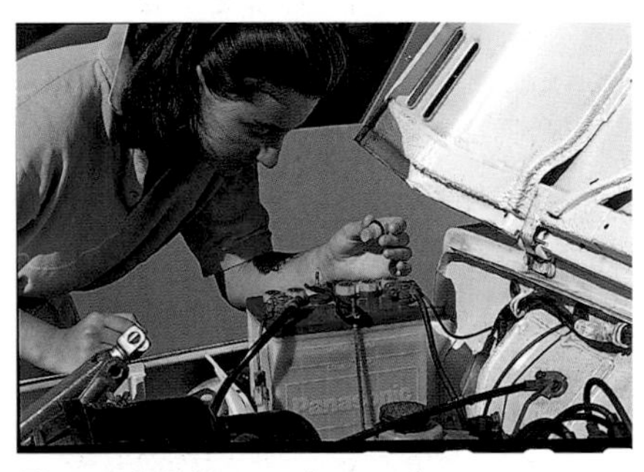

Figure 2.5 Direct electrical protection is being used to cut down the corrosion of this car's bodywork. Notice how the negative terminal of the battery is connected to the car body

Direct electrical protection

When a metal corrodes it loses electrons. Therefore, if electrons could be forced back to the metal, it would not corrode. Electrons can be supplied from the negative terminal of any direct current power supply such as a battery. This method of protection is most widely used in cars and trucks where the negative terminal of the battery is connected to the steel bodywork (see figure 2.5). Many piers and jetties and even some ocean liners when they are in port also use electrical protection to cut down on corrosion. While this method slows down corrosion considerably it unfortunately does not stop it completely.

Sacrificial protection

In Chapter 10 you read that all metals tend to lose electrons. When two metals are joined in a simple cell, electrons will flow from the metal which is higher in the electrochemical series to the one lower down in the series.

This fact is used in a method of preventing corrosion called **sacrificial protection**. For example, zinc is higher in the electrochemical series than iron. If iron is in contact with zinc then electrons will flow from the zinc to the iron and this will help prevent the iron from losing electrons and corroding. Because the zinc loses electrons, it corrodes. In other words the zinc is 'sacrificed' to protect the iron. Magnesium gives an even better protection. It is higher in the electrochemical series than zinc and so it loses electrons more readily.

Figure 2.6 shows how underground pipes such as gas pipes can be protected from corrosion by connecting them to scrap magnesium at regular intervals.

Questions

Q1 If cars were to have their bodywork connected to the *positive* terminal of their batteries, what sort of problems would this be likely to cause?

Questions

Q2 In 1824 Sir Humphry Davy suggested to the Admiralty that the corrosion of copper parts on wooden ships could be prevented by attaching blocks of zinc to the ships' hulls. Explain why this method worked well.

Q3 Galvanising gives iron two types of protection from corrosion. What are they?

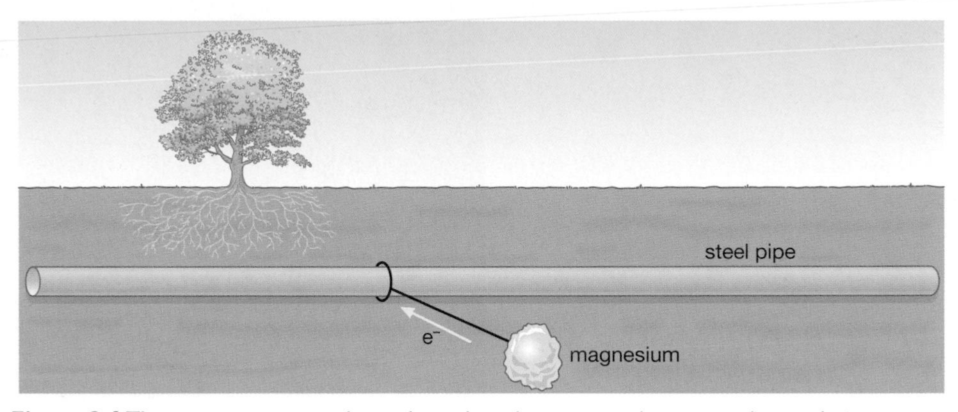

Figure 2.6 The magnesium corrodes and supplies electrons to the iron in the steel pipe, protecting the pipe from corrosion

Some ships have zinc blocks attached to their hulls to protect against the corrosive action of sea water.

SECTION 12.3 More on corrosion

You have read in section 12.1 that corrosion occurs when metals *lose electrons* and form compounds. For example, magnesium metal atoms corrode as follows:

$$Mg \rightarrow Mg^{2+} + 2e^-$$

Reactions which involve a loss of electrons are oxidation reactions. When iron rusts, the initial oxidation reaction is:

$$Fe \rightarrow Fe^{2+} + 2e^-$$

Rusting is, however, a complicated process which involves more than one step. Further oxidation of the iron(II) ions can occur:

$$Fe^{2+} \rightarrow Fe^{3+} + e^-$$

The iron(III) ions which are produced can form iron(III) oxide. This is the substance which we call rust.

When an oxidation reaction occurs there must be a corresponding reduction reaction. In the case of rusting, this involves water and oxygen molecules reacting as follows:

$$2H_2O + O_2 + 4e^- \rightarrow 4OH^-$$

This reduction reaction helps to explain why water and oxygen are needed for the rusting of iron.

Water is also required because it usually contains electrolytes. As you remember, electrolytes are compounds that dissolve in water to produce ions. For example, carbon dioxide from the air will dissolve in water to form an electrolyte. During rusting, electrolytes assist with the movement of electrons from the iron to the water and oxygen molecules.

In most cases, the higher the concentration of electrolytes present the faster the iron or steel will rust. Common salt (sodium chloride) is an electrolyte and so the presence of salt speeds up rusting.

Figure 3.1 Why does sea water speed up rusting?

Direction of electron flow during corrosion

A simple chemical cell set up as in figure 3.2 can be used to show that electrons flow *away* from iron as it corrodes.

In the iron/carbon cell, the blue colour around the iron shows that it is rusting. The milliammeter shows that the electrons are flowing away from the iron towards the carbon. These electrons are produced by the oxidation of the iron. The pink colour around the carbon is due to the formation of OH^- ions produced in the corresponding reduction reaction.

In figure 3.3 there is no blue colour around the iron. It has not rusted. This is because electrons have been flowing *towards* the iron from the magnesium. Magnesium is higher than iron in the electrochemical series. The pink colour surrounding the magnesium shows the formation of OH^- ions.

In figure 3.4 the rusting of iron is particularly rapid because iron is above tin in the electrochemical series. As a result, electrons flow from the iron to the tin.

This effect is seen with tin-plate, which is made from thin sheets of steel plated with tin. Tin-plate is used to make 'tin' cans for food. If a tin can is badly bashed or scratched the steel may become exposed. Rusting then occurs rapidly as electrons flow from the iron (the main metal in steel) to the tin. In such cases, the iron is 'sacrificed' and the tin is protected.

Questions

Q1 Look at figure 3.3. In which direction would the electrons flow if the magnesium was replaced by copper?

Q2 Which of the following would speed up the corrosion of iron by acting as electrolytes?
a) barium sulphate,
b) potassium nitrate,
c) calcium carbonate,
d) hydrochloric acid.

Q3 Imagine that a millionaire had a car in which the steel bumpers were gold-plated. What would happen if the gold was scratched and the steel was exposed?

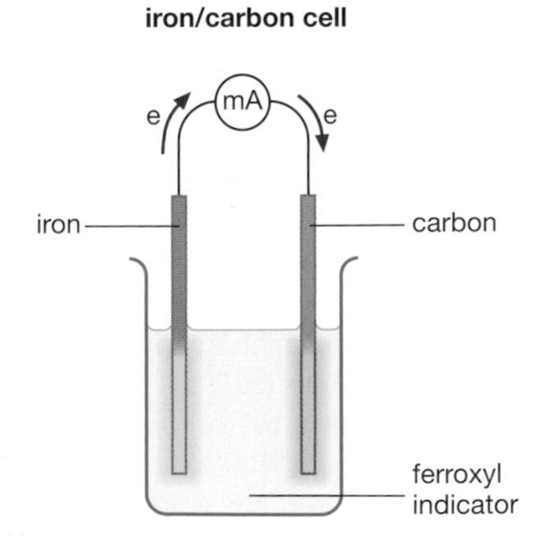

Figure 3.2 Iron in this cell rusts

iron/magnesium cell

e mA e

iron

magnesium

ferroxyl indicator

Figure 3.3 Iron in this cell does not rust

iron/tin cell

e mA e

iron

tin

ferroxyl indicator

Figure 3.4 Iron in this cell rusts badly

1 Hamish was investigating ways of protecting iron from rusting.

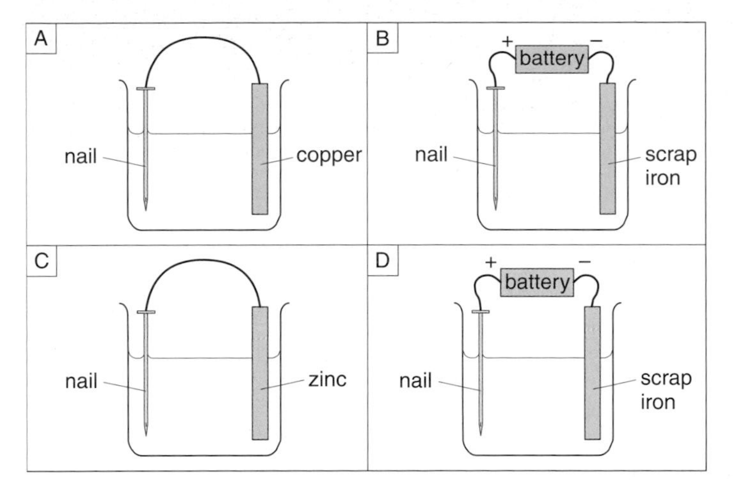

He set up four experiments using iron nails in salt water.

Identify the arrangement(s) which would protect the iron nail.

SEB GENERAL (PS)

2 Oil platforms can have metal plates fixed to the steel structure to protect them from corrosion.

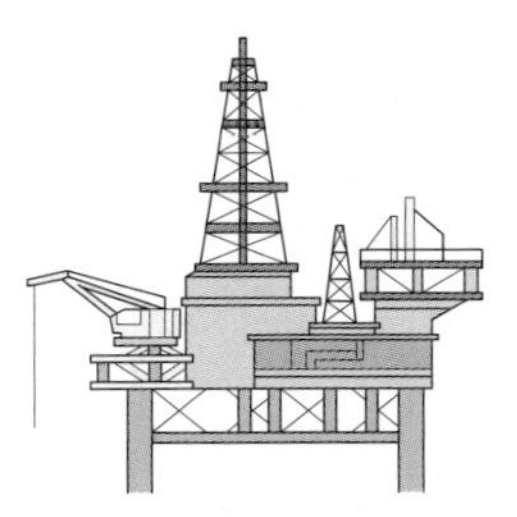

a) State what is meant by corrosion. (KU)
b) Explain **fully** why zinc could be used to make these plates.

SEB GENERAL (PS)

3 Steel can be protected from rusting by coating it with other substances. One method is by dipping it in molten plastic.

a) Why does coating steel with plastic prevent rusting?
b) Chromium-plated steel rusts quickly if scratched. What does this tell you about the reactivity of chromium?

SQA GENERAL (PS)

4 Objects made of iron are often galvanised, i.e. coated with zinc.

A	The zinc increases the rate of corrosion of iron.
B	The zinc is oxidised.
C	The zinc attracts electrons from the iron.
D	The zinc does not corrode.
E	The zinc corrodes slower than the iron.
F	The zinc is sacrificed to protect the iron.

Identify the statement(s) which can apply to a galvanised farm gate that has been scratched.

SEB CREDIT (PS)

5 Motor car manufacturers are making increased use of zinc-coated steel. This is because of the sacrificial protection given to the steel in the event of the zinc coating being broken thus exposing the steel.

a) State what happens to iron atoms when iron rusts.
b) Treating steel as if it were just iron, explain how zinc gives sacrificial protection to the steel. GENERAL (KU)

6 On some buildings the wooden roof is covered with copper sheets instead of slates.

a) Why should the copper roof not be held in position with iron nails?
b) Which metal should the nails used to hold the copper sheet in position be made of? GENERAL (KU)

7 Paint, oil, grease and plastic are said to provide 'physical protection' from corrosion when used to coat iron and steel. State what is meant by the terms:

a) corrosion
b) physical protection. GENERAL (KU)

8 Motor car manufacturers prefer to use steel which has been 'electroplated' with zinc rather than that which has been zinc-coated by dipping the steel in molten zinc. This is because the electroplated steel has the smoother surface.

a) Explain what is meant by the term 'electroplating'.
b) What name is given to the process whereby steel is coated with zinc by dipping the steel into molten zinc?

GENERAL (KU)

9 Zeenat set up the following experiments to study the rusting of iron nails under different conditions. Each of the test tubes contains ferroxyl indicator and an iron nail.

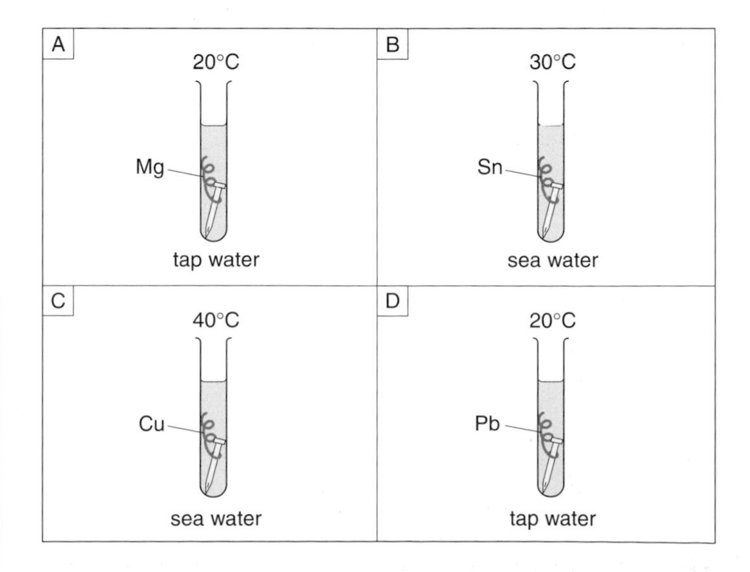

a) Identify the experiment in which the iron nail would not be expected to rust.
b) Which **two** experiments would provide a fair comparison of the ability of different metals to affect the rate of corrosion of iron? CREDIT (PS)

10 Tin-plated steel is widely used for making cans in which many types of food and drink are sold. The contents of such a can are well protected while the plating is intact because tin is quite an unreactive metal. However, once the tin coating is broken, the steel tends to rust very rapidly.

Explain why this happens. (You may assume that the steel behaves as though it were iron.) CREDIT (KU)

11 When Michael placed a drop of water containing ferroxyl indicator and common salt on a piece of sheet steel, he soon noticed the formation of colours within the drop.

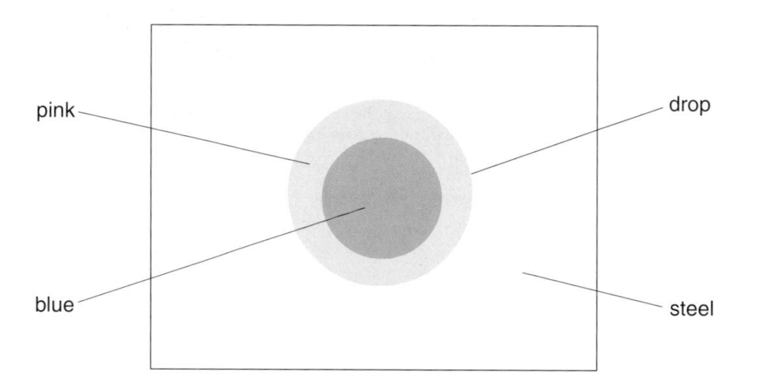

He decided that corrosion was taking place at the centre of the drop and was causing the blue colour. His chemistry teacher told him that the pink colour was due to the formation of hydroxide ions near the edge of the drop. Electrons are lost by iron atoms at the centre of the drop and move through the metal to the edge of the drop. There they join with molecules of two substances, X and Y, producing hydroxide ions.

a) Give an ion-electron equation to show the loss of electrons by iron atoms at the centre of the drop.
b) Name the two substances X and Y which accept electrons, forming hydroxide ions. CREDIT (KU)

12 For many years, 1p and 2p coins were made of an alloy containing copper, tin and zinc. However, these were replaced by coins made of copper-plated steel.

a) Treating the steel as if it were iron, explain what you would expect to happen if the copper plating on a coin was broken and it was then exposed to corroding conditions. (KU)
b) Suggest one reason for the change from an alloy coin in which the main metal is copper to a copper-plated coin in which the main metal is steel. GENERAL (PS)

13 An experiment was carried out to electroplate a steel spoon with nickel. The apparatus is shown below.

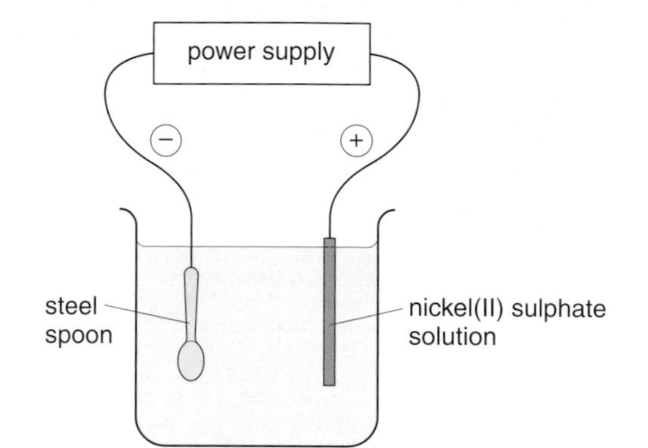

a) Explain why the object to be electroplated must be connected to the negative terminal of the power supply. (KU)
b) During the electroplating process, nickel(II) ions are converted into nickel atoms at the negative electrode. Give an ion-electron equation to represent this process. (KU)

If a piece of nickel is used as the positive electrode, the following process takes place at it while electroplating is in progress:

$$Ni(s) \rightarrow Ni^{2+}(aq) + 2e^-$$

c) What happens to the mass of the nickel positive electrode during electroplating? (PS)
d) Explain why the concentration of $Ni^{2+}(aq)$ ions in solution remains constant while electroplating takes place if a nickel positive electrode is used. CREDIT (PS)

CHAPTER THIRTEEN

Plastics

SECTION 13.1 Plastics and synthetic fibres

Figure 1.1 How many plastic objects can you see in this picture?

You read in Chapter 5 how crude oil is used to make a huge range of useful products. This section looks at two types of materials made from oil – **plastics** and **synthetic fibres**.

Think about your bedroom. Picture all the things which are made of plastic.

Everyday we use plastics in many forms, from carrier bags to CD boxes. Plastics are used for so many jobs because they have many useful **properties**. Properties of materials describe what they are like and how they behave. For example, a plastic shampoo bottle has the properties listed in table 1.1.

The plastic in the bottle is:
flexible – the shampoo can be squeezed out of the bottle *watertight* – the shampoo will not leak *shatterproof* – the bottle will not break if dropped *light* – this also makes it cheap to transport *able to be moulded* – the cap and the lid have complicated shapes which would be difficult to make using other materials.

Table 1.1 Why use plastic for shampoo bottles?

Figure 1.2 Plastic is used to make shampoo bottles

Questions

Q1 Make a list of at least four useful properties of the plastic which is used to make carrier bags.

Plastics are often used in the packaging for burgers or fish and chips. The plastic doesn't let the heat from the food escape. Plastics are **heat insulators**.

Plastics are also good **electrical insulators**. Plugs and sockets are made of plastic, and electrical wires are covered with plastic.

Plastic object	Useful properties
biscuit wrappers	flexible, airtight
hoses, 'rubber' boots	waterproof, flexible
plugs, sockets	electrical insulators, tough

Table 1.2 Some useful properties of plastics

What happens when a plastic is heated up? Plastics can be divided into two groups – those which soften when they are heated and those which do not.

Thermoplastic materials soften on heating. This can be a very useful property. It means that these plastics can be made as granules and then

Figure 1.3 What are the properties of Lycra that makes it suitable for use in swimsuits?

Natural fibres	Synthetic fibres
wool	nylon
cotton	polyester
linen	orlon

Table 1.3 Some common fibres

heated and moulded into new shapes. Nylon and PVC are both thermoplastics.

Thermosetting plastics do *not* soften on heating. Once they have been moulded into a particular form they cannot be reshaped. They are often very tough materials with high melting points. Urea-methanal, the plastic used in electrical plugs, is a thermosetting plastic. Other thermosetting plastics are used in pot handles, oven controls and kettles.

Synthetic fibres

If you look at the labels on your clothes you will probably find words such as polyamide and polyester. These are the names of the **fibres** which are woven together to make the fabrics in your clothes.

There are many different types of fibres and it is helpful to divide them into two groups, depending on where they come from. **Natural fibres** such as cotton and wool come from plants and animals. **Synthetic fibres** like nylon and polyester are made from chemicals which come from oil. The word 'synthetic' is used to describe something which is not a natural product but is man-made. Table 1.3 shows some common fibres.

Problems with plastics

Most plastics are **durable** – they last for a very long time. This is a useful property, but it can also cause problems. Natural materials usually decay after they have been thrown away – they are broken down by microorganisms into simpler substances. Anything which can be broken down in this way is said to be **biodegradable** (bio = living; degradable = able to be broken down). Most plastics do not decay, therefore they are **non-biodegradable**.

A discarded plastic bottle can remain in the environment for years and years. This causes a problem with litter. It is also difficult for some countries to cope with the huge amount of plastic refuse which is being produced. In Britain, a lot of plastic waste is buried in landfill sites, however it is not always easy to find new landfill sites once an existing one has filled up. Some plastics are burned in local authority incinerators. This disposes of the plastics, but needs to be very carefully controlled to prevent highly dangerous fumes from being produced.

There are several ways to tackle the problem of plastic waste:

1 Re-use plastic objects such as carrier bags.

2 Recycle plastics. Many areas have plastic recycling points. The labels on plastic bottles indicate which type of plastic they are made from. Most bin-liners are made from recycled plastic.

3 Use biodegradable plastics. These only became available quite recently and not everyone agrees that using them helps the environment. They decay when they come into contact with soil and this reduces the litter problem.

4 Use less plastic. A weekly trip to a supermarket means you take home huge amounts of plastic packaging, all of which has to be disposed of. Not all of the plastic packaging is necessary; most people agree that manufacturers could use less.

Question

Q2 Of the four methods given, which one do you think is the best for cutting down on plastic litter on the streets? Give reasons for your answer.

Question

Q3 Make a table with two columns, one for the advantages of using plastics and one for the disadvantages of using them. List as many advantages and disadvantages as you can.

Burning plastics

Figure 2.1 Some furniture well alight – it is the fumes that will kill you, not the flames

Figure 2.2 Furniture safety label

Would you buy a box of chemicals to put in your living room, knowing it would give off deadly fumes if it caught fire? Probably not, but that is precisely what you are doing if you buy sofas or chairs with fillings and covers made from plastics. The soft, sponge-like material which you find in cushions is polyurethane, and seat covers are made from poly(chloroethene), also known as PVC.

The problem with these and other plastics is that they can be set on fire quite easily by a match or a cigarette. When they burn they produce huge amounts of toxic fumes, some of which are invisible gases while others form clouds of thick, choking smoke. It is often these toxic fumes which kill people trapped in house fires.

The types of fumes given off depend on the plastic which is burning. Most plastics will produce carbon monoxide. This gas has no colour or smell and stops red blood cells from carrying oxygen round the body. Victims can die from oxygen starvation. Because of the danger from toxic fumes, fire-fighters normally wear special breathing apparatus when going into a burning house.

Fortunately, chemists have now developed a way of treating polyurethane foam to prevent it from burning. The treated foam will smoulder for about half an hour, after which time any burning stops. New furniture has to have a special label to show that it contains only treated foam (see figure 2.2). However, the dangers from fires involving older furniture will remain for many years.

Smoke detectors can save lives by giving an early warning that fumes are being produced. This gives people vital extra time to get out of the house – flames from a burning sofa can raise the temperature of a room to around 800°C in just a few minutes.

Questions

Q1 Look at the furniture in your living room at home and see if it contains the new treated foam or not.

Most of the poisonous fumes which burning plastics produce are made when elements in the plastics join with oxygen in the air to form new compounds.

The plastics used in the home all contain carbon and hydrogen. In many cases, carbon monoxide is formed due to the incomplete combustion of carbon in the plastic. Polyurethane plastics contain the element nitrogen. When these burn, the nitrogen can join with carbon and hydrogen in the plastic to form hydrogen cyanide (HCN), a deadly gas. Poly(chloroethene), used in covers, vinyl wallpaper and in some articles of outdoor clothing, contains chlorine. On burning, this plastic gives off a choking, acidic gas called hydrogen chloride (HCl). Table 2.1 shows some of the toxic gases given off by some commonly used plastics.

Plastic	Toxic gas produced
all plastics	carbon monoxide
polyurethane plastics	hydrogen cyanide
poly(chloroethene)	hydrogen chloride

Table 2.1 Some of the toxic gases produced when plastics burn

Questions

Q2 All plastics contain the elements carbon and hydrogen. Which two substances do you think will be produced by these elements during complete combustion? *Note*: the substances produced are not toxic fumes.

Q3 Name three toxic gases which could be given off when armchairs covered with poly(chloroethene) and containing polyurethane are burned.

In some circumstances, burning plastics will also give off a compound called dioxin. This is one of the most poisonous substances known, and it does not break down or decay easily.

Comparing natural and synthetic material

So far you have met several advantages of using plastics and synthetic fibres, as well as some of the problems connected with them. Of course, there are also advantages and disadvantages in using natural materials. Comparing natural and synthetic materials is not always simple. For example, some people believe that we could save oil and energy by using paper bags in supermarkets instead of plastic ones. Paper, being made from wood, is a natural material. However, plastic manufacturers claim that quite thick paper would be needed, and this would take a lot of energy to make. Also, because goods wrapped in thick paper would be heavier than those wrapped in plastic, more petrol would be used up in transporting them. Table 2.2 gives a summary of the advantages and disadvantages of natural and synthetic materials.

Questions

Q4 Make a table of the possible advantages and disadvantages of replacing all plastic soft-drink bottles. with glass ones.

Advantages	Disadvantages
Natural materials, e.g. wool, wood, paper:	
◆ are often biodegradable	◆ can be expensive
◆ are made from renewable resources, e.g. trees	◆ often do not last for very long
Synthetic materials, e.g. plastics, synthetic fibres:	
◆ are long-lasting	◆ are non-biodegradable
◆ can be mass-produced cheaply	◆ are made from oil – a finite resource
◆ have a range of useful properties, e.g. lightness	◆ may burn easily and give off toxic fumes

Table 2.2 Comparing natural and synthetic materials

SECTION 13.3 Making plastics

Figure 3.1 A poly(ethene) shopping bag

The molecules which make up plastics are extremely large. For example, the molecules which make up the plastic poly(ethene) contain around 50 000 atoms arranged in long chains. The basic structure of a plastic is often quite simple. Figure 3.2 shows part of a molecule of poly(ethene).

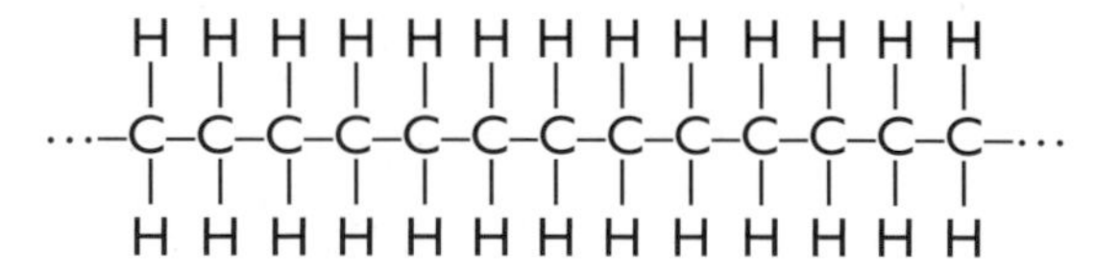

Figure 3.2 Part of a poly(ethene) molecule

Notice that the molecule contains a repeating pattern of carbon and hydrogen atoms. A molecule which is made up of many repeating units is called a **polymer** (poly = many; mer = a part). All plastics are made of many repeating parts, so all plastics are polymers.

Polymers are made when many small molecules join together. Each small molecule becomes a part of the polymer. They are called **monomers** (mono = one). The process of making a polymer by joining monomers together is called **polymerisation**.

The plastic poly(ethene) is made by joining molecules of ethene (C_2H_4). Figure 3.3 gives a simplified picture of how ethene monomers join. It shows how four molecules of ethene join to form part of a molecule of poly(ethene). More and more ethene molecules join until the reaction stops. It takes thousands of ethene monomers to make one molecule of poly(ethene).

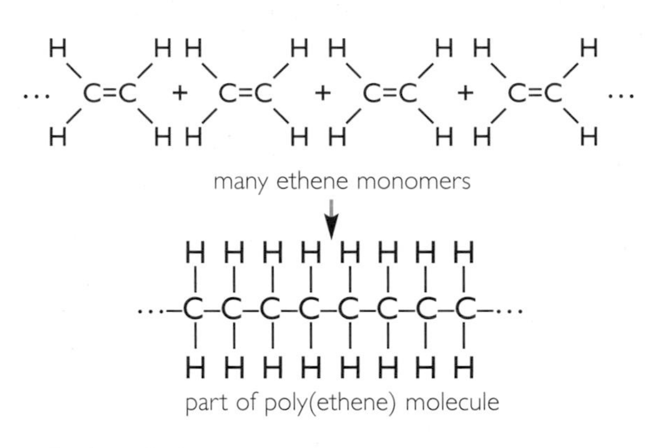

Figure 3.3 Polymerisation of ethene

Poly(ethene) is the most widely used plastic. The monomer ethene belongs to the group of chemicals called alkenes. Many alkenes and related compounds are used to make polymers. For example, propene can be polymerised to give the plastic poly(propene). This is used for hose pipes, carpets, car bumpers, biscuit wrappers, etc. All the monomers used in this kind of polymerisation must contain C=C double bonds.

As mentioned in section 13.1, plastics are made from chemicals which come from crude oil. In one part of the oil refining process, alkane molecules are broken up in a process known as **cracking**. Cracking produces alkenes, and it is these which are used as the monomers for many plastics. This is covered in detail in Chapter 6.

The name of a polymer made from an alkene or similar monomer is obtained by putting brackets round the name of the monomer and putting the word 'poly' in front. For example, the monomer chloroethene produces the polymer poly(chloroethene). This polymer is also known by its older name of PVC. Similarly, poly(ethene) is also called polythene.

Questions

Q1 Name the monomers from which the following polymers are made:

a) poly(propene),

b) poly(phenylethene).

Q2 What happens to the C=C bond in ethene monomers when they join to form a poly(ethene) polymer?

Figure 3.4 Wool and cotton are both polymers

Fibres are polymers

Natural fibres include cotton and wool. Cotton is made from the polymer cellulose. Wool contains keratin, a polymer made from several different monomers.

Synthetic fibres include Terylene and nylon, both of which are polymers.

SECTION 13.4 Addition polymerisation I

Figure 4.1 Addition polymerisation takes place at plants like this one where poly(ethene) is made

Addition polymerisation is an important type of polymerisation. Polythene and polystyrene are addition polymers. This section looks at how the monomers join together in addition polymerisation and what the polymers are used for. You may find it useful to refer back to section 6.4 which deals with addition reactions in alkenes.

Questions

Q1 Addition polymers can only be made from molecules containing a carbon-to-carbon double bond (C=C).
a) What would be the result of testing such a monomer with bromine solution?
b) What would happen if you repeated the test on a piece of poly(ethene)?

Poly(ethene)

Addition polymerisation is a complicated process. However, it always involves the joining together of monomers which contain the C=C double bond. All alkenes contain this bond. The polymerisation starts with the opening of this C=C bond. A catalyst is required for this to take place. Figure 4.2 shows how ethene monomers join to form a poly(ethene) polymer. All addition polymers are formed in this way, with thousands of monomer molecules linked together. If the letter *n* is used to represent the number of monomer molecules joining together, then the process can be written as shown in figure 4.3.

The equation can be simplified further by using molecular formulae:

$$nC_2H_4 \rightarrow (C_2H_4)_n$$

Carrier bags, basins, buckets, fertiliser sacks, water pipes and milk crates are all made of poly(ethene).

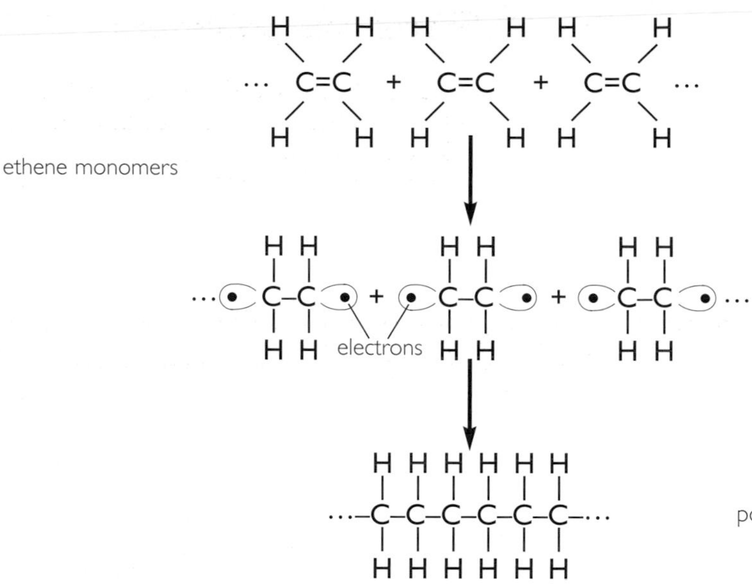

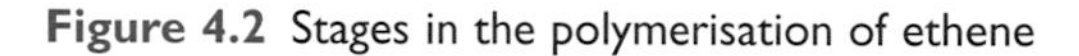

Figure 4.2 Stages in the polymerisation of ethene

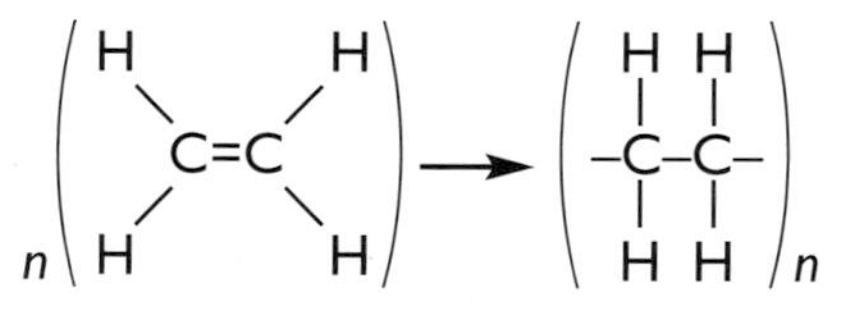

Figure 4.3 Equation for addition polymerisation

Poly(propene)

This is the second most widely used polymer after poly(ethene). It has been said that manufacturers of plastic products try poly(ethene) first, and if it is not suitable turn next to poly(propene).

Poly(propene) is produced by the addition polymerisation of propene monomers (see figure 4.4).

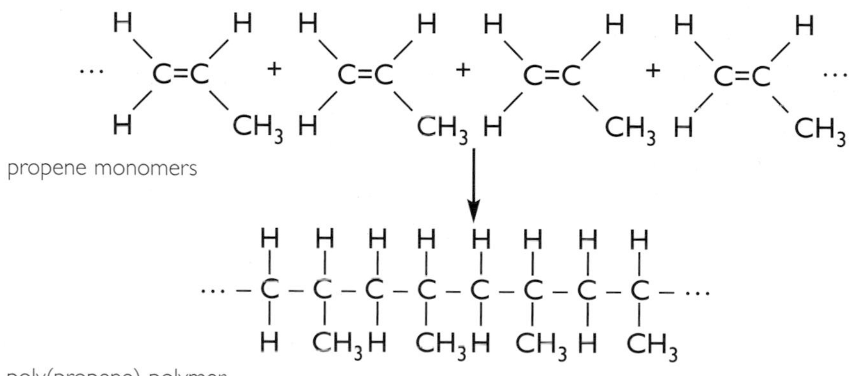

Figure 4.4 Polymerisation of propene

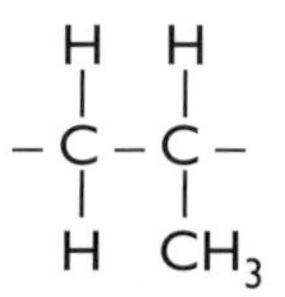

Figure 4.5 The repeating unit in poly(propene)

In common with all polymers, a group of atoms in the structure of poly(propene) is repeated over and over again. This group of atoms is called the **repeating unit**. Figure 4.4 shows four repeating units in the structure of poly(propene). The repeating unit is shown in figure 4.5.

This polymerisation process can be represented by simpler equations, as shown in figure 4.6.

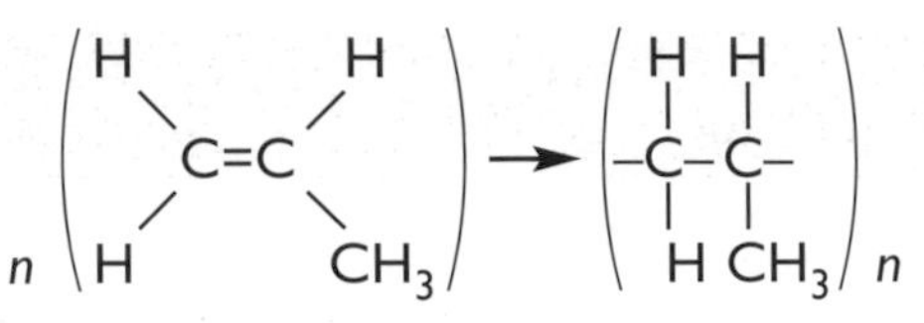

Figure 4.6 Equation for polymerisation of propene

Using molecular formulae, the equation becomes:

$$nC_3H_6 \rightarrow (C_3H_6)_n$$

Poly(propene) is often referred to by its older name, polypropylene.

Questions

Q2 Orlon and Acrylan are names for fibres which are made of the polymer shown below.

a) How many repeating units are shown in this diagram?
b) Draw a structural formula for the repeating unit.
c) Draw a structural formula for the monomer unit from which the polymer is made.

Figure 4.7 Poly (propene) is used for the synthetic grass of this hockey pitch, as well as the car bumper. What properties make it suitable for both?

Addition polymerisation 2

An increasing number of addition polymers are now made using monomers called **substituted alkenes**. These can be thought of as alkenes in which one or more hydrogen atoms have been replaced by a different type of atom or group of atoms (see figure 5.1).

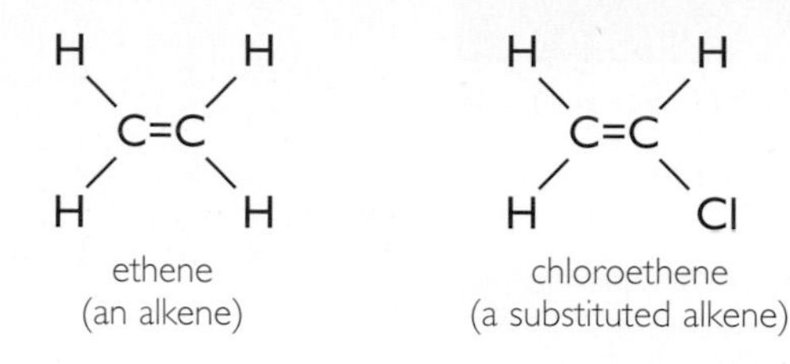

Figure 5.1

PVC (polyvinyl chloride)

The term 'vinyl' is used to describe a range of materials which includes records, wallpapers and floor coverings. These materials are made from the polymer poly(chloroethene), also known as polyvinyl chloride (PVC). This is made from a monomer which is similar to ethene, but in which one hydrogen atom has been replaced by a much larger chlorine atom. The effect of this is to produce a polymer which is rigid because the molecules are not able to flex freely.

PVC is tough as well as rigid. It is formed as shown in figure 5.2.

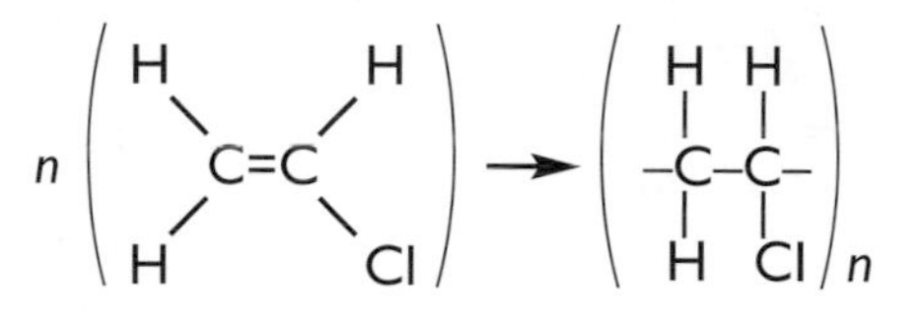

Figure 5.2 Equation for the formation of PVC

Polystyrene

Polystyrene is the very light material which is used for packaging and insulation. It is made in two very different forms. One is transparent and rigid – this is used to make the barrels of ball-point pens. The other form is called 'expanded polystyrene'. It is produced when granules of polystyrene containing a little pentane are heated. The liquid pentane boils inside each granule and, as it forces its way out, makes the granules expand.

The monomer from which polystyrene is made is called styrene. This can be thought of as an ethene molecule in which one of the hydrogen atoms has been replaced by a large hydrocarbon group called a phenyl group (C_6H_5) (see figure 5.3).

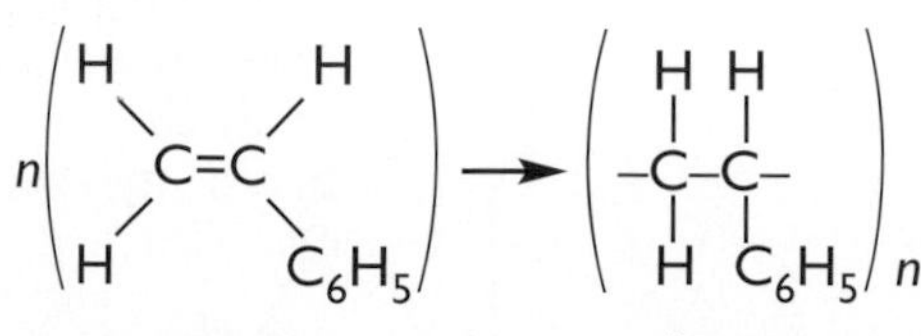

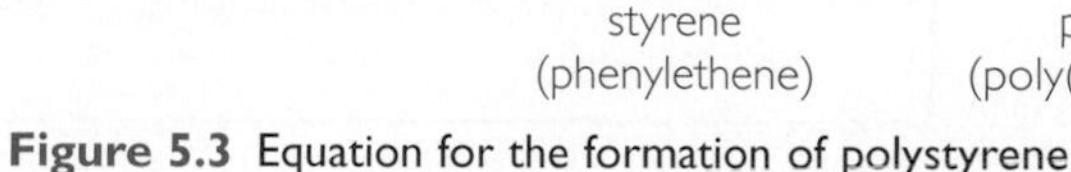

Figure 5.3 Equation for the formation of polystyrene

Questions

Q1 Polystyrene softens in boiling water and can then be reshaped. Is polystyrene a thermoplastic or a thermosetting plastic?

Figure 5.4 CD and DVD boxes are made of polystyrene

Figure 5.5 Why is normal polystyrene used for CD boxes but the expanded form is used for packaging apples?

PTFE (poly(tetrafluoroethene))

PTFE is better known by one of its brand names, 'Teflon'. It is widely used as a 'non-stick' material on frying pans and other kitchen utensils. PTFE is very unreactive.

Tetrafluoroethene is the monomer from which PTFE is made. You can imagine this molecule as an ethene molecule in which each hydrogen atom has been replaced by a fluorine atom (see figure 5.6).

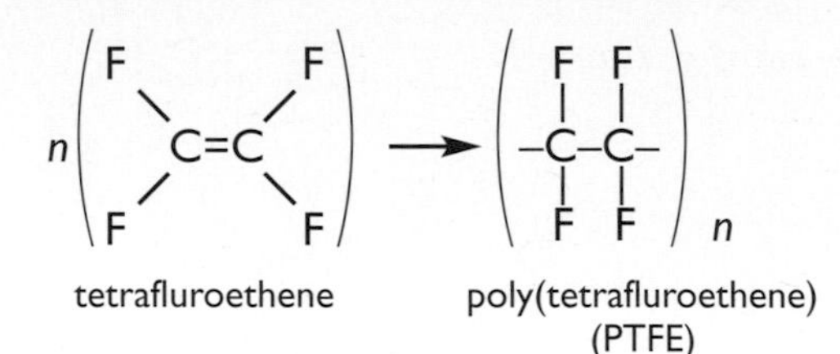

Figure 5.6 Equation for the formation of PTFE

Perspex

The brand name 'Perspex' is much better known than the chemical name for this plastic – poly(methyl-2-methylpropenoate). Bulky groups on the polymer chain make it hard and stiff, but it can also be made crystal clear with a very shiny surface. Because of this, Perspex can be used to make contact lenses, spectacle lenses and laminated windscreens for cars and aeroplanes. Perspex can also he used to make baths and washbasins.

The equation for the polymerisation process is shown in figure 5.7.

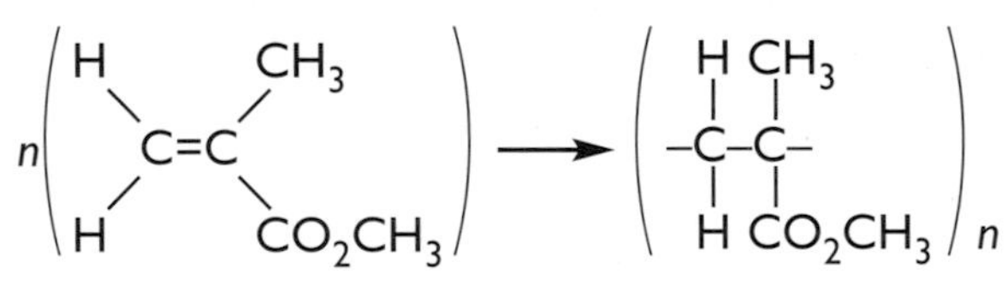

Figure 5.7 Equation for the formation of Perspex

Addition polymer 'backbones'

All addition polymers have a 'backbone' made up entirely of carbon, atoms (see figure 5.8).

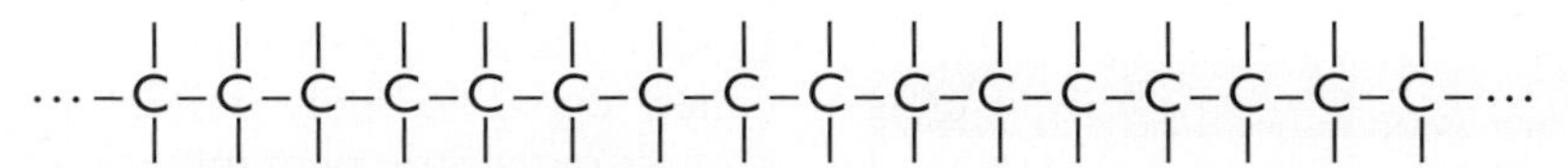

Figure 5.8 Addition polymer 'backbone'

Questions

Q2 In section 13.3 you learnt how propene molecules join to form a poly(propene) polymer. Repeat this process for

a) PVC,
b) polystyrene,
c) PTFE.

In each case show the joining of four monomer molecules using structural formulae.

1

A wood	**B** coal	**C** oil
D natural gas	**E** straw	**F** peat

Identify the substance in the grid above from which most plastics and synthetic fibres are made. GENERAL (KU)

2 A label on a shirt gave the composition of the fibres used in its manufacture as follows:

Terylene	60%
cotton	40%

a) Draw a pie chart to show the proportions of Terylene and cotton in the mixed fibre used. (PS)

b) Terylene is a synthetic fibre whereas cotton is a natural fibre. State what is meant by *synthetic* as used in the term 'synthetic fibre'. GENERAL (KU)

3 Some years ago the wreck of the liner *Lusitania* was located off the coast of Ireland. Underwater photographs showed that it was covered in torn nylon fishing nets, some of which were thought to have been lost more than forty years earlier. Nylon, in common with many plastics, is not biodegradable.

State what is meant by 'biodegradable'. GENERAL (KU)

4 Many fillers used for car body repairs contain the *monomer* phenylethene. Immediately prior to use, the filler is mixed with a small quantity of a very reactive substance which is provided in a small tube along with the filler. This results in the phenylethene undergoing a process of *polymerisation* and the filler changes from a paste to a solid in about half an hour. When the paste hardens, the phenylethene is changing into the *polymer* poly(phenylethene).

State what is meant by the terms 'monomer', 'polymer' and 'polymerisation'. GENERAL (KU)

5 About two million tonnes of ethene are produced each year in the UK. Most of this useful alkene is converted into poly(ethene), which is more familiar to people as polythene.

a) Draw the full structural formula for ethene.

b) Explain, using full structural formulae, how ethene forms poly(ethene). In your answer you should show how at least three ethene molecules join together.

c) Poly(ethene) is a thermoplastic. Explain what this means. GENERAL (KU)

6 In older houses, light fittings such as plugs, sockets and switches are often made of the dark brown thermosetting plastic Bakelite. In more modern houses, the same fittings are usually made of the white plastic urea-methanal, which is also thermosetting.

a) Explain what is meant by the term 'thermosetting plastic'.

b) Give another example of a situation where a thermosetting plastic must be used because it would not be possible to use a thermoplastic. GENERAL (KU)

7 Monomer X can be converted into polymer Y.

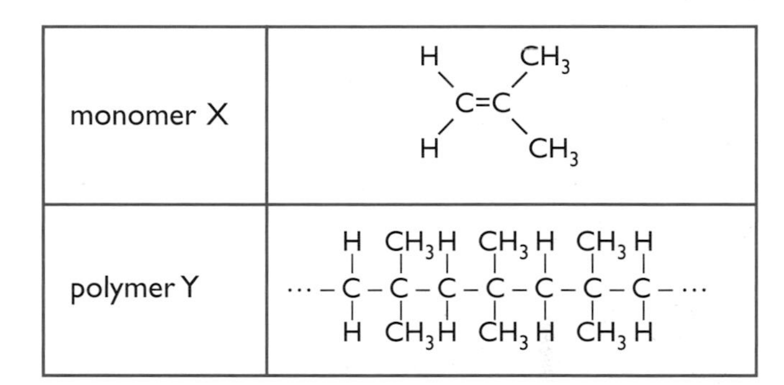

a) Which of the molecules X and Y would you classify as saturated and which as unsaturated?

b) Describe the effects of testing monomer X and polymer Y with a solution of bromine in water. GENERAL (KU)

8 Polymer P is a versatile plastic which can be made into a film for packaging, into fibres for carpets, and for the 'synthetic grass' used for some hockey and football pitches. Its structure is shown below:

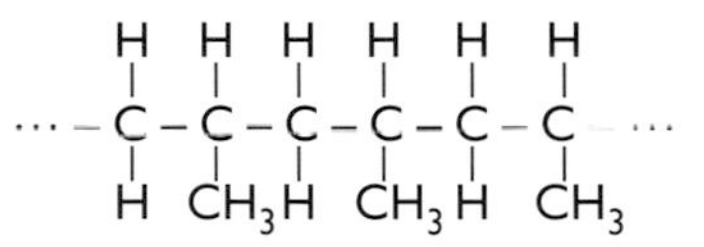

The repeating unit in the polymer is:

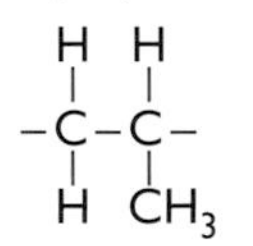

The monomer M from which the polymer is made is:

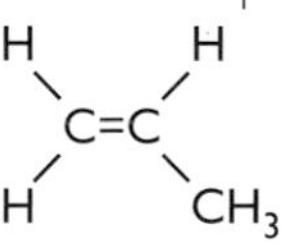

a) Name the monomer M and the polymer P.

b) To which homologous series does M belong? GENERAL (KU)

9 PVC is perhaps the most versatile plastic. It can be extremely rigid, making it ideal for window frames. If a plasticiser is added to it, it can be very flexible. In this form PVC makes an excellent insulating cover for electrical cables. A part of the polymer is shown overleaf.

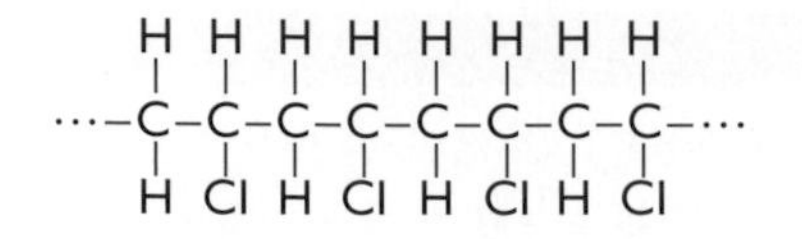

a) Draw the repeating unit present in the polymer.
b) Draw the full structural formula of the monomer used to produce the polymer PVC.
c) One unfortunate feature of plastics is that some produce very toxic gases when they burn,

Name **two** toxic gases which could be produced by PVC on combustion. CREDIT (KU)

10 When superglue sets a polymer is formed. The polymer has the following structure.

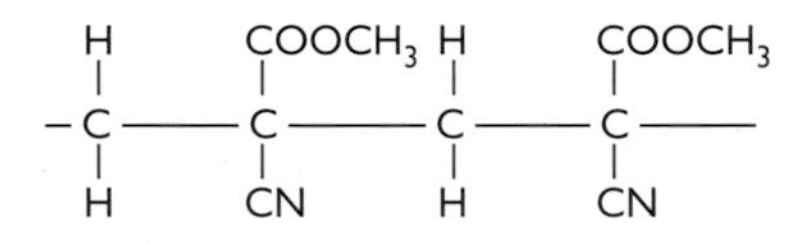

a) Draw the structural formula for the monomer in superglue.
b) Name a toxic gas given off when superglue burns.

SQA CREDIT (KU)

11 a) The graph shows the production of PVC in Western Europe.

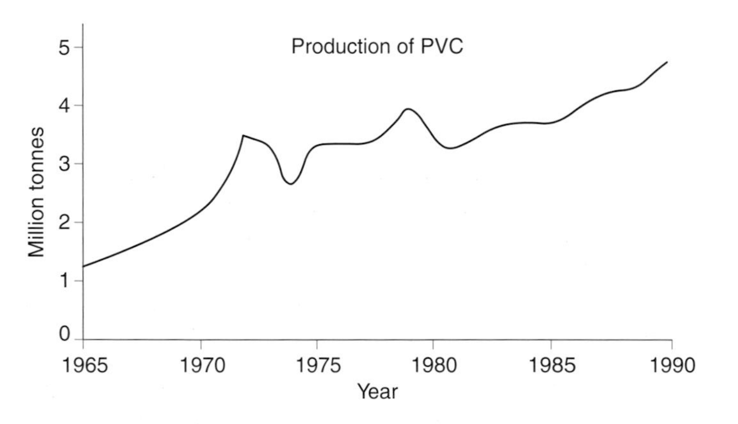

Describe the general trend in the production of PVC from 1965 to 1990.

b) PVC has many uses because of its different properties. Its strength makes it ideal for bottles. It is used for protective clothing because of its water resistance, and its chemical resistance makes it ideal for food containers. It is used for plugs and cables because it is an electrical insulator.

Present the information shown above as a table with suitable headings.

SQA GENERAL (PS)

12 Polyvinylidene chloride is an addition polymer. It is added to carpet fabrics to reduce flammability. It is made from the monomer vinylidene chloride.

Part of the polymer chain is shown below.

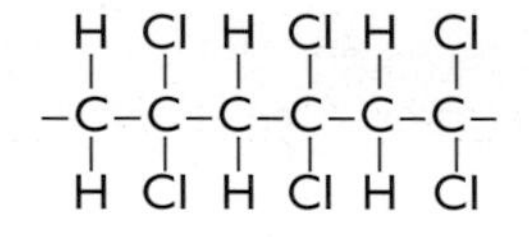

a) What is meant by an **addition** polymer?
b) Draw the full structural formula of the vinylidene chloride monomer.
c) State a problem associated with the burning of polymers. (KU)
d) Many polymers are **not** biodegradable. Why might this be an advantage?

SQA CREDIT (PS)

13 Addition polymers, like poly(ethene), are made from small unsaturated monomers.

a) Draw a section of poly(ethene), showing 3 monomer units joined together.

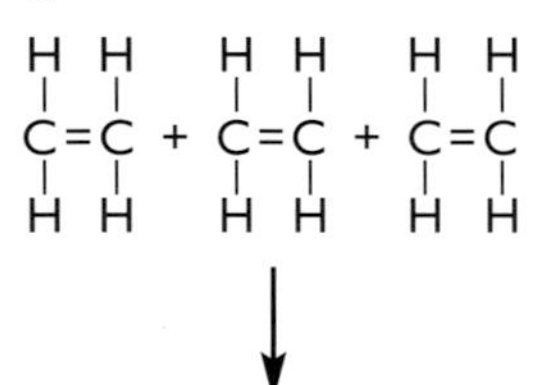

b) Poly(ethene) has many uses because it can be easily shaped when warmed. What term is used to describe this property?
c) The table shows some of the uses of poly(ethene).

Use	%
pipes	10
films	14
blow moulding	8
injection moulding	28
other uses	40

Draw a bar chart to show this information.

SQA GENERAL (PS)

CHAPTER FOURTEEN

Fertilisers

SECTION 14.1 Growing food

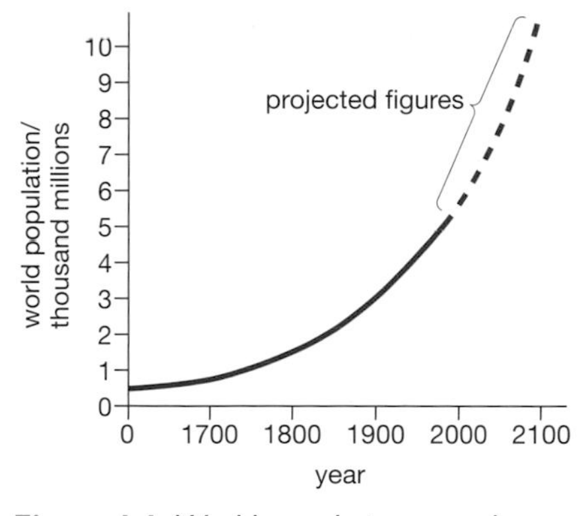

Figure 1.1 World population growth

Every quarter-second a baby is born somewhere in the world. Over the past hundred years the world population has increased five times to around 6000 million people. It is predicted that over the next 50 years the population will grow to 7000 million.

We will have to find ways of producing enough food for everyone. Most food comes from farms. Farms give us fruit, vegetables, cereals and most kinds of meat. All of the food which we eat is made from animal or plant products.

What do plants need in order to grow? Table 1.1 shows the chemical elements which plants must have.

Element	Obtained from
carbon	air
oxygen	air and water
hydrogen	water
potassium	soil
nitrogen	soil and air
phosphorus	soil

Table 1.1

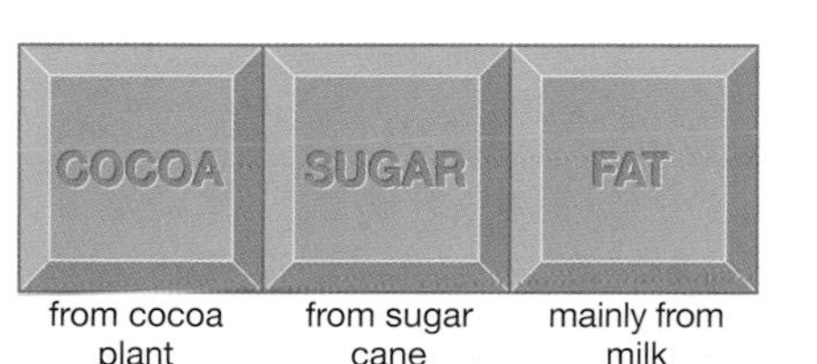

Figure 1.2 Ingredients of a chocolate bar

Plants take substances containing potassium, nitrogen and phosphorus out of the soil. These substances are called **nutrients**. If a farmer grows a crop such as potatoes in the same field every year, then the nutrients in the soil that potatoes need become used up.

Farmers can replace the lost nutrients by adding **fertilisers**. A fertiliser is a substance which contains the chemical elements needed by plants. There are two main types of fertilisers:

- **Natural fertilisers** – these are made directly from animal or plant waste. They are good sources of nitrogen. Animal manure is a natural fertiliser, as is compost.
- **Artificial fertilisers** – these are man-made fertilisers made mostly from soluble ionic compounds. Once in the soil, these compounds dissolve. Their ions can then be taken up through the roots of plants.

By using fertilisers, farmers are able to increase the amount of a crop which can be grown on a field. This means that more food can be produced.

Farmers use artificial fertilisers to supply potassium, nitrogen and phosphorus. These fertilisers are usually made up of soluble ionic compounds containing these elements. Commonly used compounds include ammonium phosphate $(NH_4)_3PO_4$ and potassium sulphate K_2SO_4. Table 1.2 shows the ones that are commonly used.

Figure 1.3 Which elements are present in this fertiliser?

Element needed by plants	Compounds present in fertiliser
potassium	potassium salts
phosphorus	phosphates
nitrogen	nitrates and ammonium salts

Table 1.2

Questions

Q1 All the main food constituents of a cheeseburger can be traced back to plants. Cheese and beef come from cattle, which eat grass. The bun is made from wheat. Do the same for the main food constituents of a bacon, lettuce and tomato sandwich.

Q2 **a)** Give the formula of the following ions:
(i) ammonium,
(ii) phosphate,
(iii) nitrate. Use the table on page 4 in the SQA Data Booklet to help you.
b) Use the solubility table on page 5 in the SQA Data Booklet to predict which of the following would be suitable as soluble fertilisers:
(i) calcium phosphate,
(ii) potassium nitrate,
(iii) ammonium sulphate.

Q3 Make your own list of the advantages and disadvantages of using artificial fertilisers.

Different types of plants require different amounts of each of these three elements. For example, cereals require more nitrogen than potatoes. Manufacturers produce fertilisers with different amounts of each of the three main elements for use with different crops. It is important to know the amount of each element in fertilisers. The percentage mass of the elements present can be calculated from the formula using the method given in section 11.4.

Problems with fertilisers

In 1950, 14 million tonnes of artificial fertiliser were used in the world. By 1985, that figure had gone up to 125 million tonnes. The use of such huge quantities of fertiliser has led to some problems:

1 Nitrates in drinking water – in certain areas nitrates have seeped into the drinking water supply. Some doctors believe that high levels of nitrates in water are harmful. However, it is not certain that artificial fertilisers alone have caused these high nitrate levels. Natural fertilisers also release nitrates.

2 The nitrates and phosphates in fertilisers can seep into rivers and streams. They can also be washed out to sea. In some cases, they boost the growth of simple types of plants, called **algae**. The algae grow quickly and can form a layer called an 'algal bloom' which floats on the surface of the water. In lochs and streams, algal blooms can cut out the light reaching the plants that live on the bottom. Without light, these plants die and without the plants most other forms of life die out too.

Legumes

Figure 1.5 shows small lumps on the roots of the clover plant. These are called **nodules**. They contain **nitrifying bacteria** (**nitrogen fixing bacteria**) which can take nitrogen from the air and produce a nitrogen compound which the clover needs. Plants which have nodules like this are called **legumes**. Peas, beans and mustard are all legumes.

Farmers can use legumes to put nutrients back into the soil. For example, a field can be planted with mustard, then once the plants have grown they are ploughed back into the soil. As the plants decay, the nutrients they contain are released. This increases the amount of nutrients in the soil, so it is able to support better crops in the future.

One advantage of using legumes is that they are a cheaper source of nitrogen for the soil than artificial fertilisers.

Figure 1.4 An algal bloom

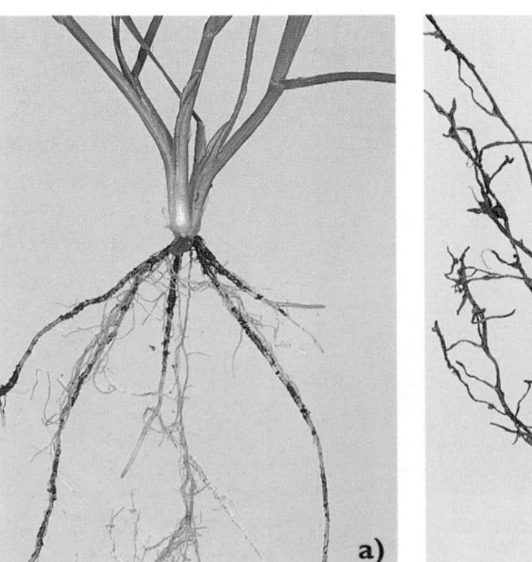

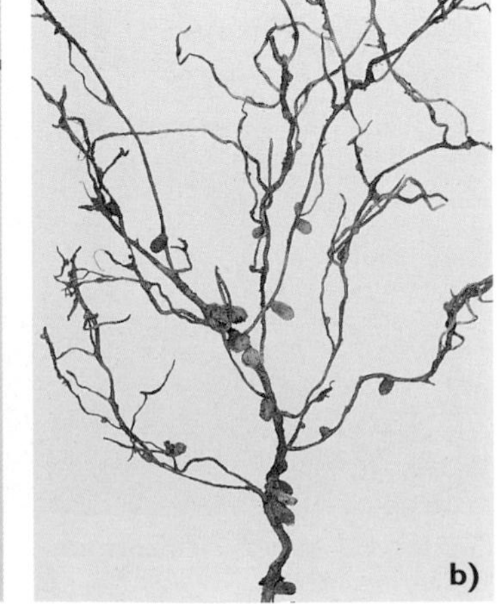

Figure 1.5 One of these plants is a legume, the other is a non-leguminous plant. Do you know which is which?

The nitrogen cycle

You have learnt that plants need nitrogen in order to grow. This section shows some of the different ways in which nitrogen can be added to and lost from the soil.

Removing nitrogen from the soil

- When plants are harvested, the nitrogen they contain is taken away and is therefore lost from the soil.
- Nitrates containing nitrogen are soluble in water and can be washed out of the soil into underground streams.
- **Denitrifying bacteria** break up nitrogen compounds in the soil and release nitrogen gas into the air.

Adding nitrogen to the soil

- Dead plants and animals all contain nitrogen. When they die, they decay and the nitrogen goes back into the soil. Compost is made mostly from decaying plants.
- Animal manure is a source of soluble nitrates. It can be spread onto fields or dug into the soil.
- Legumes (beans, peas, clover, mustard, etc.) have nodules on their roots. These take nitrogen from the air and make it into nitrates which can be used by the plant.
- During a flash of lightning, the two main gases in the air, nitrogen and oxygen, can join together to form nitrogen dioxide. This dissolves in rain water to give nitric acid, which contains the nitrate ion. When it rains, the nitrates enter the soil.
- Many artificial fertilisers add nitrogen in the form of nitrate and ammonium ions to the soil.

The continual removal and addition of nitrogen to the soil is called the **nitrogen cycle** (see figure 2.1).

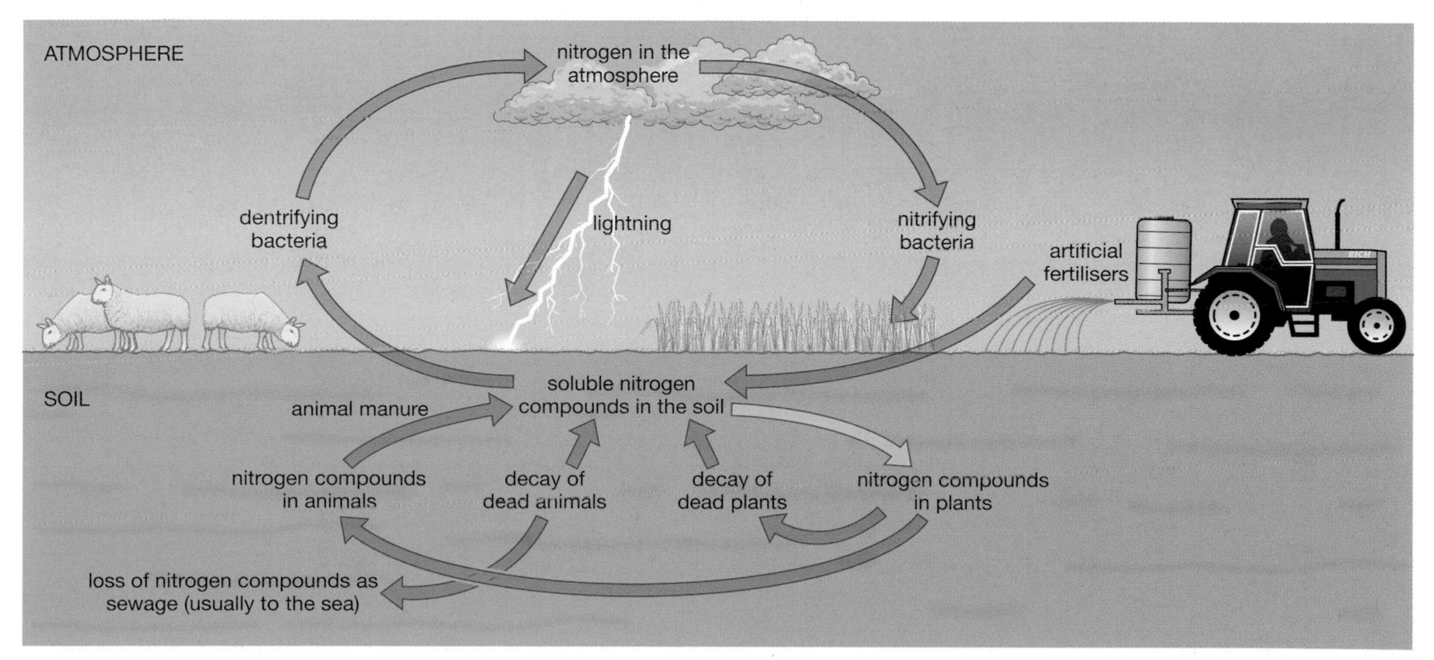

Figure 2.1 The nitrogen cycle

Questions

Q1 Why are nitrate fertilisers easily washed out of the soil?

Q2 Describe how lightning is able to add nitrates to the soil.

Q3 Describe how the crops which are known as legumes are able to add nitrates to the soil.

Figure 2.2 This farmer is spreading animal manure onto the fields to increase the nitrogen content of the soil

Figure 2.3 Lightning helps add nitrogen to the soil

SECTION 14.3 Making ammonia: the Haber process

Farmers throughout the world use millions of tonnes of nitrogen-containing fertilisers every year. This section deals with a method of making nitrogen compounds which uses nitrogen *in the air* as a starting material.

The Haber process was invented by a German chemist, Fritz Haber, along with the engineer Carl Bosch in 1908. The problem which faced them was that nitrogen gas is very unreactive and so does not form compounds easily. They were able to design a process in which nitrogen combined with hydrogen to make ammonia gas (formula NH_3). Ammonia gas can be converted easily into soluble ionic compounds that can be used in fertilisers.

Hydrogen is obtained from methane, which comes from North Sea gas. Nitrogen is obtained from the air.

In order to make nitrogen combine with hydrogen as efficiently as possible, these steps are followed:

- An iron catalyst is used to speed up the reaction.
- The gases are put under high pressure and a moderately high temperature.

Questions

Q1 a) What is the product made in the Haber process?
b) Why is it useful?

Q2 Write a word equation for the reaction between hydrogen and nitrogen in the Haber process.

- As the ammonia gas is formed, it is removed by cooling it to turn it into a liquid.
- Unchanged nitrogen and hydrogen are recycled.

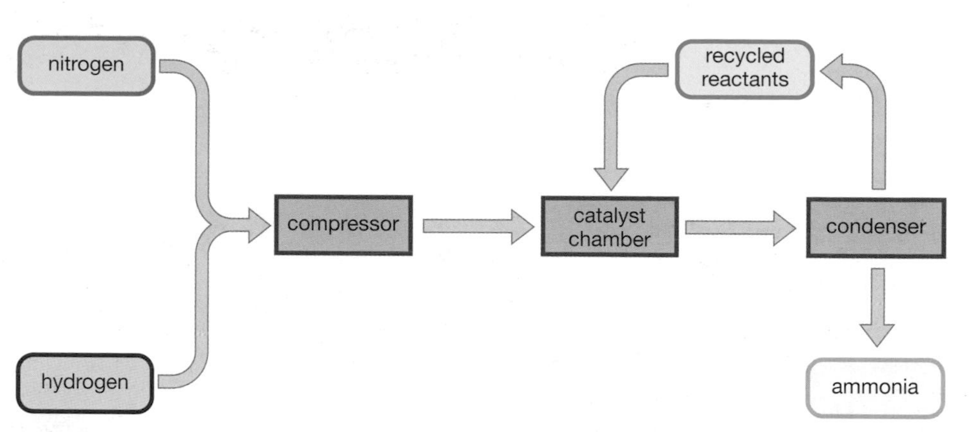

Figure 3.1 In the Haber process, nitrogen and hydrogen combine to make ammonia

More about the Haber process

The equation for the reaction between hydrogen and nitrogen to make ammonia is as follows:

$$\underset{\text{nitrogen}}{N_2} + \underset{\text{hydrogen}}{3H_2} \rightleftharpoons \underset{\text{ammonia}}{2NH_3}$$

The $\rightleftharpoons$ sign shows that the reaction is **reversible**, that is, some of the ammonia molecules decompose and turn back into hydrogen and nitrogen molecules. In practice, this means that it is impossible to change all the nitrogen and hydrogen into ammonia – some ammonia will always change back into the starting gases. However, the **yield** of ammonia (the amount produced) can be increased by controlling the temperature and pressure at which the reaction takes place. Table 3.1 shows that as the pressure increases, so does the yield of ammonia. It also shows that less ammonia is produced at higher temperatures. You would think, therefore, that the best conditions for making ammonia are high pressure and low temperature. Unfortunately, there is a problem – at low temperatures the reaction is too slow. As a compromise, a moderately high temperature of around 500°C is used.

Temperature/°C	Pressure/atmospheres		
	150	250	350
350	46.2%	57.5%	65.2%
450	22.3%	31.9%	39.3%
550	9.9%	15.6%	20.8%

Table 3.1 The final percentage yield of ammonia from the Haber process under different conditions of temperature and pressure

Questions

Q3 Why do you think the ammonia is liquefied and removed as soon as it is formed in the Haber process?

Q4 Go back to section 2.3 and suggest reasons why industrial chemists do not want the reaction for the formation of ammonia to take place slowly.

Ammonia and nitric acid

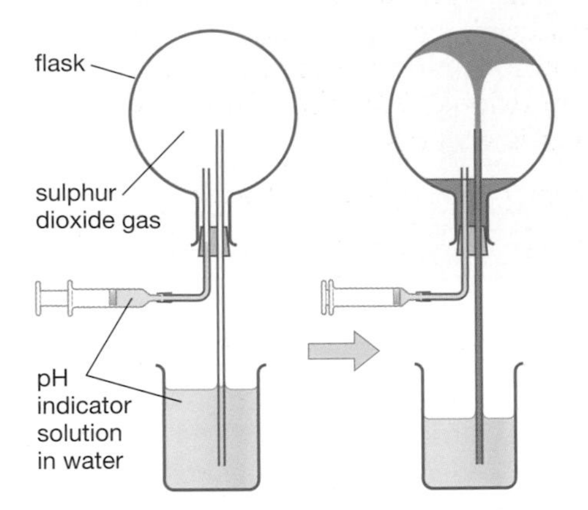

Figure 4.1 All of the ammonia gas in the flask can dissolve in just a few drops of water. When this happens, the water from the beaker rushes into the flask to fill the space

Properties of ammonia

- The formula for ammonia gas is NH_3.
- Ammonia is a colourless gas with a sharp, unpleasant smell.
- Ammonia turns damp pH paper and universal indicator solution violet.
- Ammonia gas is highly soluble in water (see section 8.1). This is shown by the fountain experiment in figure 4.1.

Reactions of ammonia

Ammonia dissolves in water to form an alkaline solution:

ammonia gas + water → ammonia solution

Ammonia reacts with acids to form salts. For example, ammonia reacts with nitric acid to form the salt ammonium nitrate:

$$\underset{\text{ammonia gas}}{NH_3} + \underset{\text{nitric acid}}{HNO_3} \rightarrow \underset{\text{ammonium nitrate}}{NH_4NO_3}$$

Ammonium salts contain the ammonium ion NH_4^+. This should not be confused with the formula for ammonia gas, which is NH_3.

Questions

Q1 a) Name the salt which is formed when ammonia reacts with sulphuric acid, (Refer to section 9.1 if you need help.)
b) Write a word equation for this reaction.

The laboratory preparation of ammonia

Ammonia is made in industry using the Haber process. In the laboratory, ammonia can be prepared by heating an ammonium compound with an alkali such as calcium hydroxide. The following reaction occurs when calcium hydroxide is warmed with ammonium sulphate.

calcium hydroxide + ammonium sulphate → calcium sulphate + ammonia gas + water

Question

Q2 Write a balanced equation for the reaction between calcium hydroxide and ammonium sulphate given above.

Making nitric acid

Nitric acid is used to make nitrates for fertilisers. One possible way to make nitric acid might be to copy the reactions that can occur during lightning storms. The energy from the lightning causes nitrogen and oxygen in the air to react to produce nitrogen dioxide:

$$\underset{\text{nitrogen}}{N_2} + \underset{\text{oxygen}}{2O_2} \rightarrow \underset{\text{nitrogen dioxide}}{2NO_2}$$

A similar reaction takes place in the air around spark plugs in car engines (see section 5.3). The nitrogen dioxide can dissolve in rain water to give nitric acid, which can then be absorbed by the soil. This increases the amount of nitrates in the soil (see section 14.2).

The reaction between nitrogen and oxygen can, therefore, be the starting point for the production of nitric acid and nitrates. However, this reaction requires huge amounts of electrical energy, making it too expensive to produce nitric acid commercially in this way.

The Ostwald process

Nitric acid is made in industry by the Ostwald process. In this process, compressed air and ammonia are heated to around 800°C and then passed through layers of platinum gauze. The platinum acts as a catalyst for the reaction which takes place between the ammonia and the oxygen in the air. Ammonia is oxidised to form nitrogen monoxide (NO), one of the oxides of nitrogen. The other product of the reaction is water:

ammonia + oxygen → nitrogen monoxide + water

The nitrogen monoxide reacts with more oxygen to produce nitrogen dioxide, a brown gas:

nitrogen monoxide + oxygen → nitrogen dioxide

Next, the nitrogen dioxide is cooled, mixed with air and passed through a flow of water. During this step nitric acid is formed:

nitrogen dioxide + oxygen + water → nitric acid

Questions

Q3 The Ostwald process needs large quantities of ammonia gas. How would this gas be made?

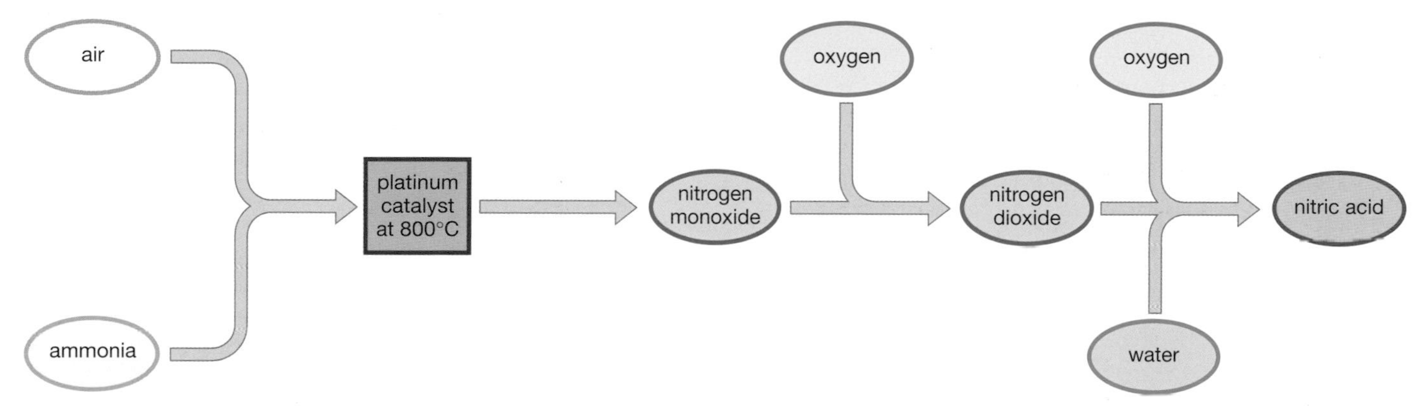

Figure 4.2 An outline of the Ostwald process

The final product contains 65 per cent nitric acid and 35 per cent water. In order to turn the nitric acid into a nitrate salt for fertilisers, it has to be neutralised. This can be carried out by reacting it with ammonia gas from the Haber process for example.

The oxidation of ammonia in the laboratory

This can be carried out by placing a hot platinum wire in a flask containing ammonia solution. A gentle stream of air is then pumped into the flask. The ammonia reacts with the oxygen in the air.

The energy to start the reaction comes from the heat in the wire. The ammonia reacts with the oxygen to form nitrogen monoxide, which then joins with more oxygen to produce nitrogen dioxide. Brown fumes of nitrogen dioxide can usually be seen around the neck of the flask.

During the reaction, the wire does not cool down; it keeps glowing red. This is because the reaction *gives out* heat. It is an **exothermic reaction**. The heat from the reaction keeps the wire glowing.

The reaction is carried out at a moderately high temperature because it is too slow at lower temperatures.

air
fumes of nitrogen dioxide
red hot platinum wire
ammonia solution

Figure 4.3 The laboratory oxidation of ammonia

Questions

Q4 During the Ostwald process, it is only necessary to heat the catalyst at the start. Why is it not necessary to keep on supplying heat to the catalyst?

Chapter 14 Study Questions

1 Nitrogen is recycled between the air, plants and animals.

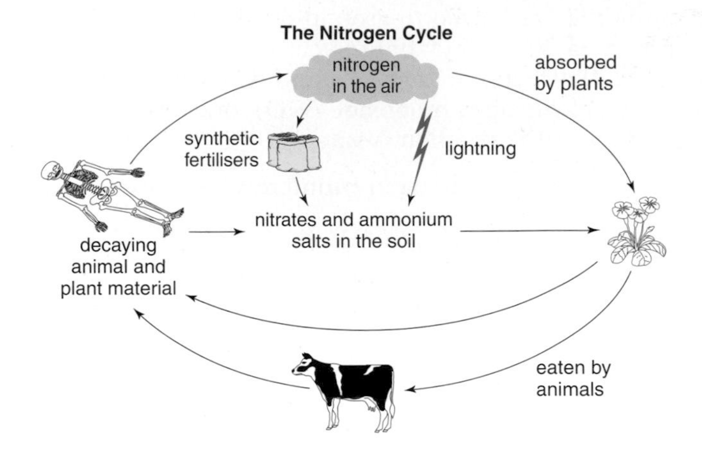

a) How are some plants able to absorb nitrogen from the air?

b) Nitrogen is an unreactive gas. Why is nitrogen able to combine with oxygen in lightning storms?

c) Fertilisers may be natural or synthetic. Why is it no longer possible to rely entirely on natural fertilisers?

d) Name the acid and alkali used to make ammonium nitrate fertiliser.

SQA GENERAL (KU)

2 The chemical industry produces large quantities of nitric acid by the catalytic oxidation of ammonia:

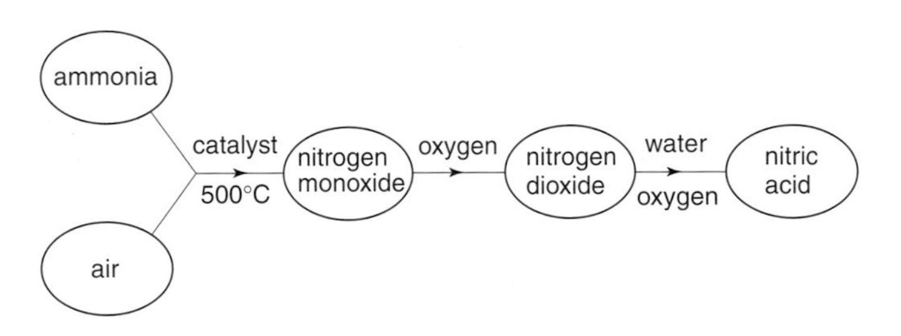

a) State the name of this industrial process.

b) Name the catalyst used to speed up the reaction.

c) The nitric acid can be used to make potassium nitrate fertiliser. Write the formula for potassium nitrate.

SEB GENERAL (KU)

3 a) Name the industrial process used to manufacture ammonia.

b) The reaction to produce ammonia is carried out at temperatures between 380°C and 450°C. Why are higher temperatures not used?

c) The graph shows the relationship between the growth of the human population and the amount of ammonia produced by industry.

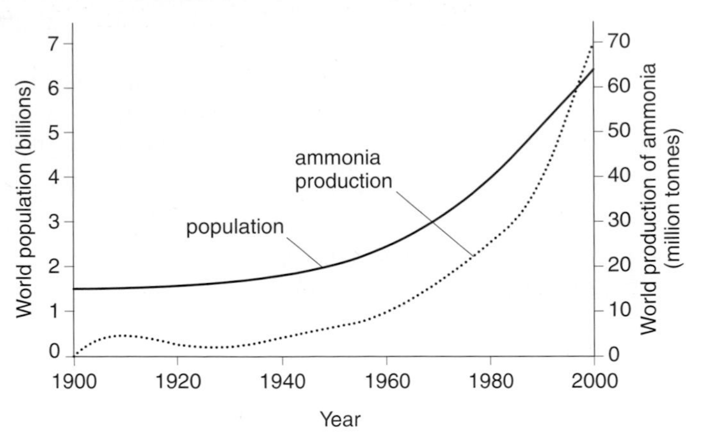

Why has the increase in world population led to an increase in ammonia production?

SQA CREDIT (PS)

4 As the world population increases, the demand for food grows. In order to meet this demand, farmers are using more and more synthetic fertilisers to improve crop yields. One of these synthetic fertilisers is Nitram. The following flow diagram shows how Nitram can be made industrially.

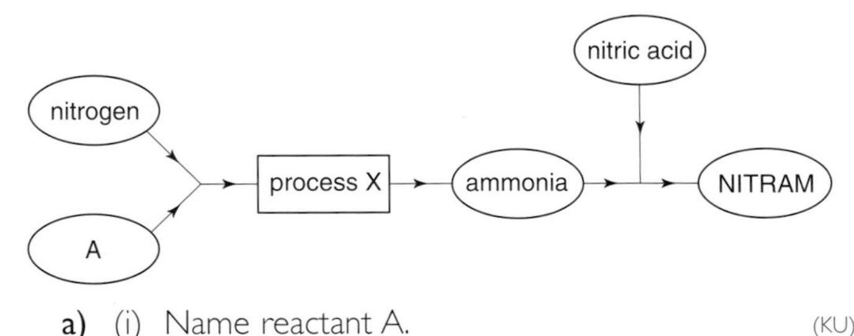

a) (i) Name reactant A. (KU)

(ii) Name industrial process X. (KU)

b) In process X, the percentage conversion of nitrogen to ammonia decreases as the temperature increases. Why, then, is process X carried out at the relatively high temperature of 450°C? (KU)

c) What is the chemical name for Nitram? (KU)

d) Nitram is very soluble in water and this allows essential elements to be taken in by the roots of crop plants very quickly. Suggest why Nitram's high solubility can also be a disadvantage in its use as a fertiliser.

SEB CREDIT (PS)

5 Study the following flow diagram which is about the manufacture of fertiliser X.

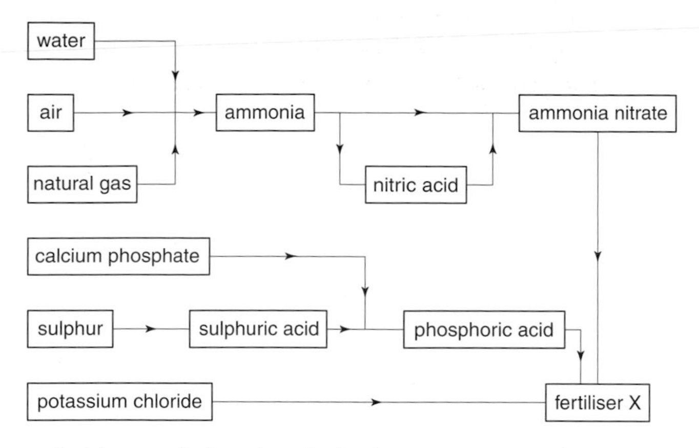

a) Name *all* the chemicals that are present in fertiliser X. (PS)
b) Calculate the formula mass of phosphoric acid, H_3PO_4. (Show your working.) CREDIT (KU)

6 Ammonia gas is converted into various solid compounds which are used as fertilisers. In one such reaction the compound urea is formed. Molecules of urea have the following structure:

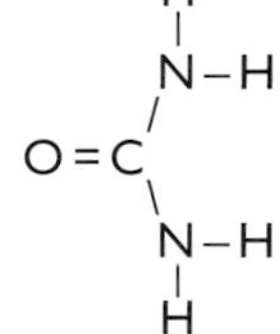

a) (i) Calculate the formula mass of urea.
(ii) Calculate the percentage by mass of nitrogen in urea.
b) Ammonia and carbon dioxide react together, using high pressure and temperature, to give urea and water. Write a balanced equation for this reaction. CREDIT (PU)

7 Calculate the percentage by mass of nitrogen in each of the following compounds which are used as fertilisers:

a) sodium nitrate
b) calcium nitrate
c) ammonium sulphate
d) ammonium nitrate. CREDIT (PS)

8 Potassium is a major plant nutrient. It promotes the rate of plant growth. Potassium is usually added to soils in the form of the salt potassium chloride.

Calculate the percentage by mass of potassium in potassium chloride. CREDIT (KU)

9 Phosphorus is an important plant nutrient as it regulates leaf development and size. Phosphorus is found as calcium phosphate $Ca_3(PO_4)_2$, but this is converted into calcium dihydrogenphosphate $Ca(H_2PO_4)_2$ which is then used as a fertiliser.

a) Calculate the percentage by mass of phosphorus in
(i) calcium phosphate
(ii) calcium dihydrogenphosphate. (KU)
b) Calcium phosphate is converted to calcium dihydrogenphosphate by reaction with sulphuric acid (H_2SO_4). The only other product of the reaction is calcium sulphate. Write a balanced chemical equation for this reaction. (KU)
c) Refer to the table of solubilities on page 5 of the SQA Data Booklet and explain why calcium dihydrogenphosphate is preferred to calcium phosphate as a fertiliser.

CREDIT (PS)

10 Rain falling during a lightning storm is often found to contain relatively large amounts of a nitrogen-containing acid X. The energy provided by lightning enables oxygen and nitrogen to combine, producing gas Y, which dissolves in water to give acid X.

a) Name the acid X and the gas Y. The acid X can be both beneficial and harmful to soil.
b) (i) State one way in which acid X is beneficial to soil.
(ii) State one way in which acid X could be harmful to soil. GENERAL (KU)

11 A test tube containing ammonia gas was placed in a beaker of water which contained universal indicator, as shown below.

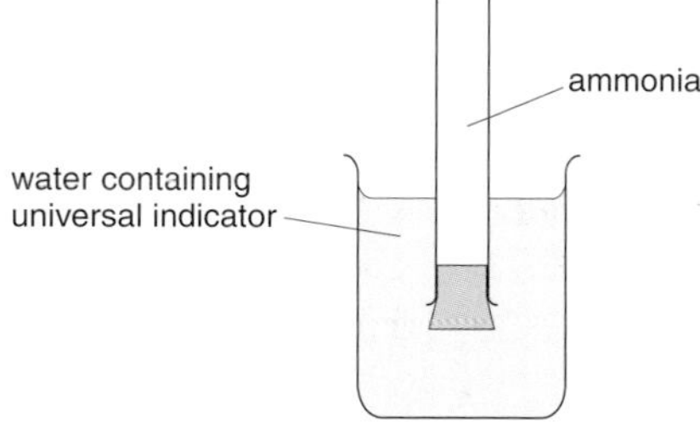

a) What would you expect to happen when the stopper was removed from the test tube? State *two* expected observations.
b) If some ammonia leaked into the laboratory, why would it be easily noticed? GENERAL (KU)

12 Crop rotation is a system used by many farmers. A field is planted with different crops each year, for example potatoes one year, perhaps oats the next and clover the year after. Why do you think this system is used? GENERAL (KU)

13 A gardener tests the pH of the soil in his garden. The soil has a pH of 5.8. The table shows the pH range in which different vegetables will grow successfully.

Vegetable	pH range
Broad beans	5.5–7.0
Carrots	6.0–7.5
Lettuce	6.5–7.5
Potatoes	5.5–6.5

a) Which vegetables will the gardener be able to grow successfully in his garden?
b) Name a substance that the gardener could add to the soil in order to grow all of the vegetables successfully.

SQA GENERAL (PS)

CHAPTER FIFTEEN

Carbohydrates and Alcohols

SECTION 15.1 Carbohydrates

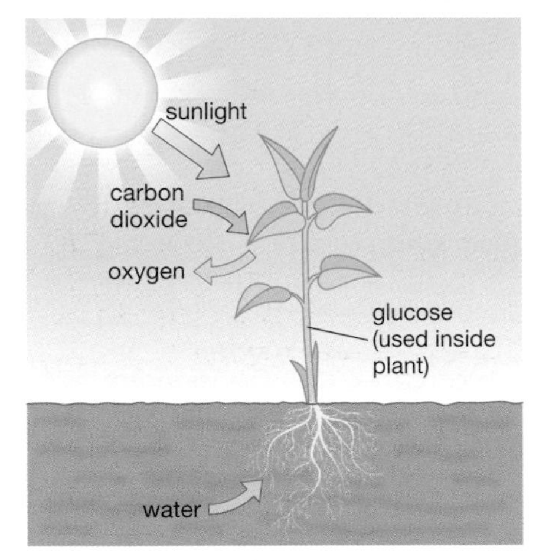

Figure 1.1 Photosynthesis

Photosynthesis

Green plants are able to make their own food. They do this by a process called **photosynthesis** (photo = light; synthesis = making something from simple materials). The plants need light, water, carbon dioxide and a special substance called **chlorophyll** in order to do this. Chlorophyll is a green pigment usually found in the leaves of plants. It traps light energy which is used to start the food-making reactions. These are very complicated. In general terms, however, they involve water and carbon dioxide combining to form compounds called **carbohydrates**. During photosynthesis oxygen is given out by the plant (see figure 1.1).

Carbohydrates are a group of compounds which contain only three elements: carbon, hydrogen and oxygen. Most of the foods which we eat to give us energy contain carbohydrates (see table 1.1). When green plants make food, they produce a carbohydrate called **glucose**. The production of glucose by photosynthesis can be shown as:

carbon dioxide + water → glucose + oxygen

The molecular formula for glucose is $C_6H_{12}O_6$, so the balanced chemical equation for this process is:

$$6CO_2 + 6H_2O \rightarrow C_6H_{12}O_6 + 6O_2$$

bread
sugar
rice
potatoes
pasta

Table 1.1 Foods that contain carbohydrates

Respiration

You may have heard of **respiration** in connection with breathing. However, scientists use the word 'respiration' to cover the whole process of taking in oxygen and combining it with glucose in order to give energy.

Humans, and other animals, need energy for warmth, movement and growth. During respiration, oxygen which has been breathed in combines with glucose in our body cells. Carbon dioxide and water are produced, and **energy** is released at the same time:

glucose + oxygen → carbon dioxide + water

The chemical equation for this is:

$$C_6H_{12}O_6 + 6O_2 \rightarrow 6CO_2 + 6H_2O$$

Foods such as potatoes and bread give us energy because they contain **starch**, a carbohydrate which our bodies can turn into glucose.

It is important to realise that respiration also occurs in plants. During photosynthesis, a plant will make food. During respiration this food is broken down to give the plant energy.

Questions

Q1 Name the substances which are
a) used up,
b) produced by photosynthesis.

Figure 1.2 Respiration supplies these athletes with the energy they need to run

Figure 1.3 Some carbohydrate-rich foods

The combustion of carbohydrates

Respiration can be thought of as a slow form of combustion without any flames. In the experiment shown in figure 1.4, a carbohydrate is burned in air and any gases which are produced are pumped into the two test tubes. Cobalt chloride paper turns from blue to pink when water is present. Lime-water turns milky white when carbon dioxide is passed through it.

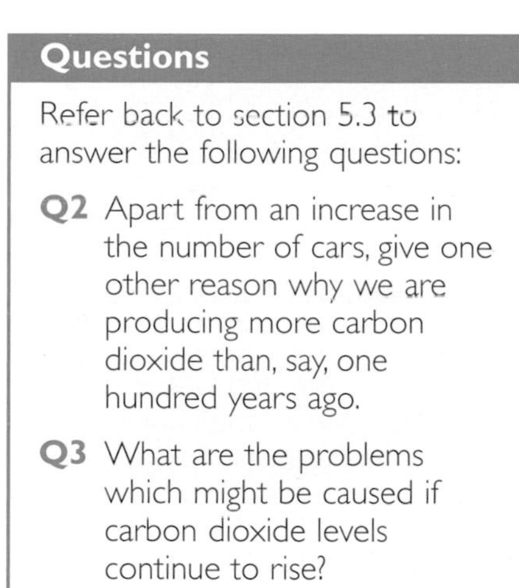

Questions

Refer back to section 5.3 to answer the following questions:

Q2 Apart from an increase in the number of cars, give one other reason why we are producing more carbon dioxide than, say, one hundred years ago.

Q3 What are the problems which might be caused if carbon dioxide levels continue to rise?

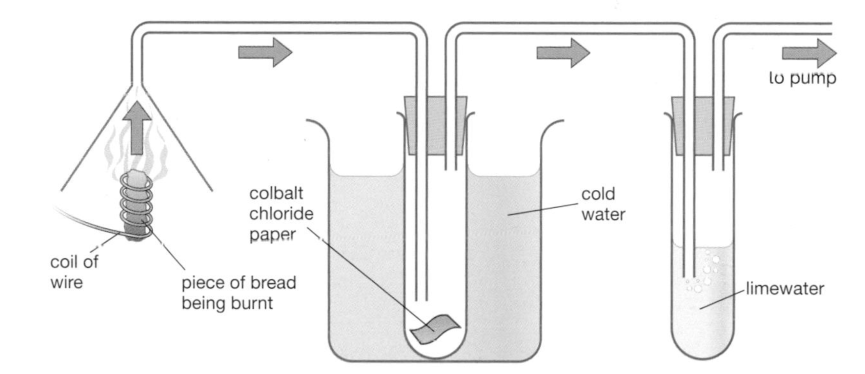

Figure 1.4 Combustion of carbohydrates experiment

The results of experiments like these show that when a carbohydrate burns, water and carbon dioxide are produced:

carbohydrate + oxygen → carbon dioxide + water

The production of carbon dioxide when a carbohydrate burns shows that there must be carbon present in the carbohydrate. Similarly, if water (H_2O) is formed then hydrogen must also be present (see section 5.3).

The air around us contains many gases. Green plants (including trees) play an important part in controlling the balance of some of these gases. All green plants produce oxygen. They also *take in* carbon dioxide. This therefore creates a balance between the levels of oxygen and carbon dioxide in the atmosphere. However, carbon dioxide levels have recently begun to increase (see figure 1.5). This suggests that the balance is no longer being maintained. One of the major reasons for this is the cutting down of tropical rainforests. This has resulted in a significant decrease in the numbers of green plants, so less carbon dioxide is being absorbed. Unfortunately, at the same time as the forests are being cut down, the number of motor vehicles producing carbon dioxide is increasing.

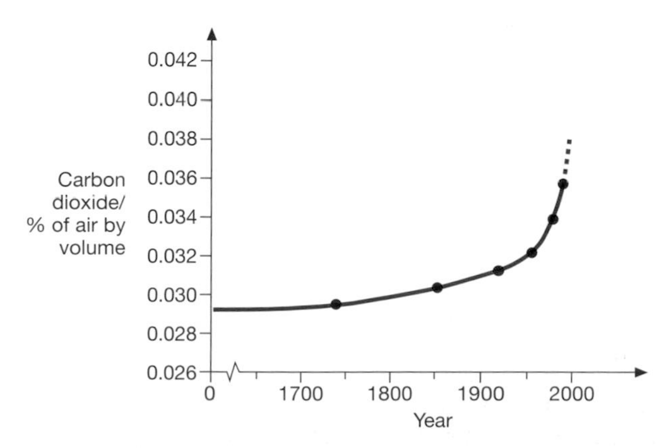

Figure 1.5 Increasing global levels of carbon dioxide

Comparing glucose and starch

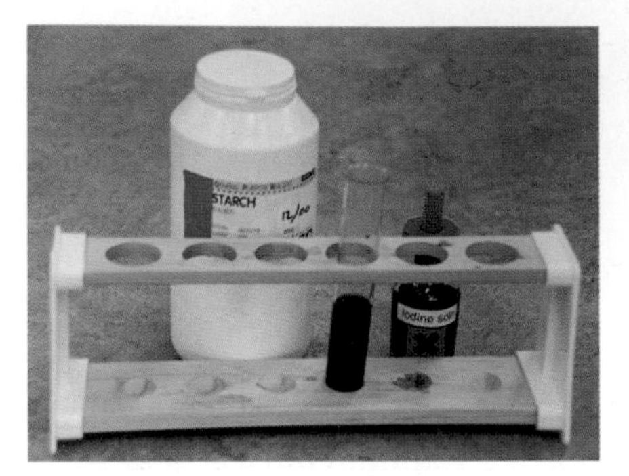

Figure 2.2 A positive test for starch, using iodine

Glucose and starch have many features in common. They are both carbohydrates and they both burn to give carbon dioxide, water and energy. However, there are important differences between them:

- Glucose tastes sweet but starch has a slightly unpleasant taste.
- Glucose dissolves easily in water. Starch does not dissolve well in water.
- A beam of light passes straight through a glucose solution with no trace of its path being left behind. A light beam is reflected back by a solution of starch so that the path of the beam can be seen. This is shown in the experiment in figure 2.1.

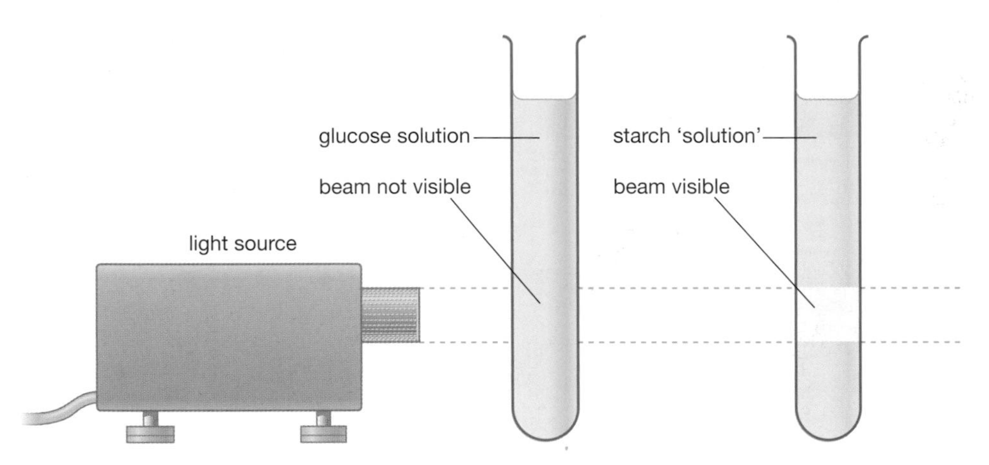

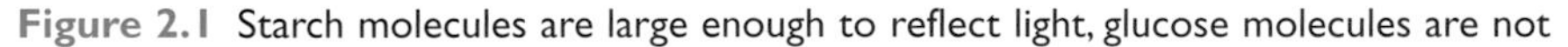

Figure 2.1 Starch molecules are large enough to reflect light, glucose molecules are not

The reason the beam of light can be seen in the starch solution is that the starch molecules are big enough to reflect the light. The glucose molecules are too small to do this. Starch molecules are therefore bigger than glucose molecules.

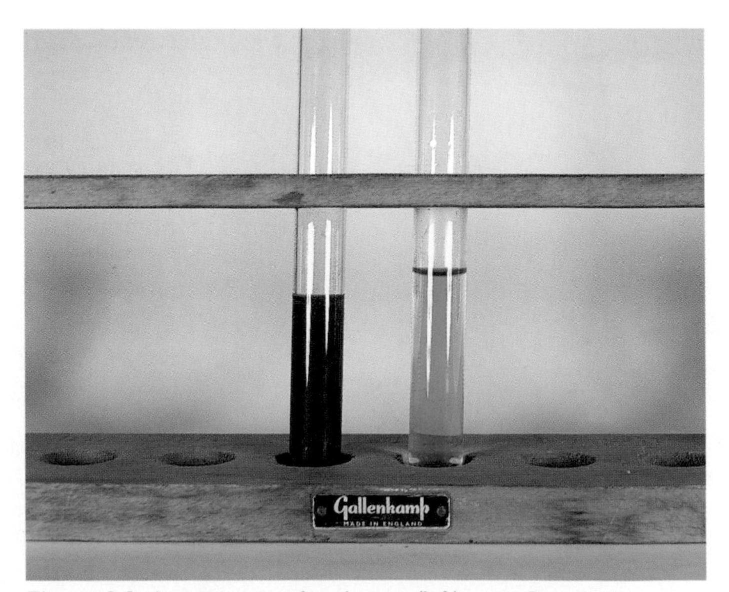

Figure 2.3 A positive test for glucose (left), using Benedict's solution (right)

Sugar	Where found
glucose	grapes, glucose drinks, chocolate
sucrose	sugar cane, sugar beet
fructose	many fruits

Table 2.1 Some common sugars

Questions

Q1 Fructose and maltose both give an orange colour on heating with Benedict's or Fehling's solution. Sucrose does *not* react with Benedict's or Fehling's solution. You are given unlabelled solutions of starch, sucrose and fructose. What tests would you carry out to see which one was which?

Figure 2.5 Honey contains a mixture of glucose and fructose

Questions

Q2 Glucose and fructose are both isomers. Give the names and the molecular formula of two other sugars which are isomers.

Chemical tests for carbohydrates

Starch is the only carbohydrate which gives a blue-black colour when iodine solution is added to it. This is the simplest chemical test for starch (figure 2.2).

Some carbohydrates are classified as **sugars**. These are carbohydrates which taste sweet and which dissolve in water. Glucose, fructose, sucrose and maltose are all sugars (see table 2.1).

Benedict's solution or Fehling's solution can be used to test for glucose. Both solutions give an orange colour when they are heated with glucose (figure 2.3). Granulated sugar, which we normally use for cooking, is called **sucrose**. It does not react with Benedict's or Fehling's solutions.

Polymerisation of glucose

Glucose is a relatively small molecule. It is formed in plants during photosynthesis. Plants turn glucose into starch by joining together many glucose molecules. This is an example of polymerisation. Glucose is a monomer and starch is the polymer that it produces.

In polymerisation, the important parts of the glucose molecules are the H atom and the –OH group. Equations for the polymerisation of glucose can be written using $H-\boxed{G}-OH$ to represent a glucose molecule. The polymerisation can then be written as in figure 2.4.

$$\cdots H-\boxed{G}-OH + H-\boxed{G}-OH + H-\boxed{G}-OH\cdots$$
$$\downarrow$$
$$\cdots\boxed{G}-\boxed{G}-\boxed{G}\cdots + {}_{n}H_2O$$

Figure 2.4 Polymerisation of glucose

Notice that a molecule of water is released each time two glucose monomers join together. Polymerisation reactions where this occurs are called **condensation polymerisation** reactions.

Chemists classify carbohydrates according to the *size* of their molecules. The simplest carbohydrates are called **monosaccharides**. They all have the molecular formula $C_6H_{12}O_6$. Glucose and fructose are both monosaccharides. Glucose and fructose are **isomers**, that is they both have the same molecular formula but they each have a different structure. (This is covered in section 6.3.)

Disaccharides all have the molecular formula $C_{12}H_{22}O_{11}$. Sucrose and maltose are disaccharides. These molecules are formed by the joining together of two monosaccharide molecules.

Polysaccharides are large polymers with a molecular formula of $(C_6H_{10}O_5)_n$ where n is a very large number. Starch and cellulose are polysaccharides. Thick strands of cellulose in the stems of plants give them support.

Type of carbohydrate	monosaccharide	disaccharide	polysaccharide
Molecular formula	$C_6H_{12}O_6$	$C_{12}H_{22}O_{11}$	$(C_6H_{10}O_5)_n$
Examples	glucose, fructose	maltose, sucrose	starch, cellulose

Table 2.2 Classification of carbohydrates

Breaking down carbohydrates

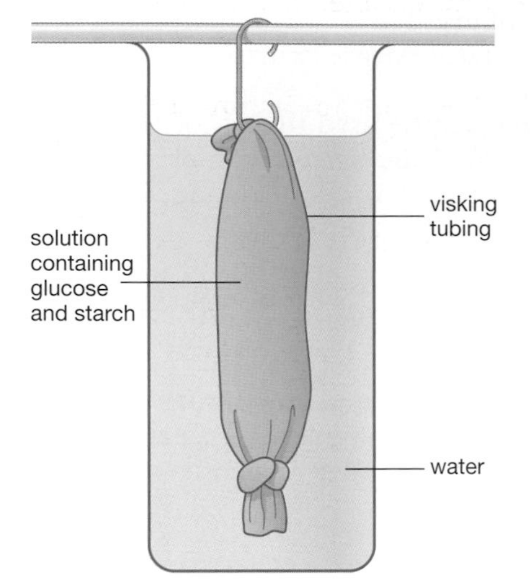

Figure 3.1

Try chewing a piece of bread for about ten minutes. Eventually it tastes of sugar! Bread contains starch. As you chew, the starch is broken down into smaller sugar molecules. During **digestion** our bodies break down large molecules such as starch, making small molecules which can then pass through the gut wall and into the bloodstream. Our bodies produce special chemicals called **enzymes** to help break down large molecules. Enzymes are biological catalysts.

In figure 3.1, starch and glucose solutions are placed inside some visking tubing. This is made from a plastic which contains tiny holes. At regular intervals the water in the beaker is tested for starch and for glucose. It is found that the glucose can pass through the tubing and go into the water but starch cannot. This is because starch molecules are too big to pass through the pores in the tubing. Similarly, in the gut, glucose molecules can pass through the gut wall into the bloodstream but starch molecules cannot.

Breaking down starch

Amylase is an enzyme found in saliva. It breaks down starch to maltose. If a mixture of starch solution and amylase is kept at around 37°C (body temperature) then the starch is quickly broken down.

Starch can also be broken down into simpler sugar molecules by heating it in a dilute solution of hydrochloric acid. In this case the starch breaks down to give molecules of glucose.

Many packet foods, such as instant mashed potato, contain starch which has been partly broken down by warming with chemicals. This makes the food cook more quickly.

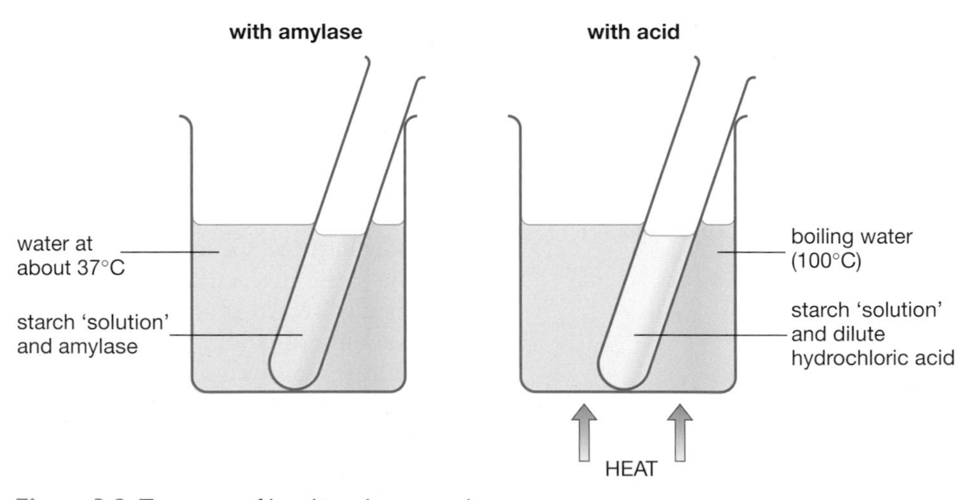

Figure 3.2 Two ways of breaking down starch

Questions

Q1 In the experiment in figure 3.2, samples of the liquid inside each test tube were taken every 5 minutes and tested with iodine. What would you notice about the results of the test that would tell you that starch was being used up?

Hydrolysis

How does starch break down into simpler molecules? It does this by reacting with water. When starch is heated with diluted hydrochloric acid, it forms glucose as follows:

$$\underset{\text{starch}}{(C_6H_{10}O_5)_n} + \underset{\text{water}}{H_2O} \rightarrow \underset{\text{glucose}}{C_6H_{12}O_6} \text{ (not balanced)}$$

A reaction like this is called a ***hydrolysis*** reaction. When hydrolysis takes place, a large molecule reacts with water and breaks down into two or more smaller molecules.

Sucrose can by hydrolysed by heating with dilute acid. In this case the sucrose molecules break down to give molecules of glucose and fructose. In nature, enzymes break down sucrose.

$$\underset{\text{sucrose}}{C_{12}H_{22}O_{11}} + \underset{\text{water}}{H_2O} \rightarrow \underset{\text{glucose}}{C_6H_{12}O_6} + \underset{\text{fructose}}{C_6H_{12}O_6}$$

Identifying the products of hydrolysis

A special technique called **paper chromatography** is used to identify carbohydrates. This is shown in figure 3.3.

Small amounts of carbohydrates as solutions to be tested are 'spotted' onto a line on special chromatography paper. In figure 3.3 we are looking at glucose, maltose and samples taken from two hydrolysis experiments. In the first, starch has been hydrolysed with the enzyme amylase, and in the second it has been hydrolysed with dilute acid.

The paper is placed in contact with a solvent, which spreads up through the paper. As the solvent moves, it carries the carbohydrates with it. When the solvent has almost reached the top of the paper, the paper is removed and dried. Carbohydrates are colourless and so a chemical called a **locating agent** is used to make them visible.

The carbohydrates can be identified by examining the distance they have been carried up the paper. In figure 3.3, the spot from the acid hydrolysis has travelled the same distance as the glucose. This indicates that they are the same, thus proving that glucose is formed when starch is hydrolysed with acid.

Questions

Q2 What does the experiment in figure 3.3 show about the product formed when starch is hydrolysed by amylase? Give a reason for your answer.

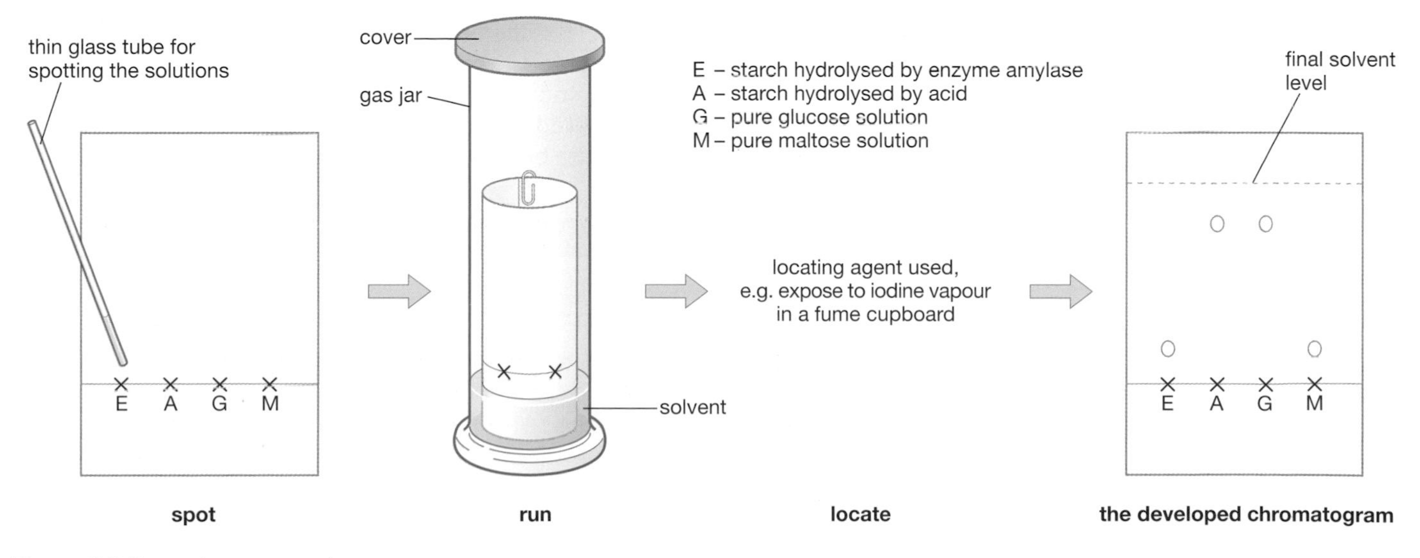

Figure 3.3 Paper chromatography

Alcoholic drinks

Figure 4.1 Each of these alcoholic drinks is made from a fruit or a vegetable

How many kinds of alcoholic drink can you name? There are beers, lagers, wines, ciders and spirits such as whisky. All of these drinks are made from fruits or vegetables. Table 4.1 shows the ones used to make some common alcoholic drinks. In fact, almost any fruit or vegetable can be used to make an alcoholic drink if there are sugars or starch present. For example, the alcohol in many wines comes from the sugar glucose which is present in grapes (see figure 4.2). In other words, an alcoholic drink can be made from any fruit or vegetable which contains starch or sugar.

Alcoholic drink	Fruit or vegetable used	Carbohydrate present
beer	barley	starch
lager	barley	starch
whisky	barley	starch
vodka	potatoes	starch
cider	apples	sugars
wine	grapes	sugars
brandy	grapes	sugars

Table 4.1 Starting materials for some common alcoholic drinks

Figure 4.2 Grapes contain glucose

Fermentation

How is alcohol produced from starch and sugar? 'Alcohol' is a general term which includes many different substances. The alcohol which is present in alcoholic drinks is called **ethanol**. Its molecular formula is C_2H_5OH. Ethanol is the second member of a chemical family called the **alkanols** (see table 4.2).

Name	Formula
Methanol	CH_3OH
Ethanol	C_2H_5OH
Propanol	C_3H_7OH
Butanol	C_4H_9OH

Table 4.2 The first four members of the alkanols

Ethanol is produced from glucose molecules by a process called **fermentation**. Carbon dioxide is also formed during fermentation:

$$\underset{\text{glucose}}{C_6H_{12}O_6} \rightarrow \underset{\text{ethanol}}{C_2H_5OH} + \underset{\text{carbon dioxide}}{CO_2} \quad \text{(not balanced)}$$

For fermentation to occur, a catalyst is needed. **Yeast**, which consists of living cells, contains enzymes – these are biological catalysts. In some cases, yeast occurs naturally in the fruit or vegetable which is fermenting. For example, a yeast grows on the surface of grapes. In other cases, such as beer-making, yeast has to be added to the fermentation mixture.

For many alcoholic drinks, such as lager or vodka, the starting material contains starch. Enzymes are used to break down the starch into glucose molecules. The glucose can then be turned into ethanol during fermentation.

Questions

Q1 Enzymes are biological catalysts. What is a catalyst?

Questions

Q2 Balance the equation given above for the fermentation of glucose.

Figure 4.3 Copper pot stills are used for whisky production

Distillation

The amount of ethanol which is produced during fermentation is fairly low. Most of the fermentation liquid is water. Lagers, for example, contain around 5 per cent alcohol.

For drinks called **spirits**, such as whisky, vodka and brandy, a higher concentration of ethanol is required, usually about 40 per cent. The production of spirits can be divided into two main stages.

First, during fermentation, a liquid with a small concentration of ethanol is produced. In the second stage the ethanol concentration is increased by a process called **distillation**. Water boils at 100°C, but ethanol has a much lower boiling point of 79°C. During distillation, the fermentation liquid is heated slowly. The ethanol boils first, and could, in theory, be separated from the rest of the watery liquid. In practice, however, the distilled product, although rich in ethanol, still contains a lot of water and other substances from the fermentation stage.

Enzymes are biological molecules which work best under certain conditions. For example, enzymes are affected by temperature. Fermentation is slow at 15°C because the enzymes in yeast do not work efficiently at that temperature. However, if the temperature is increased to around 25°C the rate of fermentation increases considerably; 25°C is the best temperature for the yeast enzymes. It is called the **optimum temperature**. If the temperature rises above 25°C, the reaction slows down.

Enzymes are also affected by the pH of their surroundings. Most of the enzymes in the human body work best in a neutral pH of about 7.

Yeast cells can be poisoned by ethanol. This means that fermentation stops after the concentration of ethanol becomes too high for the yeast cells to survive. Usually, fermentation cannot produce an ethanol concentration of more than about 14 per cent. This explains why distillation has to be used to increase the ethanol concentration of certain drinks.

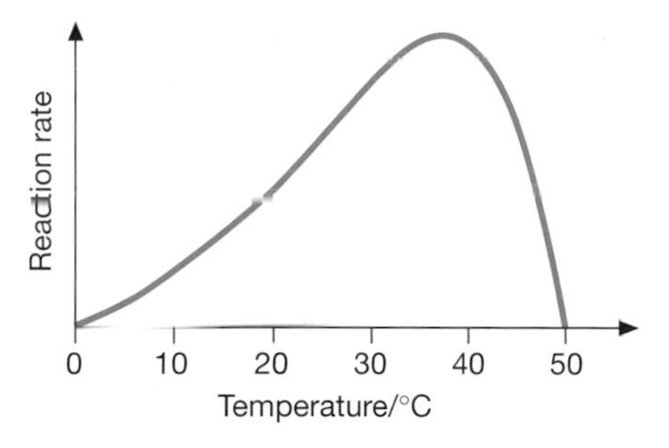

Figure 4.4 Relationship between reaction rate and temperature for an enzyme-catalysed reaction

Questions

Q3 What is the optimum temperature for the enzyme shown in figure 4.4?

Q4 The enzyme pepsin operates in the stomach where conditions are very acidic. Is its optimum pH likely to be 2, 5, 9 or 13?

Chapter 15 Study Questions

1 This diagram shows the main stages in the making of malt vinegar.

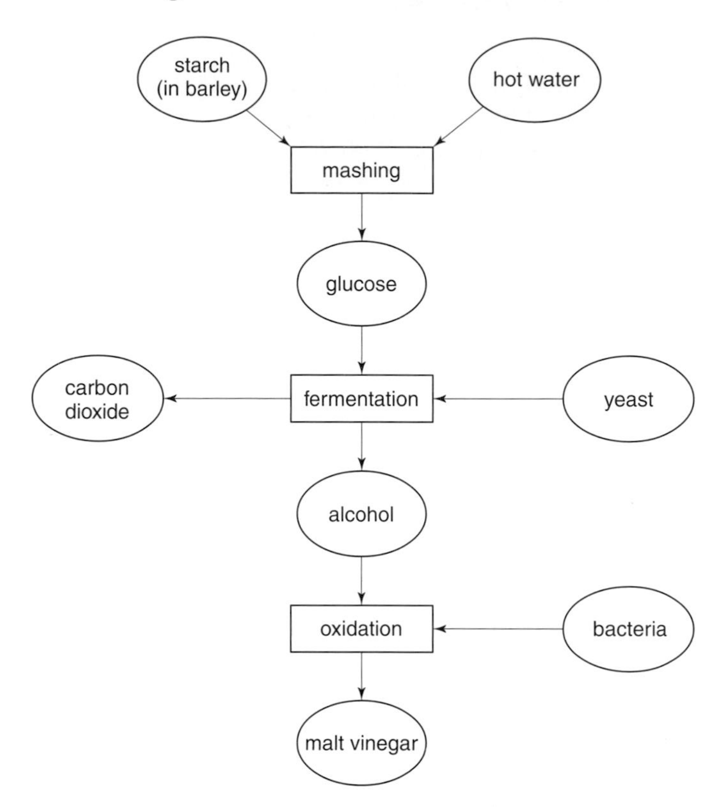

a) In the mashing process, some of the starch is broken down to glucose. Starch and glucose are carbohydrates. Name the elements present in a carbohydrate.

b) Describe how you could test the mash to show it still contained some **starch**.

SQA GENERAL (KU)

2 Carbohydrates are formed in plants.

A	fructose
B	glucose
C	maltose
D	starch
E	sucrose

a) Identify the carbohydrate which does not dissolve well in water.

b) Identify the **two** carbohydrates with the formula $C_{12}H_{22}O_{11}$.

c) Identify the cararbohydrate which is a condensation polymer.

d) Identify the **two** carbohydrates which **cannot** be hydrolysed.

SQA CREDIT (KU)

3 Visking tubing can be used to model the gut wall.

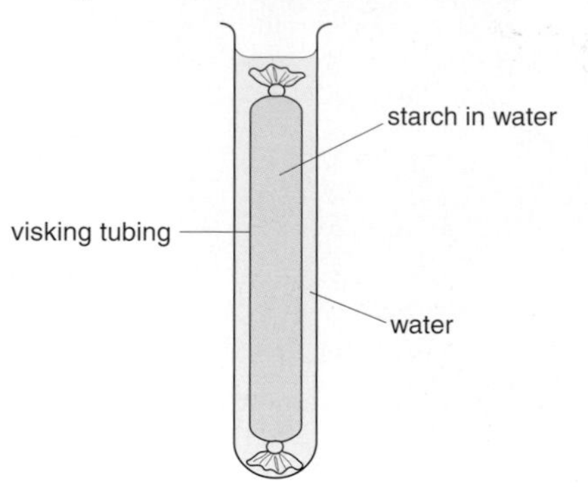

a) Describe how you would show that starch molecules are too large to pass through the visking tubing. (PS)

b) During digestion starch is hydrolysed by amylase.

(i) What is meant by 'hydrolysed'? (KU)

(ii) Using all the chemicals and apparatus describe the experiment you would carry out to show that hydrolysed starch can pass through the visking tubing. (You may wish to draw a diagram.)

You may use other apparatus if required.
Chemicals and apparatus.

SQA CREDIT (PS)

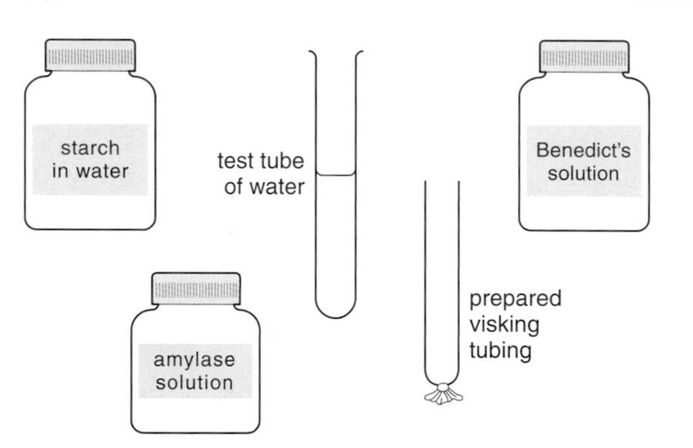

4 a) Yeast can be used to convert carbohydrates to ethanol. What name is given to this process? (KU)

Name of alkanol	Heat released when one mole of alkanol is burned (kJ)
methanol	726
ethanol	1367
propanol	2017
butanol	2665

b) Ethanol is the second member of the alkanol family. The combustion of an alkanol releases heat energy.

(i) Make a general statement linking the amount of heat released and the number of carbon atoms in the alkanol molecule.

(ii) Predict the amount of heat released, when 1 mole of pentanol burns.

SQA CREDIT (PS)

5 The following gases are present in the air, some in very small amounts:

A ammonia	**B** carbon dioxide	**C** argon
D oxygen	**E** nitrogen	**F** methane

a) Identify the gas which is given out during the fermentation of glucose.
b) Identify the gas which is given out by plants during photosynthesis.
c) Identify the gas which is taken in by human beings to be used in the process of respiration. GENERAL (KU)

6 Give details of chemical tests that could be carried out to distinguish between the following carbohydrates:

a) starch and sucrose
b) sucrose and glucose.
Remember to describe the results of each test. GENERAL (KU)

7 Give details of chemical tests that could be carried out to distinguish between the following carbohydrates:

a) maltose and sucrose
b) starch and fructose.
Remember to describe the results of each test. CREDIT (KU)

8 a) Which elements are present in all carbohydrates?

b) Name the *two* products when a carbohydrate burns completely. GENERAL (KU)

9 a) Write a balanced equation for the complete combustion of glucose.
b) Calculate the mass of oxygen that is required for the complete combustion of 360 g of glucose. CREDIT (KU)

10 In plants, glucose molecules undergo a process of condensation polymerisation to produce starch and cellulose.

a) Explain what happens during a condensation reaction.
b) Explain the meaning of the term condensation polymerisation. CREDIT (KU)

11 A reaction catalysed by an enzyme was studied and the following graph of reaction rate against temperature was obtained:

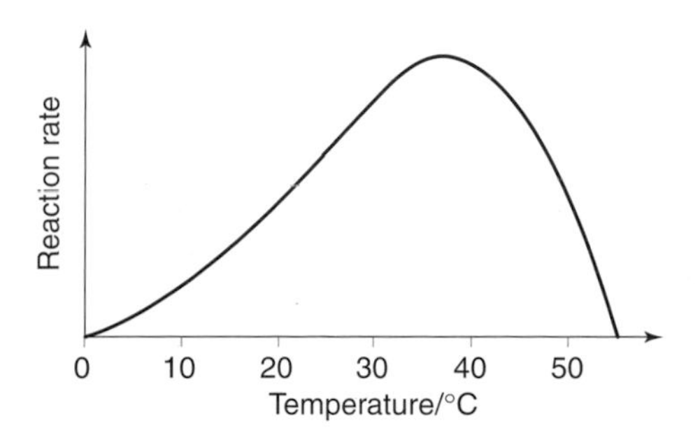

a) What is the optimum temperature for the enzyme?
b) The enzyme was obtained from an animal. Explain whether the enzyme was obtained from a 'warm-blooded' animal, such as a human being, or a 'cold-blooded' animal, such as a fish. CREDIT (PS)

12 Hilary added 1 g quantities of glucose and starch to separate beakers, each containing 100 cm^3 of boiling water. She stirred both thoroughly and then allowed them to cool.

a) One solution was found to be slightly cloudy. Which was it likely to be?
b) Explain clearly any effects that might be seen when Hilary shone a beam of light through each solution. GENERAL (KU)

13 Recent measurements have shown that throughout recent decades the amount of carbon dioxide in the air has been increasing. Many scientists claim that this could prove to be very harmful to the Earth.

a) Explain how it should be possible for the amount of carbon dioxide in the air to stay at a certain level and neither rise nor fall.
b) Explain why the clearing of forests would be expected to lead to a rise in carbon dioxide levels.
c) Name one harmful effect that scientists have predicted may take place if carbon dioxide levels continue to rise. GENERAL (KU)

14 Lactose is a disaccharide which can be hydrolysed to give two monosaccharides, glucose and galactose. Lactose, glucose and galactose are all reducing sugars.

a) Write molecular formulae for (i) lactose and (ii) glucose. (KU)
b) Glucose and galactose are compounds with the same molecular formulae but different molecular structures. What word describes the relationship between glucose and galactose? (KU)
c) State what is meant by the term 'hydrolysis'. (KU)
d) Explain why testing at intervals with Benedict's solution (or Fehling's solution) would *not* help to show that hydrolysis had taken place in this case. (PS)
e) Explain how chromatography could be used to show that hydrolysis to glucose and galactose had taken place. (You may assume that a suitable locating agent is available.) CREDIT (KU)

15 The percentage of alcohol in some drinks is shown.

Identify the **two** drinks which are made by fermentation followed by distillation.

A beer (5%)	**B** cider (8%)	**C** rum (45%)	**D** whisky (40%)	**E** wine (11%)

SEB GENERAL (KU)

CHAPTER SIXTEEN

More about Ions in Formulae and Equations

Ionic formulae

You will sometimes be asked to write **ionic formulae** for compounds. This section describes a method for doing this.

In simple compounds involving only two elements, you can assume that a compound formed between a metal and a non-metal will probably be ionic. The metal atoms lose electrons and form positively charged ions, whereas atoms of non-metals gain electrons and form negatively charged ions.

Ionic formulae are obtained by considering the electron arrangements of the atoms concerned, and how these could be altered, by loss or gain of electrons, to give the electron arrangement of the nearest noble gas.

For example, to predict the ionic formula for magnesium chloride:

step 1	element	Mg	Cl
step 2	electron arrangement	2,8,2	2,8,7
step 3	electron arrangement of nearest noble gas	2,8 (neon)	2,8,8 (argon)
step 4	number of electrons to be lost or gained	2 (lost)	1 (gained)
step 5	ions present	Mg^{2+}	Cl^-
step 6	number of ions needed to balance charges	1 Mg^{2+}	$2Cl^-$
step 7	ionic formula	$Mg^{2+}(Cl^-)_2$	

Questions

Q1 Use the technique shown above to write ionic formulae for: (a) lithium oxide, (b) calcium sulphide, (c) aluminium oxide.

Where the charge on the metal ion is shown by a roman numeral, for example I or II, there is no need to consider its electron arrangement. For example, any copper(II) compound will contain the ion Cu^{2+} and any iron(III) compound will contain the ion Fe^{3+}.

Some ionic compounds contain more than two elements. This is the case, for example, in compounds where an ion such as the carbonate ion CO_3^{2-} is present. In such cases, the procedure used can be started at step 5.

For example, to predict the ionic formula for lead(II) nitrate:

ions present	Pb^{2+}	NO_3^-
number of ions needed to balance charges	$1Pb^{2+}$	$2NO_3^-$
ionic formula	$Pb^{2+}(NO_3^-)_2$	

Questions

Q2 Write ionic formulae for:
a) copper(I) oxide,
b) iron(II) sulphate,
c) ammonium carbonate,
d) chromium(III) sulphate.

Ionic equations

If you are asked to write an ionic equation, or an ion-electron equation, you should appreciate that not all of the substances present may be ionic. For many elements and compounds, their formula will be the same in both ionic and non-ionic equations.

It should also be noted that, if an ionic compound is present as a solid, then the ionic formula is written in the normal way, but with the state symbol (s) placed after it. For example, the solid silver(I) carbonate would be written as $(Ag^+)_2CO_3^{2-}(s)$. Similarly, if an ionic compound is present dissolved in water, then the ions are separated in a special way and the state symbol (aq) is placed after each ion. Thus, sodium sulphate solution would be written as $2Na^+(aq) + SO_4^{2-}(aq)$, whereas solid sodium sulphate would appear as $(Na^+)_2SO_4^{2-}(s)$.

For example, consider the displacement reaction in which magnesium metal is added to silver(I) nitrate solution. This reaction results in the formation of silver metal and magnesium nitrate solution. Using simple formulae and state symbols, we can write the following equation:

$$Mg(s) + 2AgNO_3(aq) \rightarrow 2Ag(s) + Mg(NO_3)_2(aq)$$

However, the two compounds present are ionic and soluble in water, so the following ionic equation can be written for the reaction:

$$Mg(s) + 2Ag^+(aq) + 2NO_3^-(aq) \rightarrow 2Ag(s) + Mg^{2+}(aq) + 2NO_3^-(aq)$$

In this equation, the nitrate ions appear on both sides and can therefore be omitted as they are spectator ions. The equation then becomes:

$$Mg(s) + 2Ag^+(aq) \rightarrow 2Ag(s) + Mg^{2+}(aq)$$

The displacement reaction taking place between the magnesium atoms and the silver ions is a redox process in which electrons are lost by the magnesium atoms and gained by the silver ions. The reaction therefore involves both oxidation (loss of electrons) and reduction (gain of electrons) and can be written as two ion-electron equations:

$$Mg(s) \rightarrow Mg^{2+}(aq) + 2e^- \text{ (oxidation of Mg atoms)}$$

$$2e^- + 2Ag^+(aq) \rightarrow 2Ag(s) \text{ (reduction of Ag}^+\text{ ions)}$$

Questions

Q3 Write the following as an ionic equation:

$$CuO(s) + H_2SO_4(aq) \rightarrow CuSO_4(aq) + H_2O(l)$$

Q4 Zinc reacts with hydrochloric acid to give zinc chloride solution and hydrogen.

a) Write an ionic equation for this reaction.

b) Identify any spectator ions present.

c) Write ion-electron equations to show that oxidation and reduction take place during the reaction.

CHAPTER SEVENTEEN

Types of Formulae for Covalent Molecular Substances

Molecular formula

C_2H_6

Figure 1.1 Molecular formula for ethane

This shows the number of atoms of different elements which are present in one molecule of a substance. It is so commonly used that it is often referred to simply as 'the formula' for a given substance, although there are in fact several types of formula. The molecular formula for ethane, for example, is C_2H_6 (figure 1.1).

Empirical formula

CH_3

Figure 1.2 Empirical formula for ethane

This shows the simplest *ratio* of atoms in a compound. Thus for ethane, molecular formula C_2H_6, the empirical formula is CH_3, meaning that the ratio of carbon atoms to hydrogen atoms is 1:3 (figure 1.2). This is obtained by simplifying the ratio 2:6 by dividing both numbers by 2.

Full structural formula

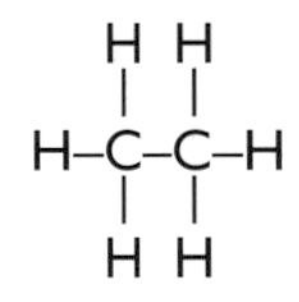

Figure 1.3 Full structural formula for ethane

This shows all of the bonds present in a molecule, although it does not always show the actual positions of the atoms. (See 'perspective formula' below.) The full structural formula for ethane is shown in figure 1.3.

Shortened structural formula

CH_3CH_3

Figure 1.4 Shortened structural formula for ethane

This shows the sequence of groups of atoms in a molecule. In ethane, for example, two CH_3 groups are joined together, so the shortened structural formula is CH_3CH_3 (figure 1.4).

Perspective formula

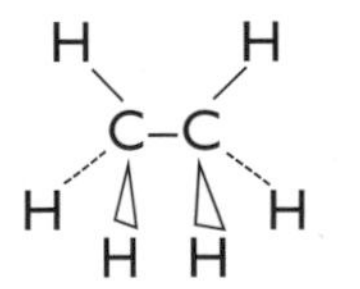

Figure 1.5 Perspective formula for ethane

This is simply the formula which shows the true arrangement of the atoms in a molecule, and the bonds as well. In ethane, for example, the four single covalent bonds around each carbon atom are arranged in a tetrahedral shape, as they are in all alkane molecules. The perspective formula for ethane is shown in figure 1.5, but you will probably be more familiar with the tetrahedral shape of methane, which is shown in figure 1.6.

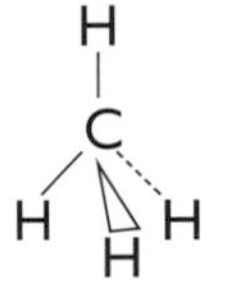

Figure 1.6 Perspective formula for methane

Chapter 17 Study Questions

1 Vinegar is a dilute solution of ethanoic acid, which has the following full structural formula:

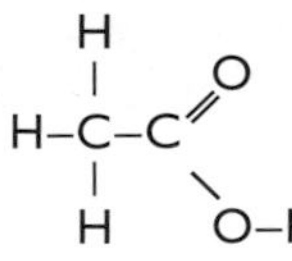

a) Give the molecular formula for ethanoic acid.
b) Give the empirical formula for ethanoic acid.

2 Give empirical formulae for the following compounds. In each case the molecular formula is given after the name:

a) glucose, $C_6H_{12}O_6$
b) sucrose, $C_{12}H_{22}O_{11}$
c) octane, C_8H_{18}
d) carbon dioxide, CO_2

3 Aluminium chloride is unusual in that although it forms ions in aqueous solution, it normally exists as molecules. The shape of these molecules (that is, the perspective formula) is shown below:

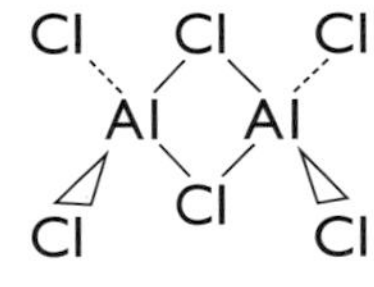

a) Give the molecular formula for aluminium chloride.
b) Give the empirical formula for aluminium chloride.

4 Give the perspective formula for water, that is, the formula which shows the shape of a water molecule.

5 For the hydrocarbon butane give:

a) the full structural formula
b) the shortened structural formula
c) the molecular formula
d) the empirical formula.

6 Trichloromethane (chloroform), which has molecular formula $CHCl_3$, was used many years ago as an anaesthetic.

a) Draw a full structural formula for trichloromethane.
b) Draw a perspective formula for trichloromethane.

7 For each of the following, draw (i) the molecular formula, (ii) the empirical formula, (iii) the full structural formula and (iv) the shortened structural formula.

a) Hexane
b) Propene
c) Cyclobutane

CHAPTER EIGHTEEN

Identification tests

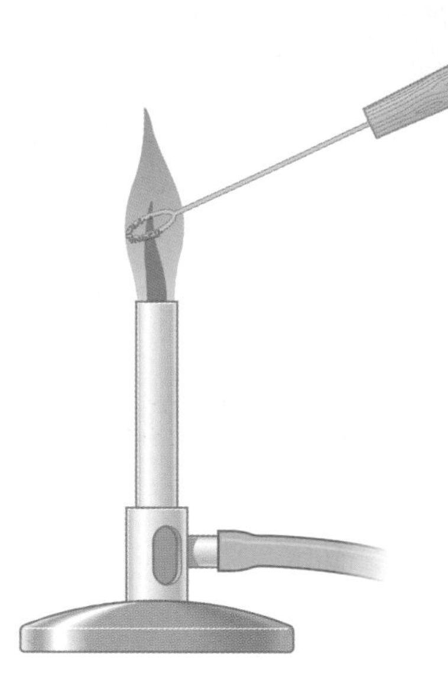

Figure 1.1 Flame test

Flame tests

Certain metals may be identified by the characteristic colours they produce when heated in a bunsen flame. The tests are usually carried out on compounds which contain the metals as ions.

Element	Ion	Flame colour
barium	Ba^{2+}	green
calcium	Ca^{2+}	orange-red
copper	Cu^{2+}	blue-green
lithium	Li^{+}	red
potassium	K^{+}	lilac
sodium	Na^{+}	yellow
strontium	Sr^{2+}	red

Identification of gases

Gas	Test	Result
carbon dioxide	pass into lime water	solution turns milky white
hydrogen	apply a burning taper	gas burns with a 'pop'
oxygen	apply a glowing taper	taper is re-lit

Test for a carbonate

Test	Result
add dilute hydrochloric acid (or similar acid)	carbon dioxide gas is given off (which turns lime water milky)

Testing for rust

Rust may be detected using ferroxyl indicator solution. When iron rusts, Fe^{2+}(aq) and OH^{-}(aq) ions are formed. These ions produce characteristic colours with ferroxyl indicator:

Ion	Colour produced with ferroxyl indicator
Fe^{2+}(aq)	blue
OH^{-}(aq)	pink

Testing for acidic, alkaline and neutral solutions

The pH of a solution may be determined using a pH meter or, less accurately, with pH paper or universal indicator. Although there can be differences in the results obtained, these methods usually produce the following colours:

Nature of solution	pH	Colour produced
quite strongly acid	much less than 7	red
neutral	7	green
quite strongly alkaline	much greater than 7	violet

Tests for carbohydrates

Carbohydrate	Test	Result
reducing sugars e.g. glucose	heat with Benedict's (or Fehling's) solution	orange-red colour
starch	iodine solution	dark-blue colour

Test for unsaturated hydrocarbons

An unsaturated hydrocarbon is one which contains a C=C bond. For example, ethene, $CH_2=CH_2$, and propene, $CH_3CH=CH_2$, or any alkene.

Test	Result
add bromine solution and shake	rapid decolorisation

This test is *not* given by saturated hydrocarbons such as alkanes and cycloalkanes, which contain only C–C bonds.

Test for water

Water is identified by measuring its freezing point and boiling point:

freezing point 0°C
boiling point 100°C

(No two substances have the same freezing points and boiling points.)

Chapter 18 Study Questions

In an examination, you may be asked to suggest a way of distinguishing between two substances. You will be expected to describe a chemical test in which the two substances behave differently. It is important to describe the result of the test.

Example:
Question: The labels have fallen off two bottles containing colourless liquids in the chemical store of a school, but it is known that one is pentene and the other is pentane. Describe a chemical test, together with the results, by which the bottles could be correctly relabelled.

Answer: Add bromine solution to both. The one which decolourises the bromine solution is pentene.

It may be possible to give more than one correct answer to questions of this type.

Now answer the following questions:

Describe chemical tests which could be carried out to distinguish between the following. Also give the results of the tests.

1 sodium chloride and potassium chloride

2 lithium carbonate and magnesium sulphate

3 barium oxide and calcium oxide

4 dilute hydrochloric acid and dilute sodium hydroxide

5 hydrogen gas and helium gas

6 copper(II) oxide and carbon (both are black solids)

7 iron(II) sulphate solution and dilute nickel(II) sulphate solution (both are pale green in colour)

8 glucose and sucrose

9 sucrose and starch

10 glucose and starch

11 propene gas and cyclopropane gas

12 heptane and heptene.

CHAPTER NINETEEN

Gas Production and Collection

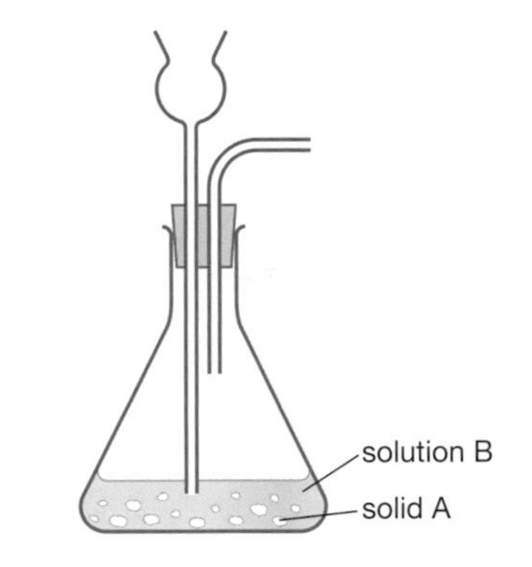

Figure 1.1

Gas production

Several gases can be produced using a solid and a solution in the apparatus shown in Figure 1.1. Solid A is placed in the flask and solution B is added via the thistle funnel.

Solid A	Solution B	Gas produced
zinc	dilute sulphuric acid	hydrogen
marble chips ($CaCO_3$)	dilute hydrochloric acid	carbon dioxide
maganese(IV) oxide	hydrogen peroxide	oxygen

Table 1.1

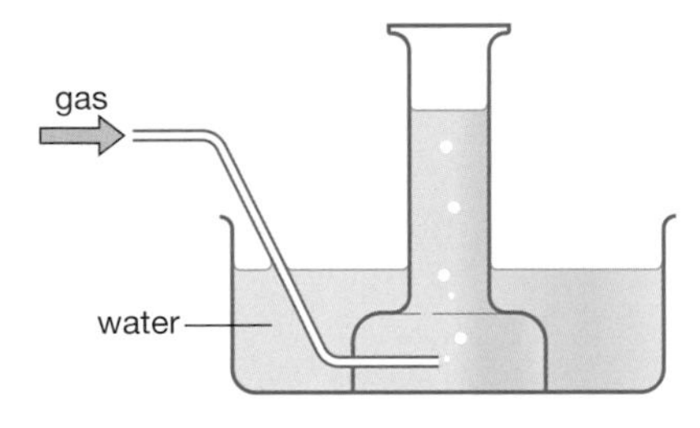

Figure 1.2

Gas collection

Collection over water

If a gas is *not* very soluble in water then it may be collected over water as shown in Figure 1.2. Gases that can be collected in this way include oxygen, nitrogen, carbon dioxide, the noble gases and all of the hydrocarbon gases such as methane.

Gases that are appreciably soluble in water and therefore *cannot* be collected over water include ammonia, sulphur dioxide and nitrogen dioxide. It should also be noted that collection over water does not provide a *dry* sample of gas.

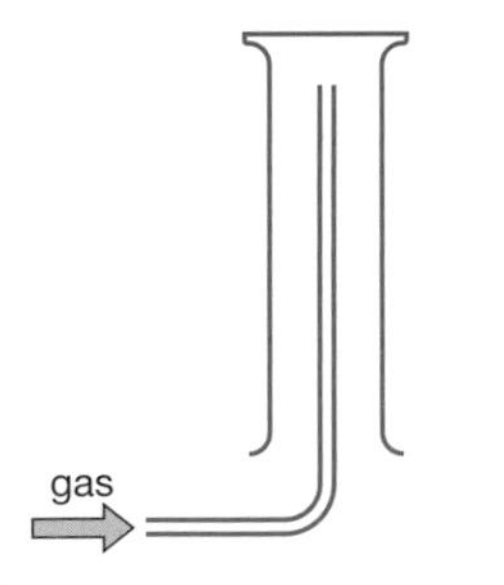

Figure 1.3

Collection by downward displacement of air

Gases that are *less* dense than air may be collected by downward displacement of air as shown in Figure 1.3. Since air is roughly 80% nitrogen, which has a formula mass of 28, and 20% oxygen, with a formula mass of 32, the average air molecule has a formula mass of about 29. If a gas has a formula mass that is less than 29 it is less dense than air and can be collected by downward displacement of air.

Gases that can be collected using this method include hydrogen, helium, ammonia and methane. Collection by this method provides a *dry* sample of gas, but since none of the lighter gases are coloured, it is difficult to know when the gas jar or test tube is full.

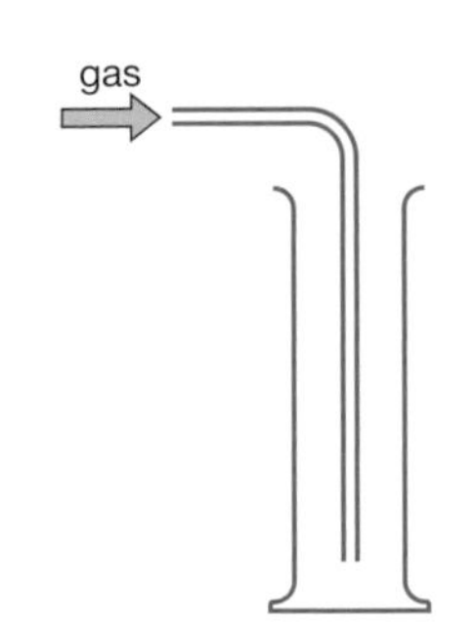

Figure 1.4

Collection by upward displacement of air

Gases that are *more* dense than air may be collected by upward displacement of air as shown in Figure 1.4. Examples of gases that have formula masses well above 29 and can therefore be collected using this method include carbon dioxide, sulphur dioxide, nitrogen dioxide and butane.

Collection by this method provides a *dry* sample of gas, but only in the case of a coloured gas, such as the reddish brown nitrogen dioxide, is it easy to tell when the gas jar or test tube is full.

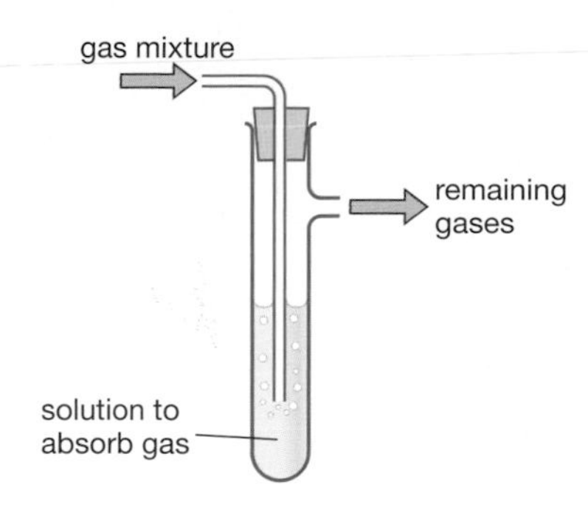

Figure 1.5

Removing acidic and alkaline gases

An acidic gas, such as carbon dioxide, can be removed from other neutral gases by passing the mixture through a solution of an alkali such as sodium hydroxide. An alkaline gas, such as ammonia, can be removed from other neutral gases by passage through an acid solution, such as dilute hydrochloric acid. Apparatus that could be used to remove acidic or alkaline gases is shown in figure 1.5.

Although ammonia is the only common alkaline gas, there are several fairly common acidic gases. Examples of acidic gases include carbon dioxide, sulphur dioxide, nitrogen dioxide and hydrogen chloride.

Chapter 19 Study Questions

In questions 1–4 choose the correct word(s) **from the following list** to complete the sentence.

downward, less, more, upward, nitrogen dioxide, nitrogen, insoluble, soluble

1 Neon is _______ dense than air and can be collected by downward displacement of air.

2 Ammonia is _______ in water and is therefore collected by downward displacement of air.

3 Carbon dioxide can be collected by _______ displacement of air.

4 Because _______ is insoluble in water it can be collected over water.

5 Which of the following gases is more dense than air?

A Ammonia
B Methane
C Helium
D Argon

6 Which of the following gases can be collected over water?

A Ammonia
B Carbon dioxide
C Nitrogen dioxide
D Sulphur dioxide

7 Which of the following can be collected by downward displacement of air?

A Krypton
B Xenon
C Argon
D Helium

8 Oxygen, nitrogen, hydrogen and carbon dioxide are common gases. Which of the following statements is true?

A All can be collected by upward displacement of air.
B All can be collected by downward displacement of air.
C All can be collected over water.
D Different methods must be used for their collection.

9 A sample of argon gas (neutral and insoluble) is contaminated with acidic hydrogen bromide gas. Copy and complete the diagram below to show how the hydrogen bromide could be removed and the argon collected over water.

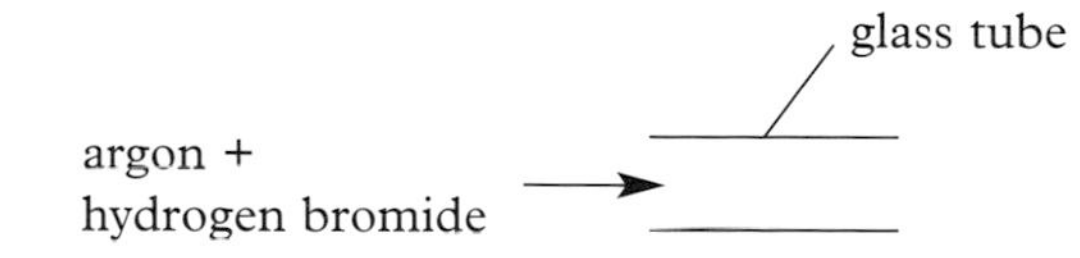

CHAPTER TWENTY

Separation Techniques

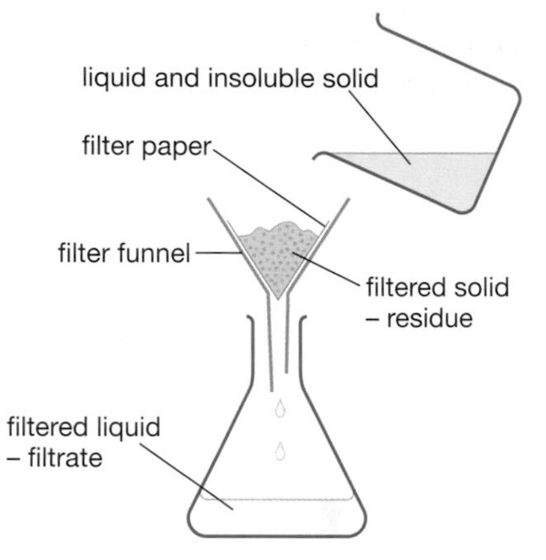

Figure 1.1 Apparatus for filtration

Filtration

- Insoluble solids can be separated from liquids by passage through filter paper.
- The solid which remains behind in the filter paper is called the **residue**.
- The liquid which passes through the filter paper is called the **filtrate**.
- Precipitates may be separated from aqueous solutions of other substances using this technique.

Evaporation

- A solid solute may be recovered from a solution by evaporating off the solvent.
- Soluble salts may be recovered from solutions using this technique.

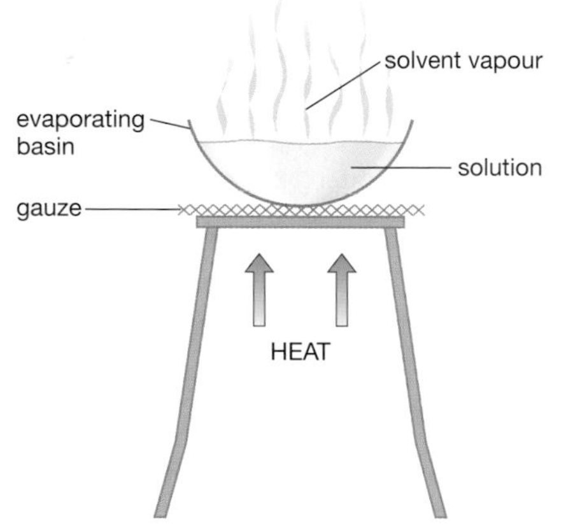

Figure 1.2 Apparatus for evaporation

Distillation

- A mixture of liquids may be separated by distillation, the one with the lowest boiling point distilling off first.
- Ethanol (boiling point 79°C) may be separated from water (boiling point 100°C) after the fermentation of sugars such as glucose.
- A solvent can be recovered from a solution containing dissolved solids.

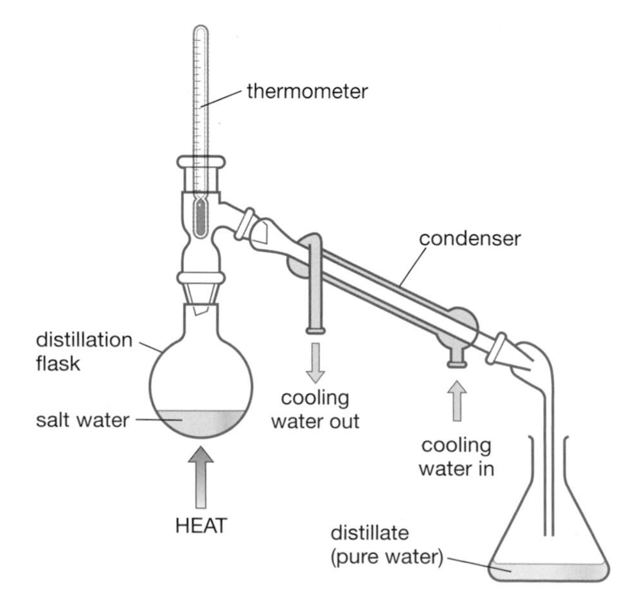

Figure 1.3 Apparatus for distillation

Chromatography

- Small quantities of soluble compounds can be separated using paper chromatography.
- The separation of coloured substances, such as food colours or coloured inks, is easily seen.
- A drop of the mixture is spotted onto chromatography paper (or filter paper) and the bottom of the paper is placed in a suitable solvent (water can often be used).
- If the substances used are colourless, such as sugars or amino acids, a locating agent is used to show up their presence.

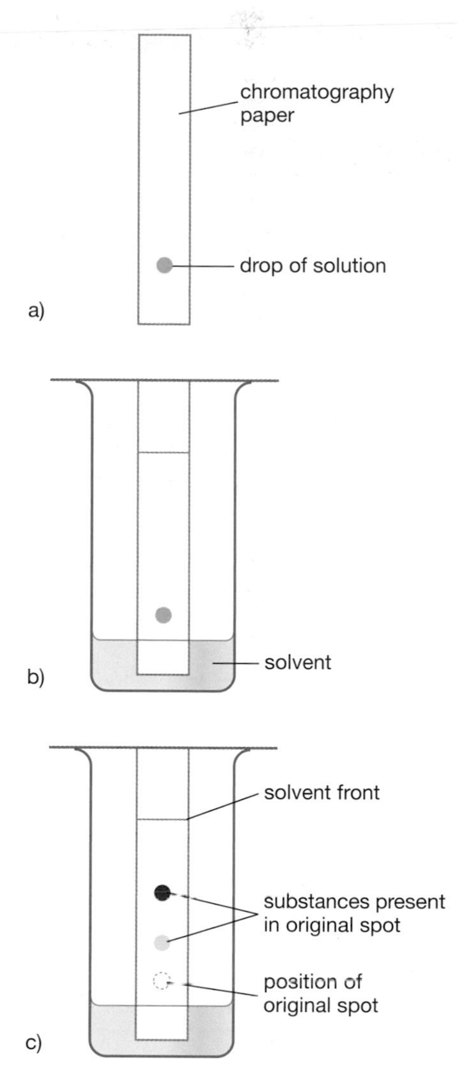

Figure 1.4 Paper chromatography can work using only a single drop of a solution

CHAPTER TWENTY ONE
Data

Name, symbols and relative atomic masses of selected elements simplified for calculations

Relative atomic masses are also known as the average atomic masses.

Element	Symbol	Relative atomic mass	Element	Symbol	Relative atomic mass
aluminium	Al	27	magnesium	Mg	24.5
argon	Ar	40	mercury	Hg	200.5
bromine	Br	80	neon	Ne	20
calcium	Ca	40	nickel	Ni	58.5
carbon	C	12	nitrogen	N	14
chlorine	Cl	35.5	oxygen	O	16
copper	Cu	63.5	phosphorus	P	31
fluorine	F	19	platinum	Pt	195
gold	Au	197	potassium	K	39
helium	He	4	silicon	Si	28
hydrogen	H	1	silver	Ag	108
iodine	I	127	sodium	Na	23
iron	Fe	56	sulphur	S	32
lead	Pb	207	tin	Sn	118.5
lithium	Li	7	zinc	Zn	65.5

Formulae of selected ions containing more than one kind of atom

One positive		One negative		Two negative		Three negative	
Ion	Formula	Ion	Formula	Ion	Formula	Ion	Formula
ammonium	NH_4^+	ethanoate	CH_3COO^-	carbonate	CO_3^{2-}	phosphate	PO_4^{3-}
		hydrogencarbonate	HCO_3^-	chromate	CrO_4^{2-}		
		hydrogensulphate	HSO_4^-	dichromate	$Cr_2O_7^{2-}$		
		hydrogensulphite	HSO_3^-	sulphate	SO_4^{2-}		
		hydroxide	OH^-	sulphite	SO_3^{2-}		
		nitrate	NO_3^-				
		permanganate	MnO_4^-				

Periodic table of the elements showing dates of discovery

Atomic number Symbol
Electron arrangement
Name
Date of discovery

1	2											3	4	5	6	7	0
1 H 1 hydrogen 1766																	2 He 2 helium 1862
3 Li 2,1 lithium 1817	4 Be 2,2 beryllium 1789											5 B 2,3 boron 1808	6 C 2,4 carbon prehistoric	7 N 2,5 nitrogen 1772	8 O 2,6 oxygen 1774	9 F 2,7 fluorine 1771	10 Ne 2,8 neon 1898
11 Na 2,8,1 sodium 1807	12 Mg 2,8,2 magnesium 1775	Transition metals										13 Al 2,8,3 aluminium 1827	14 Si 2,8,4 silicon 1823	15 P 2,8,5 phosphorus 1669	16 S 2,8,6 sulphur prehistoric	17 Cl 2,8,7 chlorine 1774	18 Ar 2,8,8 argon 1894
19 K 2,8,8,1 potassium 1807	20 Ca 2,8,8,2 calcium 1808	21 Sc 2,8,9,2 scandium 1879	22 Ti 2,8,10,2 titanium 1791	23 V 2,8,11,2 vanadium 1830	24 Cr 2,8,13,1 chromium 1797	25 Mn 2,8,13,2 manganese 1774	26 Fe 2,8,14,2 iron prehistoric	27 Co 2,8,15,2 cobalt 1735	28 Ni 2,8 16,2 nickel 1751	29 Cu 2,8,18,1 copper prehistoric	30 Zn 2,8,18,2 zinc 16th century	31 Ga 2,8,18,3 gallium 1875	32 Ge 2,8,18,4 germanium 1866	33 As 2,8,18,5 arsenic ~1250	34 Se 2,8,18,6 selenium 1817	35 Br 2,8,18,7 bromine 1826	36 Kr 2,8,18,8 krypton 1898
37 Rb 2,8,18,8,1 rubidium 1861	38 Sr 2,8,18,8,2 strontium 1790	39 Y 2,8,18,9,2 yttrium 1794	40 Zr 2,8,18,10,2 zirconium 1789	41 Nb 2,8,18,11,2 niobium 1801	42 Mo 2,8,18,13,1 molybdenum 1778	43 Tc 2,8,18,14,1 technetium 1937	44 Ru 2,8,18,15,1 ruthenium 1844	45 Rh 2,8,18,16,1 rhodium 1803	46 Pd 2,8,18,18,0 palladium 1803	47 Ag 2,8,18,18,1 silver prehistoric	48 Cd 2,8,18,18,2 cadmium 1817	49 In 2,8,18,18,3 indium 1863	50 Sn 2,8,18,18,4 tin prehistoric	51 Sb 2,8,18,18,5 antimony ~1450	52 Te 2,8,18,18,6 tellurium 1782	53 I 2,8,18,18,7 iodine 1811	54 Xe 2,8,18,18,8 xenon 1898
55 Cs 2,8,18,18,8,1 caesium 1860	56 Ba 2,8,18,18,8,2 barium 1808	•	72 Hf 2,8,18,32,10,2 hafnium 1923	73 Ta 2,8,18,32,11,2 tantalum 1802	74 W 2,8,18,32,12,2 tungsten 1783	75 Re 2,8,18,32,13,2 rhenium 1925	76 Os 2,8,18,32,14,2 osmium 1804	77 Ir 2,8,18,32,17,0 iridium 1804	78 Pt 2,8,18,32,17,1 platinum 16th century	79 Au 2,8,18,32,18,1 gold prehistoric	80 Hg 2,8,18,32,18,2 mercury prehistoric	81 Tl 2,8,18,32,18,3 thallium 1861	82 Pb 2,8,18,32,18,4 lead prehistoric	83 Bi 2,8,18,32,18,5 bismuth 1753	84 Po 2,8,18,32,18,6 polonium 1898	85 At* 2,8,18,32,18,7 astatine 1940	86 Rn 2,8,18,32,18,8 radon 1900
87 Fr* 2,8,18,32,18,8,1 francium 1939	88 Ra 2,8,18,32,18,8,2 radium 1898	••	104 Rf* rutherfordium 1964	105 Db* dubnium 1970	106 Sg* seaborgium 1974	107 Bh* bohrium 1981	108 Hs* hassium 1984	109 Mt* meitnerium 1982									
Alkali metals																Halogens	Noble gases

•	57 La 2,8,18,18,9,2 lanthanum 1839	58 Ce 2,8,18,20,8,2 cerium 1803	59 Pr 2,8,18,21,8,2 praseodymium 1885	60 Nd 2,8,18,22,8,2 neodymium 1885	61 Pm 2,8,18,23,8,2 promethium 1947	62 Pm 2,8,18,24,8,2 samarium 1879	63 Eu 2,8,18,25,8,2 europium 1896	64 Gd 2,8,18,25,9,2 gadolinium 1880	65 Tb 2,8,18,27,8,2 terbium 1843	66 Dy 2,8,18,28,8,2 dysprosium 1886	67 Ho 2,8,18,29,8,2 holmium 1879	68 Er 2,8,18,30,8,2 erbium 1843	69 Tm 2,8,18,31,8,2 thulium 1879	70 Yb 2,8,18,32,8,2 ytterbium 1907	71 Lu 2,8,18,32,9,2 lutetium 1907
••	89 Ac 2,8,18,32,18,9,2 actinium 1899	90 Th 2,8,18,32,18,10,2 thorium 1828	91 Pa 2,8,18,32,20,9,2 protactinium 1917	92 U 2,8,18,32,21,9,2 uranium 1789	93 Np* 2,8,18,32,22,9,2 neptunium 1940	94 Pu* 2,8,18,32,24,8,2 plutonium 1940	95 Am* 2,8,18,32,25,8,2 americium 1944	96 Cm* 2,8,18,32,25,9,2 curium 1944	97 Bk* 2,8,18,32,26,9,2 berkelium 1944	98 Cf* 2,8,18,32,28,8,2 californium 1952	99 Es* 2,8,18,32,29,8,2 einsteinium 1953	100 Fm* 2,8,18,32,30,8,2 fermium 1955	101 Md* 2,8,18,32,31,8,2 mendelevium 1957	102 No* 2,8,18,32,32,8,2 nobelium 1961	103 Lr* 2,8,18,32,32,9,2 lawrencium 1969

* element not found in nature but has been made by scientists

CHAPTER TWENTY TWO
Chemical Dictionary

A

acid A substance which produces $H^+(aq)$ ions.

acid solution One which contains a greater concentration of $H^+(aq)$ ions than pure water.

activity series A list of metals in order of reactivity (also called a 'reactivity series').

addition polymerisation A process whereby many small molecules (monomers) join to form one large molecule (a polymer) and nothing else.

addition reaction One in which two (or more) molecules join to produce a single larger molecule and nothing else.

alcohol Common name for the alkanol called ethanol, C_2H_5OH.

alkanols A homologous series with general formula $C_nH_{2n+1}OH$. (The simplest is methanol, CH_3OH.)

alkali A substance which dissolves in water to give a solution with a pH greater than 7.

alkaline solution One which contains a greater concentration of $OH^-(aq)$ ions than pure water.

alkali metal Metal in Group 1 of the periodic table (Li, Na, K, Rb, Cs, Fr).

alkanes A homologous series with general formula C_nH_{2n+2}. (The simplest is methane, CH_4.)

alkenes A homologous series with general formula C_nH_{2n}. (The simplest is ethene, C_2H_4.)

alloy A mixture of metals or of metals and non-metals. (The mixture must be melted together.) For example, bronze – copper/tin, steel – iron/carbon.

anode The positive electrode during electrolysis where oxidation takes place.

atom The smallest part of an element that can exist. It has a nucleus of protons and neutrons surrounded by moving electrons.

atomic number The number of protons in the nucleus of an atom.

B

balanced equation One with the same number of atoms of each element on both sides of the equation.

base A substance which reacts with an acid and neutralises it giving water as a product.

battery A series of chemical cells joined together.

Benedict's solution A solution which turns from blue to orange-red when warmed with a reducing sugar such as glucose.

biodegradable Able to rot away by natural biological processes.

C

catalyst A substance which speeds up a reaction without itself being changed.

cathode The negative electrode during electrolysis where reduction takes place.

cell In a chemical cell, chemical energy is changed into electrical energy. In an electrolytic cell, electrical energy is used to produce chemical changes.

chemical equation Uses chemical formulae to show the reactants and products during a chemical reaction, for example:

$$C + O_2 \rightarrow CO_2$$

chemical formula Shows the number of atoms of each element in a molecule or the ions in a compound. For example, CO_2 and $Ca^{2+}(Cl^-)_2$.

chemical reaction A chemical process in which one or more new substances are formed (usually accompanied by an energy change and a change in appearance).

chromatography A method of separating and identifying similar substances, for example glucose and fructose by paper chromatography.

combustion The burning of a substance during which it combines with oxygen.

compound A substance in which two or more elements are joined together chemically.

concentration The amount of solute dissolved in a given volume of solution. The usual units are moles per litre (mol/l).

condensation polymerisation A process whereby many small molecules (monomers) join to form a large molecule (a polymer), with water or another small molecule formed at the same time.

condensation reaction One in which two (or more) molecules join to produce a single larger molecule, with water or another small molecule formed at the same time.

corrosion A chemical reaction in which the surface of a metal changes from an element to a compound. This is an oxidation process.

covalent bond Bond formed between two non-metal atoms by the sharing of a pair of electrons.

cracking The breaking up of larger hydrocarbon molecules (usually alkanes) to produce a mixture of smaller molecules (usually alkanes and alkenes). The use of a catalyst (in catalytic cracking) allows the process to be carried out at a lower temperature.

cycloalkanes A homologous series of ring molecules with general formula C_nH_{2n}. (The simplest is cyclopropane, C_3H_6.)

D

decomposition The breaking down of a compound into two or more substances (usually by heating). For example,

$$2HgO \rightarrow 2Hg + O_2$$

diatomic molecule Molecule which contains only two atoms, for example H_2 and HCl.

disaccharide A carbohydrate which is formed when two monosaccharide molecules join in a condensation reaction, for example sucrose, $C_{12}H_{22}O_{11}$.

displacement Formation of a metal from a solution containing its ions by reaction with a metal higher in the electrochemical series.

distillate The liquid which is collected as a result of distillation.

distillation A process of separation or purification dependent on differences in boiling point. The changes of state involved are:

liquid → gas → liquid.

ductile The ability to be drawn out into wires – a physical property of metals.

E

electrochemical series A list of metals (and hydrogen) in order of their ability to lose electrons and form ions in solution. (Similar to the activity series of metals.)

electrode A conductor which is used to pass electricity into and out of solutions or melts.

electrolyte A compound which will conduct electricity when dissolved in water or melted.

electrolysis The process which occurs when a current of electricity is passed through a molten electrolyte (resulting in decomposition) or an electrolyte solution (which *in some cases* results in decomposition).

electron A particle which moves around the nucleus of an atom. It has a single negative charge but its mass is negligible compared to that of a proton or neutron.

electroplating Process whereby a layer of metal is deposited on an object during electrolysis. The object is used as the negative electrode in a solution containing ions of the metal being deposited.

element A substance which cannot be broken down into simpler substances by chemical means. All of its atoms have the same atomic number.

empirical formula Shows the simplest ratio of atoms in a compound, for example CH_3 is the empirical formula for ethane (molecular formula C_2H_6).

endothermic process Process in which heat energy is taken in.

enzyme A biological catalyst.

equilibrium State attained when forward and reverse reactions are taking place at the same rate.

exothermic process Process in which heat energy is given out.

F

fermentation The breakdown of glucose to form alcohol (ethanol) and carbon dioxide, brought about by the presence of yeast.

ferroxyl indicator Used to show the production of Fe^{2+}(aq) ions (by the formation of a blue colour) and OH^-(aq) ions (by the formation of a pink colour) during corrosion of iron experiments.

filtrate The liquid which passes through filter paper during filtration.

filtration The separation of an insoluble solid from a liquid by passage through filter paper.

finite Limited.

formula mass The sum of the relative atomic masses of all the atoms present in a formula.

fossil fuel One which has been formed from the remains of living things, for example coal, oil and natural gas.

fraction A mixture of hydrocarbons with similar boiling points obtained by fractional distillation of crude oil.

fractional distillation A means of separating crude oil into groups of hydrocarbons with similar boiling points (fractions).

fuel A substance which is used as a source of energy. This is released when the fuel burns.

G

galvanising Process by which iron is coated with a protective layer of zinc (by dipping into molten zinc).

group A column of elements in the periodic table.

H

Haber process The industrial production of ammonia from nitrogen and hydrogen, using high pressure and moderately high temperature, with iron as a catalyst.

halogen An element in Group 7 of the periodic table (F, Cl, Br, I, At).

homologous series A group of chemically similar compounds which can be represented by a general formula. Physical properties change gradually through the series, for example the alkanes, general formula C_nH_{2n+2}.

hydrocarbon A compound containing the elements carbon and hydrogen only.

hydrolysis reaction Reaction in which a large molecule is broken down into two (or more) smaller molecules by reaction with water.

I

indicator A substance whose colour is dependent on pH.

ionic bond Bond formed as a result of attraction between positive and negative ions.

ion bridge Used to complete the circuit in a chemical cell by allowing a flow of ions through it.

ion-electron equation Equation which shows either the loss of electrons (oxidation) or the gain of electrons (reduction).

ionic equation Equation which shows any ions that may be present among the reactants and products.

ions Atoms or groups of atoms which possess a positive or negative charge due to loss or gain of electrons, for example Na^+ and CO_3^{2-}.

isomers Compounds which have the same molecular formula but different structural formulae.

isotopes Atoms of the same element which have different numbers of neutrons. They have the same atomic number but different mass numbers.

L

lime-water Calcium hydroxide solution. Used to test for carbon dioxide, which turns it milky white.

M

malleable The ability to be beaten out into thin sheets – a physical property of metals.

mass number The total number of protons and neutrons in the nucleus of an atom.

metals Shiny, malleable and ductile elements found on the left of the periodic table. They all conduct electricity.

mixture Two or more substances mixed together but not joined chemically, for example air, which is a mixture of gases.

mole A formula mass expressed in grams.

molecular formula Formula which shows the number of atoms of the different elements which are present in one molecule of a substance.

molecule A group of atoms held together by covalent bonds.

monatomic Existing as single atoms. For example, the noble gases.

monomers Relatively small molecules which can join together to produce a very large molecule (a polymer) by a process called polymerisation.

monosaccharide Carbohydrate with molecular formula $C_6H_{12}O_6$, for example glucose and fructose.

N

neutral solution Solution in which the concentrations of $H^+(aq)$ and $OH^-(aq)$ ions are equal (pH = 7).

neutralisation reaction Reaction that moves the pH of a solution towards 7.

neutron A particle found in the nucleus of an atom. It has the same mass as a proton but no charge.

noble gas An element in Group 0 of the periodic table (He, Ne, Ar, Kr, Xe, Rn).

non-metals Non-conducting elements (except carbon as graphite) found on the right of the periodic table.

nucleus The extremely small centre of an atom where the neutrons and protons are found.

nutrient Something that helps a plant or animal to grow.

O

Ostwald process The industrial production of nitric acid from ammonia by a process which includes catalytic oxidation.

oxidation reaction One in which electrons are lost.

oxidising agent An electron acceptor.

P

period A horizontal row in the periodic table.

periodic table An arrangement of the elements in order of increasing atomic number, with chemically similar elements occurring in the same main vertical columns (groups).

pH A number which indicates the degree of acidity or alkalinity of a solution. Acidic solutions pH <7; neutral solutions pH = 7; alkaline solutions pH >7.

pH indicator See universal indicator.

photosynthesis A process whereby green plants convert carbon dioxide and water into carbohydrates such as glucose, and release oxygen into the air.

pollutant Something that harms the environment.

polymer A very large molecule which is formed by the joining together of many smaller molecules (monomers).

polymerisation The process whereby a polymer is formed. (See *addition polymerisation* and *condensation polymerisation*.)

polysaccharide A carbohydrate which is formed when many monosaccharide molecules join by condensation, for example starch.

precipitate An insoluble solid which is formed on mixing certain solutions.

precipitation reaction One in which a precipitate is formed.

proton A particle found in the nucleus of an atom. It has a single positive charge and the same mass as a neutron.

R

reactivity series See *activity series*.

redox reaction Reaction in which reduction and oxidation take place. Electrons are lost by one substance and gained by another.

reducing agent An electron donor.

reducing sugar Sugar which gives a positive test with Benedict's (or Fehling's) solution, for example glucose.

reduction reaction One in which electrons are gained.

relative atomic mass The average mass of one atom of an element on a scale where one atom of $^{12}_{6}C$ has a mass of 12 units exactly. It is the average of the mass numbers of the isotopes present, taking into account the proportion of each.

residue Solid left behind in the filter paper after filtration.

renewable energy source One which will not run out in the forseeable future e.g. solar, wind and tidal power.

respiration A process whereby carbohydrates are broken down by reaction with oxygen to release energy. Carbon dioxide and water are formed in this process, which takes place in both plants and animals.

reversible reaction One which proceeds in both directions, for example:

$$N_2 + 3H_2 \rightleftharpoons 2NH_3$$

rusting The corrosion of iron. It is caused by oxygen and water (which contains a dissolved electrolyte).

S

sacrificial protection A method for protecting a metal from corrosion by attaching it to a metal which is higher in the electrochemical series.

salt A compound which is formed when hydrogen ions in an acid are replaced by metal ions or ammonium ions.

saturated hydrocarbon Hydrocarbon in which all carbon-to-carbon covalent bonds are single bonds.

saturated solution Solution in which no more solute will dissolve at a given temperature.

solute A substance which dissolves in a liquid to give a solution.

solution A liquid with something dissolved in it.

solvent A liquid in which a substance dissolves.

spectator ion Ion which is present in a reaction mixture but takes no part in the reaction.

standard solution Solution of known concentration.

starch A polysaccharide, with molecular formula $(C_6H_{10}O_5)_n$.

state symbols Symbols used to indicate the state of atoms, ions or molecules: (s) = solid; (l) = liquid; (g) = gas; (aq) = aqueous (dissolved in water).

structural formula Formula which shows the arrangement of atoms in a molecule or ion. A *full structural formula* shows all of the bonds, for example propane:

```
  H H H
  | | |
H-C-C-C-H
  | | |
  H H H
```

A *shortened structural formula* shows the sequence of groups of atoms, for example propane: $CH_3CH_2CH_3$.

sugars Sweet-tasting, water-soluble carbohydrates, e.g. glucose and sucrose.

synthetic Man-made.

T

thermoplastic Plastic which softens on heating and can be reshaped.

thermosetting plastic Plastic which does not soften on heating.

titration An experiment in which volumes of reacting liquids are measured. In acid/alkali titrations, it is normal practice to use a pipette for measuring out the alkali and to add the acid from a burette.

toxic Poisonous.

transition metals The elements which form a 'bridge' in the periodic table between Groups 2 and 3, for example iron and copper.

U

universal indicator A solution containing several indicators which can be used to give the approximate pH of a solution. It gives a range of colours depending on the pH of the solution.

unsaturated hydrocarbon Hydrocarbon in which there is a carbon-to-carbon double bond, C=C, for example ethene.

V

valency A number which indicates the 'combining power' of atoms or ions.

variable Something that can be changed in a chemical reaction e.g. temperature, particle size, concentration, etc.

viscosity A description of how 'thick' a liquid is, for example engine oil is 'thicker' (more viscous) than petrol.

volatile Having a low boiling point and therefore easily vaporised.

W

word equation Use of words to show the reactants and products during a chemical reaction, for example:

magnesium + oxygen → magnesium oxide

CHAPTER TWENTY THREE

Whole course questions

Grid questions – General

1 This diagram shows part of the periodic table. The letters do **not** represent the symbols for the elements.

			C				
	B					E	
A							F
			D				

a) Identify the alkali metal.
b) Identify the element which exists as diatomic molecules.
c) Identify the **two** elements which are in the same group. (KU)

2 Hydrocarbons can be useful both as fuels and as starting materials for the production of plastics. The grid shows the names of some hydrocarbons.

A butane	**B** propene	**C** propane
D ethene	**E** methane	**F** ethane

a) Which hydrocarbon makes up about 95% of natural gas?
b) Which **two** hydrocarbons would quickly decolourise bromine water?
c) The structure of part of a polymer molecule is shown below.

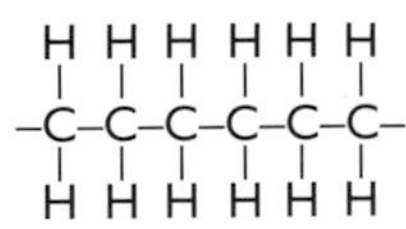

Which hydrocarbon is used to prepare this polymer? (KU)

3 The periodic table on page 8 of the SQA Data Booklet shows the names of the elements.

A chlorine	**B** copper	**C** oxygen
D lithium	**E** sulphur	**F** bromine

a) Identify the **two** elements in the same group as fluorine.
b) Identify the element which is a transition metal.
c) Identify the **two** elements which were discovered in 1774. (PS)

4 Hydrocarbons are compounds containing carbon and hydrogen only.

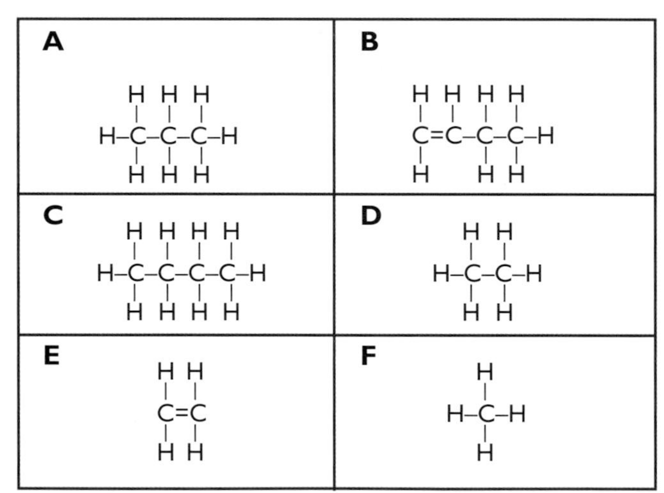

a) Identify the **two** hydrocarbons which can be used to make plastics.
b) Identify butene. (KU)

5 Compounds are formed when elements react together.

A sodium fluoride	**B** potassium sulphite
C potassium sulphide	**D** ammonium carbonate

a) Identify the **two** compounds which contain only two elements.
b) Identify the compound which contains nitrogen. (KU)

6 Many substances can be represented by simple models.

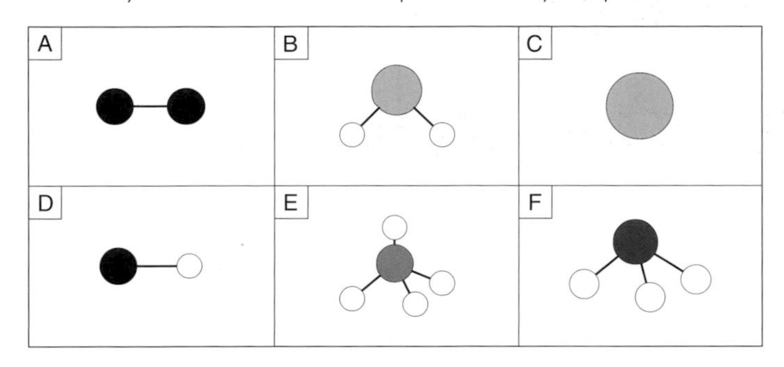

a) Identify the model which could represent water. (PS)
b) Identify the **two** models which represent diatomic molecules.
c) Identify the **two** models which represent elements. (KU)

7 Many terms are used to describe different chemical reactions.

A addition	**B** combustion	**C** cracking
D neutralisation	**E** polymerisation	**F** respiration

Identify the reaction(s) in which oxygen is used up. (KU)

8 Electricity can be produced using electrochemical cells.

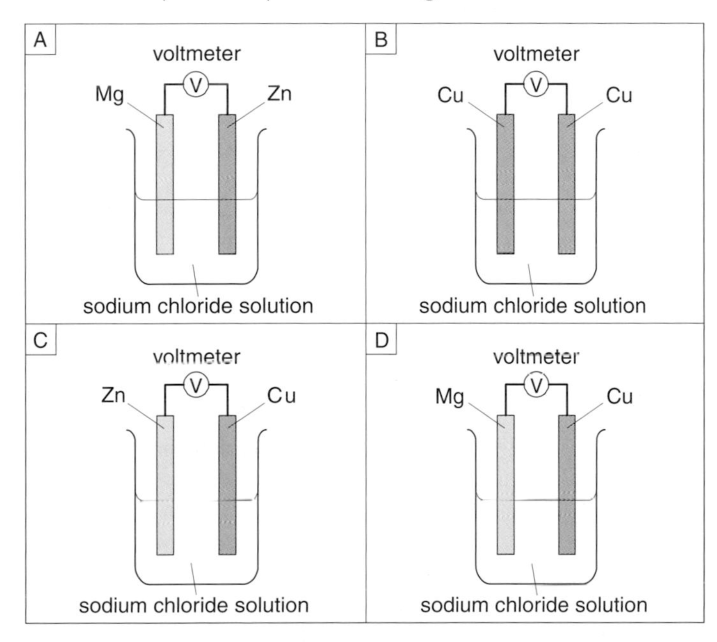

a) Identify the arrangement which would **not** produce electricity. (KU)
b) Identify the arrangement which would produce the greatest voltage.

(You may wish to refer to page 7 in the SQA Data Booklet to help you.) (PS)

9 Substances can be grouped as conductors or insulators.

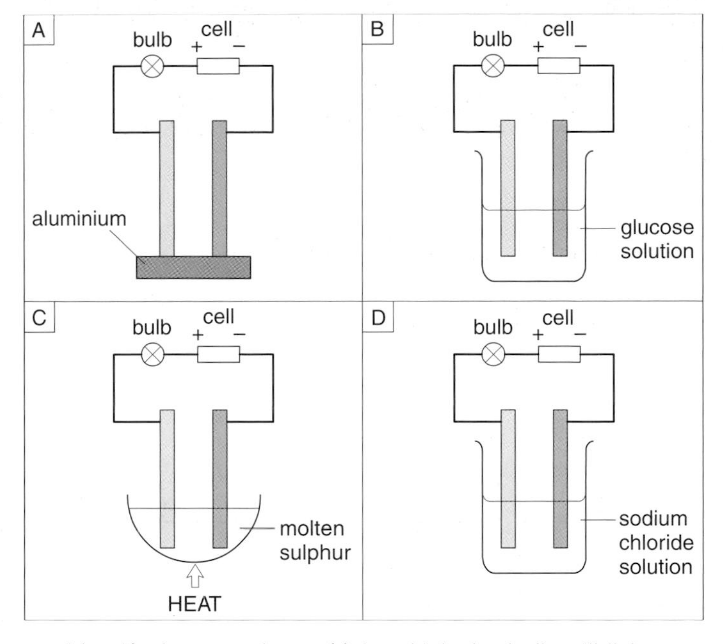

Identify the experiment(s) in which the bulb will light. (KU)

Extended answer questions – General

10 Lumps of calcium carbonate react with nitric acid. One of the products is a gas.

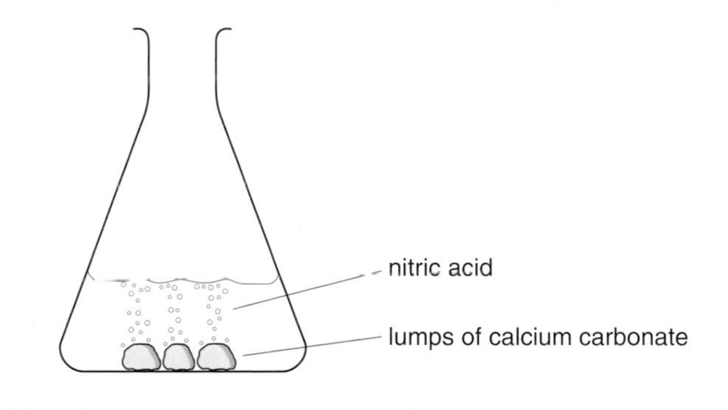

Identify the correct statement(s). (PS)

A	The gas produced is hydrogen
B	The reaction also produces calcium nitrate
C	No gas will form if the acid used is hydrochloric acid
D	The same gas will form if the carbonate used is copper carbonate
E	The gas will be produced at the same rate if powdered calcium carbonate is used.

11 The following flow diagram shows some of the processes carried out by the petroleum industry.

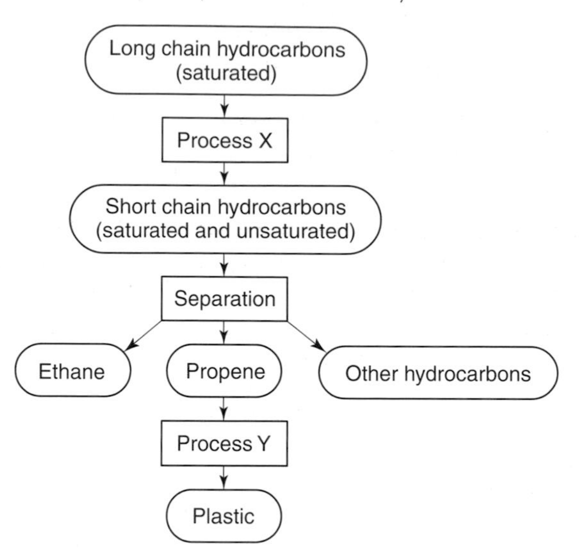

a) State the name of:
   (i) process X,
   (ii) process Y.

b) (i) Draw the full structural formula for propene.
   (ii) Name the plastic made from propene.
   (iii) Describe a test that can be used to show that propene is unsaturated.

c) Ethane is a short chain hydrocarbon. A substitution reaction occurs when a mixture of ethane and chlorine is exposed to light. Hydrogen chloride, an acidic gas, is produced. The other product is chloroethane.
   (i) Write a word equation for this reaction. (KU)
   (ii) Suggest why light is needed for the reaction to occur. (PS)

12 The analysis of salts is important in forensic science. Salts connected with certain occupations are shown below.

Occupation	**Salt(s)**
plasterer	calcium sulphate
farmer	ammonium nitrate ammonium sulphate

a) A forensic scientist carried out a flame test on some powder scraped from a plasterer's work clothes. What colour of flame would have been seen?

You may wish to use page 4 of the SQA Data Booklet. (PS)

b) A crime suspect was thought to have been in a field on which a farmer had recently sprayed ammonium fertiliser.
The forensic scientist heated mud from the suspect's shoe with an alkali called soda lime. She tested to see if ammonia gas was given off (see diagram top right).
How would she know if ammonia gas was produced? (KU)

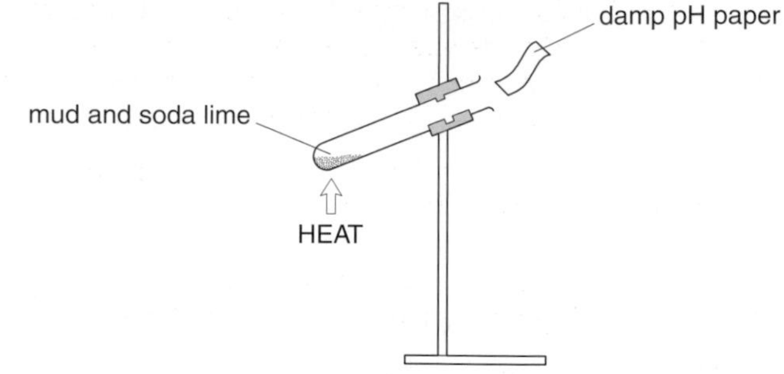

13 Class 4B were studying fermentation.

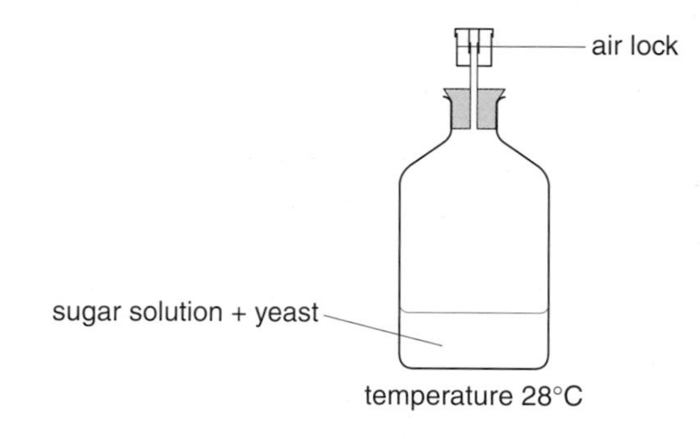

a) Why are sugar solutions fermented? (KU)

b) The class tested some of the gas which was produced by fermentation. Copy and complete the table to show the results. (PS)

Test	**Result**
burning splint	
lime water	

14 Paul connected a low voltage supply to carbon electrodes in a copper chloride solution. He wrote some notes in his jotter.

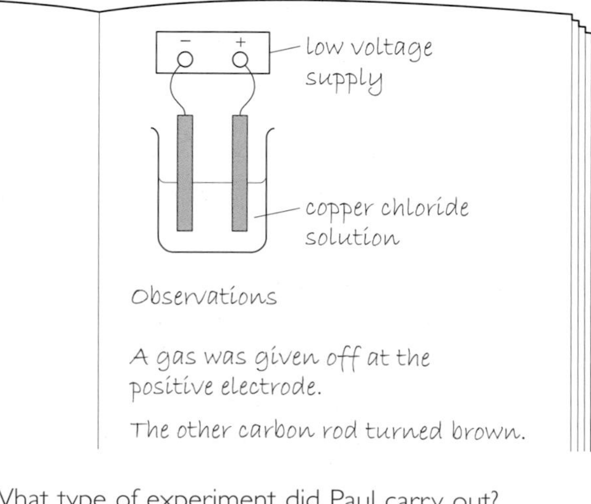

a) What type of experiment did Paul carry out?

b) Name the gas given off at the positive electrode. (KU)

c) Explain **fully** why the negative electrode 'turned brown'. (PS)

15 During the Apollo 13 space flight, the astronauts discovered that the carbon dioxide level inside the spacecraft was increasing. They used lithium hydroxide to remove the carbon dioxide. Lithium carbonate and water were produced.

Write an equation using formulae for the reaction. There is no need to balance the equation. (KU)

16 Wire gauzes in laboratories can be made to last longer by using either galvanised iron or an alloy of iron.

a) State what is meant by:
(i) galvanised iron,
(ii) an alloy. (KU)

b) Suggest **two** properties that make iron a good choice for a wire gauze. (PS)

17 Mary was investigating the reaction of marble chips with excess dilute hydrochloric acid. Her results for two experiments are shown below.

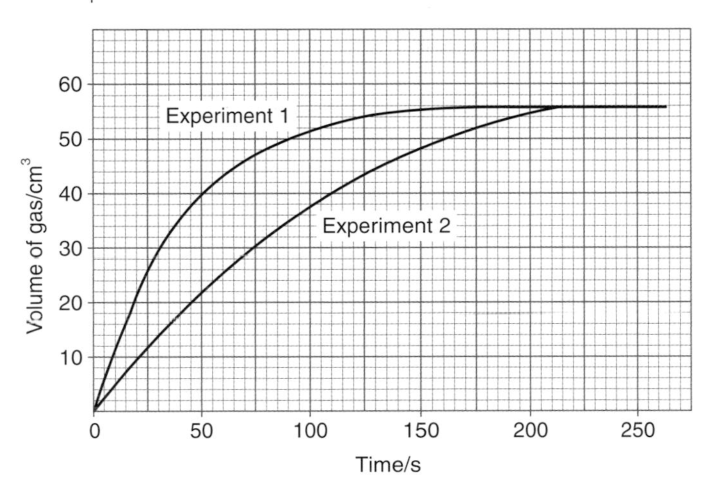

a) What volume of gas was produced in Experiment 1 during the first 90 s?

b) In the initial stages, the reaction in Experiment 2 was slower than in Experiment 1.
(i) How can this be concluded from the graph? (PS)
(ii) Suggest **two** changes in conditions which would have resulted in the slower reaction. (KU)

c) Mary used the same mass of marble chips in each experiment. How can this be concluded from the graph? (PS)

18 A plastic bottle has the following label:

NON-BIODEGRADABLE

To dispose of this plastic bottle, simply fill with hot water. When soft, empty and flatten.

a) What is meant by non-biodegradable? (KU)

b) What type of plastic is suggested by the instructions on the label? (PS)

c) Part of the structure of the plastic is shown (top right).

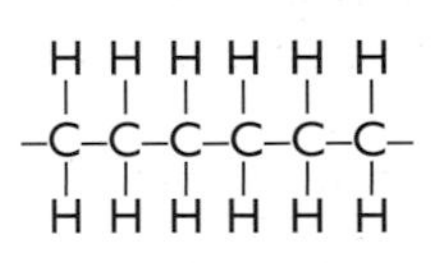

Draw the **full structural formula** for the monomer which is used to make the plastic. (KU)

19 Crude oil contains a mixture of chemicals. This table compares the composition of a sample of crude oil from the North Sea with one from an oilfield in the Middle East.

Chemicals	**% of chemicals in two samples of oil**	
	North Sea crude	**Middle East crude**
gases and gasoline	7	6
petroleum spirit	20	14
kerosene diesel	30	25
residue chemicals	43	55

a) Copy and label the pie chart below. It shows the composition of one of the above samples of crude oil.

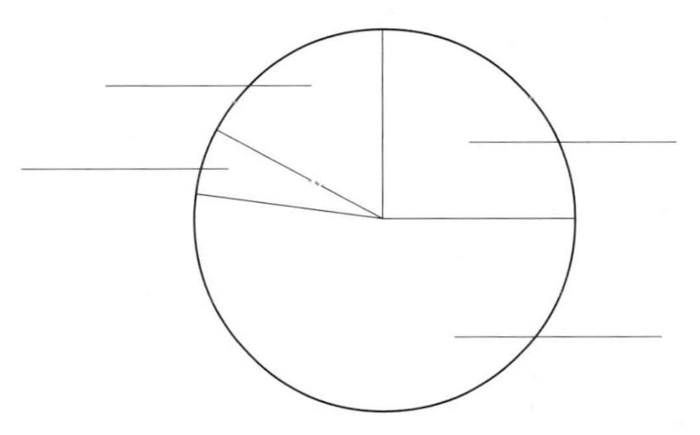

b) Which sample of crude oil has the composition shown in the pie chart? (PS)

c) (i) Name the process used to separate the different chemicals in crude oil. (KU)
(ii) Use the information in the table to suggest one reason why North Sea crude oil might be more useful than Middle East crude oil for modern day needs. (PS)

20 Jack and Iona measured the pH of some fizzy drinks.

Drink	**pH**
Just Fizz	5
Fizz Alive	3
Jupiter	4

a) Describe how you would use Universal Indicator or pH paper to measure the pH of a fizzy drink.

b) The more acidic the drink the more likely it is to increase tooth decay.
Name the fizzy drink which would be most likely to increase tooth decay.

c) Some fizzy drinks also contain a sugar called fructose.

(i) Suggest why fructose is added to some fizzy drinks. (PS)

(ii) Fructose is a carbohydrate.

Name the **three** elements present in fructose. (KU)

Grid questions – Credit

21 Iron can be coated with different materials which provide a physical barrier against corrosion.

A	tin
B	grease
C	paint
D	plastic
E	zinc

a) Identify the coating which also provides sacrificial protection.

b) Identify the coating which, if scratched, would cause the iron to rust faster than normal. (KU)

22 Frank and Dave carried out several experiments with metals and acids.

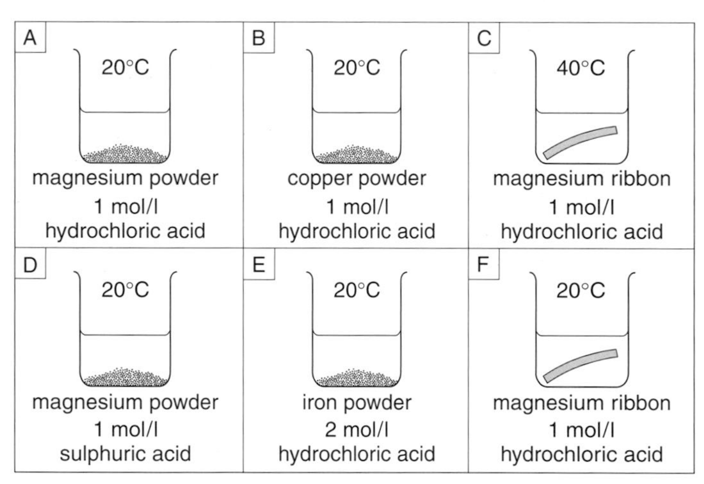

a) Identify the **two** experiments which should be compared to show the effect of particle size on reaction rate. (PS)

b) Identify the experiment in which no reaction would take place. (KU)

23

A	B	C
$^{24}_{11}Na$	$^{14}_{6}C$	$^{19}_{9}F$
D	**E**	**F**
$^{24}_{12}Mg^{2+}$	$^{19}_{9}F^{-}$	$^{12}_{6}C$

a) Identify the **two** particles with the same number of neutrons.

b) Identify the **two** atoms which are isotopes. (KU)

c) Identify the **two** particles with the same electron arrangement as neon. (PS)

24 The grid shows the names of some chemical compounds.

A	B	C
sodium hydroxide	potassium nitrate	sodium chloride
D	**E**	**F**
lithium carbonate	sodium phosphate	barium sulphate

a) Identify the **two** bases.

b) Identify the compound which could be prepared by precipitation.
You may wish to refer to page 5 of the SQA Data Booklet. (PS)

25 Several conductivity experiments were carried out using the apparatus shown below.

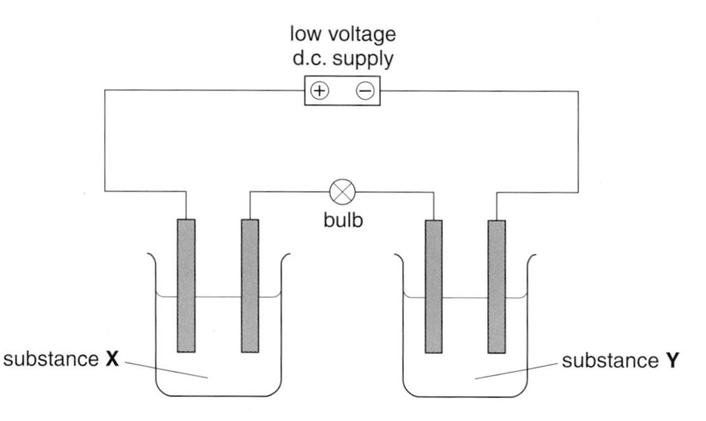

Identify the experiment(s) in which the bulb would light. (PS)

Experiment	Substance X	Substance Y
A	glucose solution	sodium chloride solution
B	molten tin	liquid mercury
C	sodium chloride solution	hexane
D	nickel bromide solution	molten sodium chloride
E	solid potassium nitrate	copper sulphate solution

26 The table below shows the names and colours of some common ions.

Ion	Formula	Colour
copper	Cu^{2+}	blue
nickel	Ni^{2+}	green
zinc	Zn^{2+}	colourless
lithium	Li^{+}	colourless
magnesium	Mg^{2+}	colourless
nitrate	NO_3^{-}	colourless
sulphate	SO_4^{2-}	colourless
permanganate	MnO_4^{-}	purple
dichromate	$Cr_2O_7^{2-}$	orange

Identify the true statement(s) based on the information in the table. (PS)

A	Copper nitrate is blue.
B	Coloured ions contained transition metals.
C	Ions containing oxygen are colourless.
D	All transition metal ions are coloured.
E	All lithium compounds are colourless.

27 There are different types of chemical reaction.

A displacement	**B** hydrolysis	**C** fermentation
D condensation	**E** addition	**F** redox

a) Identify the type of reaction that occurs when glucose molecules join to form starch. (KU)

b) Identify the type(s) of reaction represented by the following equation.

$Fe(s) + Cu^{2+}SO_4^{2-}(aq) \rightarrow Cu(s) + Fe^{2+}SO_4^{2-}(aq)$ (PS)

28 The grid shows the formulae for a number of hydrocarbons.

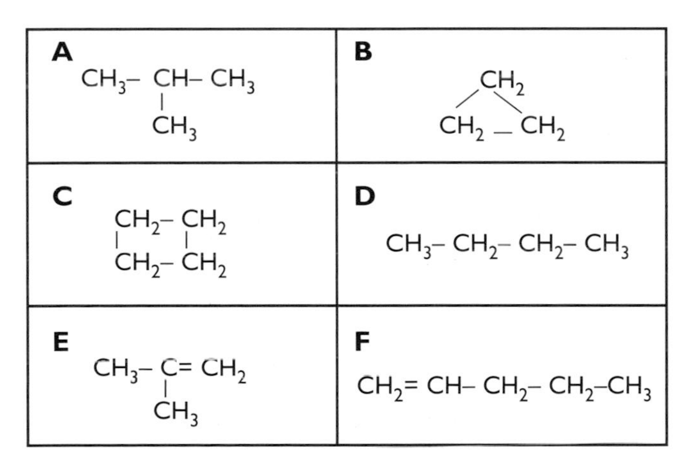

a) Identify the isomer(s) of

$$CH_2 = CH - CH_2 - CH_3$$ (ku)

b) Identify the **two** compounds with general formula C_nH_{2n}, which do **not** react quickly with bromine solution. (PS)

Extended answer questions – Credit

29 The diagram (top right) shows how sulphur dioxide is removed from the gases given off in a coal-fired power station.

The gases given off are passed through powdered limestone (calcium carbonate) in water.

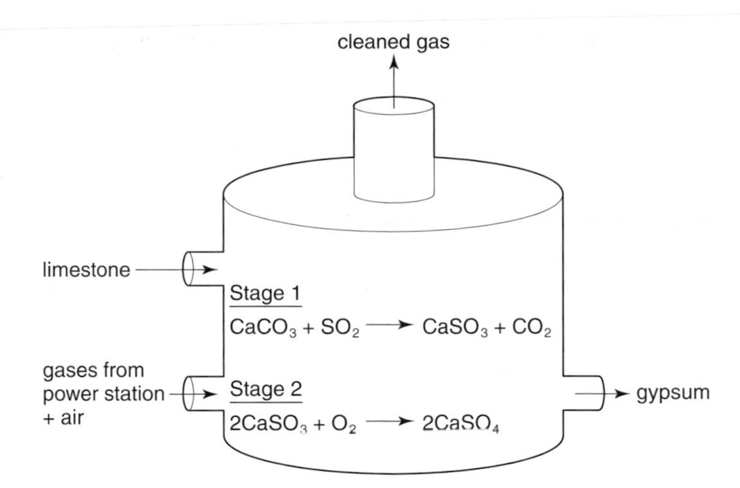

a) Why should sulphur dioxide be removed? (KU)

b) Limestone is insoluble in water. A soluble carbonate would be more efficient. Suggest why limestone is used.

c) Give the chemical name for gypsum. (PS)

d) A power station produces 580 tonnes of sulphur dioxide per day. Calculate the mass of calcium carbonate required to remove this completely.

Show your working clearly. (KU)

30 Tetrafluoromethane is a covalent compound. Its formula is CF_4.

a) Draw a diagram to show the **shape** of a molecule of tetrafluoromethane.

b) The atoms in a hydrogen molecule are held together by a covalent bond. A covalent bond is a shared pair of electrons.

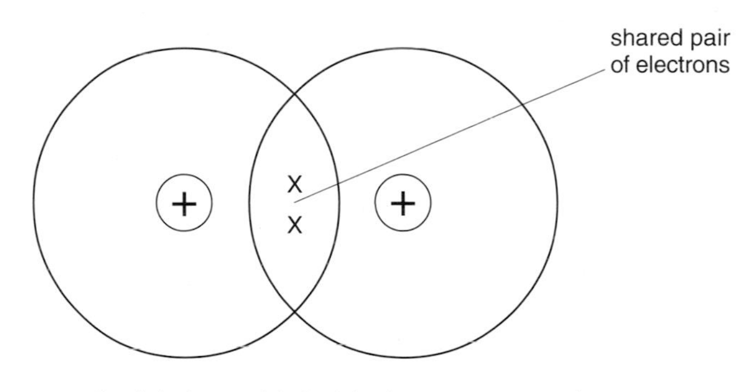

Explain how this holds the atoms together. (KU)

31 $^{14}_{6}C$ is an isotope of carbon. It occurs naturally in carbon dioxide and is present in plants as a result of photosynthesis.

a) What are isotopes?

b) Copy and complete the table to show the numbers of protons, neutrons and electrons in an atom of $^{14}_{6}C$.

Particle	Number
protons neutrons electrons	

c) Sunlight is essential for photosynthesis. Name the substance in green plants which can absorb sunlight. (KU)

32 Stuart wanted to prepare ammonium sulphate. He carried out a titration using 0.5 mol/l sulphuric acid and 0.5 mol/l ammonia solution.

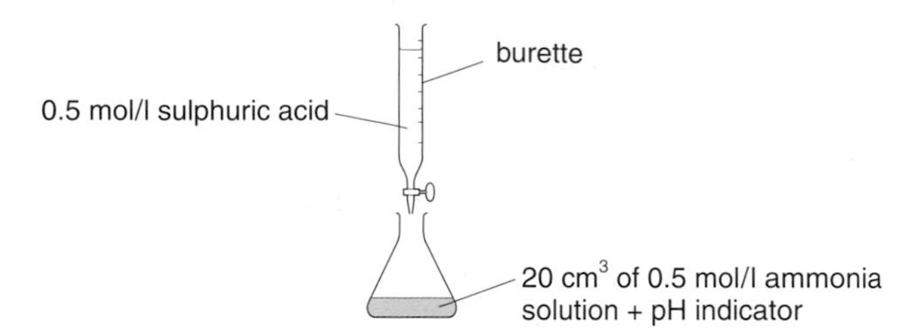

The equation for the reaction is

$$2NH_3(aq) + H_2SO_4(aq) \rightarrow (NH_4)_2SO_4(aq)$$

a) Calculate the volume of sulphuric acid Stuart used to neutralise the ammonia solution. (KU)
b) The indicator was removed from the ammonium sulphate solution by filtering the solution through charcoal. How would Stuart then obtain a sample of solid ammonium sulphate from the solution? (PS)

33 When chlorine gas is prepared in a fume cupboard it contains acid fumes.

The chlorine gas is bubbled through water to remove the acid fumes. It is then bubbled through concentrated sulphuric acid to dry the gas. The dried gas is collected in a gas jar.

Copy, complete and label the diagram to show how a sample of dry gas could be obtained. (PS)

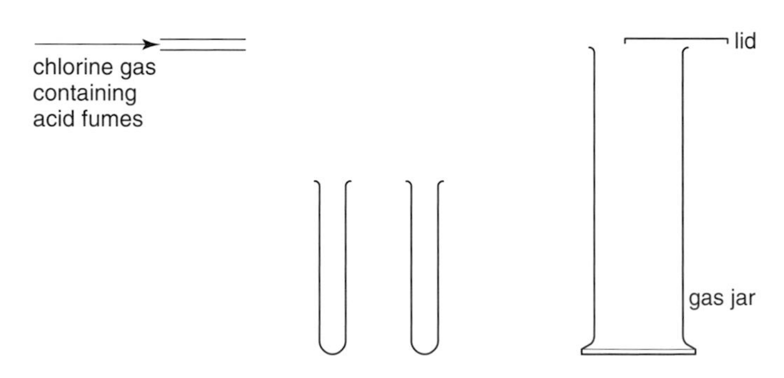

34 Urea (NH_2CONH_2) solution is broken down when the enzyme urease is added to it. During the reaction, ammionia and carbon dioxide are produced.

$$NH_2CONH_2(aq) + H_2O(l) \rightarrow 2NH_3(aq) + CO_2(g)$$

a) What is an enzyme? (KU)
b) The enzyme activity can be determined by removing a sample from the solution and measuring the concentration of ammonia. Suggest how the concentration of ammonia could be found.
c) Enzyme activity was determined at different temperatures (see top right).

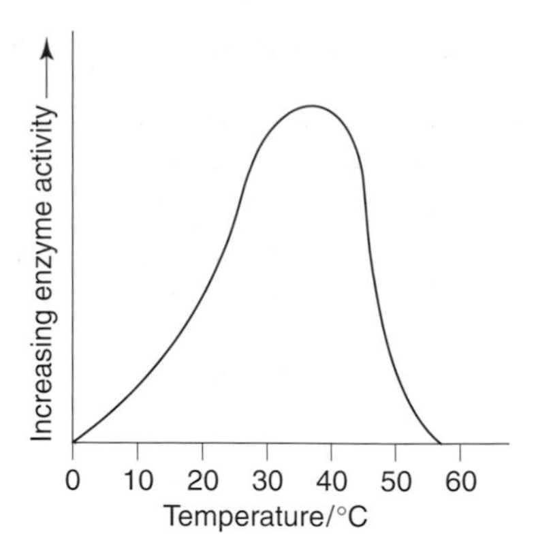

What effect does increasing temperature have on the activity of the enzyme? (PS)

35 Pupils in a class were trying to show that a green mineral in a rock was a copper ore.

Firstly they reacted the rock with sulphuric acid.

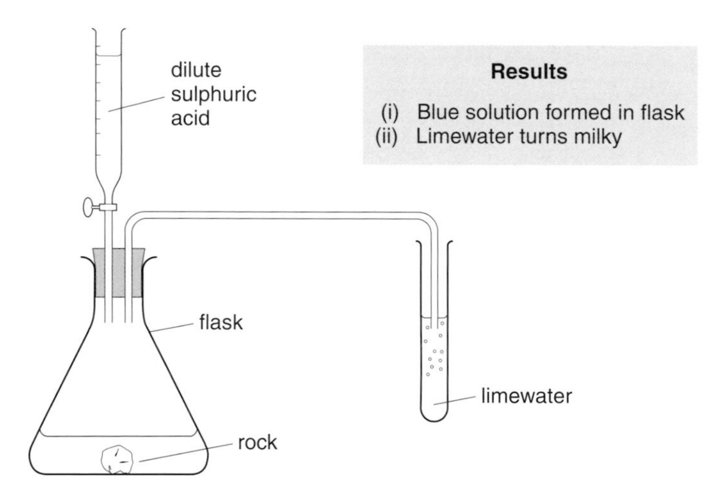

a) During the experiment, the limewater turned milky. Which **ion** does this indicate is present in the mineral?
b) The class carried out a second experiment using the blue solution formed in the flask.
Describe an experiment that could have been carried out by the class and explain how this could help them decide that the mineral was a copper ore. (You may wish to draw a diagram.) (PS)

36 Siobhan carried out some experiments with four metals (**W**, **X**, **Y** and **Z**) and some of their compounds. She made the following observations.

> When each metal was placed in cold water, only metal Y reacted.
>
> Only metal W was obtained from its oxide by heating.
>
> When metal X was placed in a solution containing ions of metal Z, metal X dissolved and solid metal Z was formed.

a) Name the gas formed when metal **Y** reacts with water. (KU)

b) Suggest names for metals **W** and **Y**.

c) Place the four metals (**W**, **X**, **Y** and **Z**) in order of reactivity (most reactive first). (PS)

d) Name the type of chemical reaction which takes place when a metal is extracted from its oxide. (KU)

37 A compound of sulphur and phosphorus is used to make matchheads.

A sample weighing 11 g is found to contain 4.8 g sulphur. Work out the empirical (simplest) formula of the compound.

Show your working clearly. (KU)

38 During hip replacement operations the new hip joint is fixed in place using bone cement.

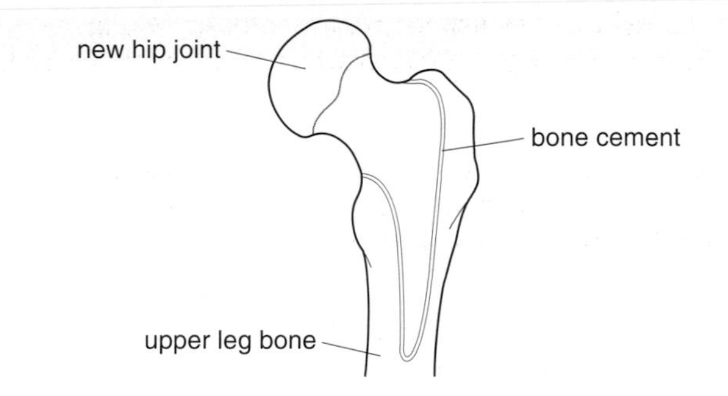

The most popular bone cement is a synthetic polymer formed from the monomer methyl methacrylate.

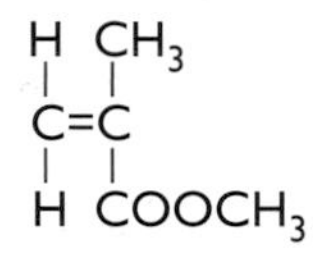

methyl methacrylate

a) Draw a section of this polymer, showing 3 monomer units joined together.

b) As the polymer forms, it releases large amounts of heat, which can damage surrounding bone. What term is used to describe a reaction releasing heat?

c) Calculate the percentage by mass of carbon in methyl methacrylate. (KU)

Chapter 1

Section 1.1

Q1 Chemical reactions take place in the burning fireworks.
Q2 The silvery surface of the nail becomes covered with a reddish-brown substance (rust).
Q3 **a)**, **b)** and **c)** In each case something burns producing mainly heat energy. Light and sound energy are also produced, but the light produced within the car engine would not be seen.

Section 1.2

Q1 **a)** I
b) K
c) Hg
Q2 The Latin name for iron was 'ferrum'.
Q3 **a)** hydrogen and sulphur;
b) zinc, nitrogen and oxygen;
c) potassium and chlorine;
d) barium and nitrogen;
e) magnesium, chlorine and oxygen;
f) sodium, nitrogen and oxygen.
Q4 Heat the solution to evaporate off the water leaving the dry salt.
Q5 **a)** and **b)** You can check the meaning of these and many other terms by referring to the chemical dictionary (Chapter 22).
Q6 In the compound iron sulphide the elements iron and sulphur are joined chemically, whereas in a mixture of iron and sulphur they are not.

Section 1.3

Q1 The solvent is water. The solutes are carbon dioxide, sugar, citric acid, flavourings, preservative (E211), caffeine, colours (E110 and E124), ammonium ferric citrate.
Q2 A concentrated solution contains a large quantity of solutes (dissolved substances) in a small volume of solvent (water in this case). A dilute solution contains a small quantity of solutes in a large volume of solvent. (Note that in apple juice there will be more than one solute.)
Q3 About 51 g of copper sulphate per 100 g of water.

Chapter 2

Section 2.1

Q1 Rusting (slowest), glue setting, apple turning brown, gas catching fire (fastest).
Q2 **a)** See graph top right.
b) The reaction is fastest at the beginning.
Q3 Take two washing-up basins each containing the same volume of water at the same temperature. Add a given volume of brand X to one basin and an equal volume of brand Y to the other. Wash equally dirty dishes of the same size and type in each basin, noting how many are cleaned in each case.

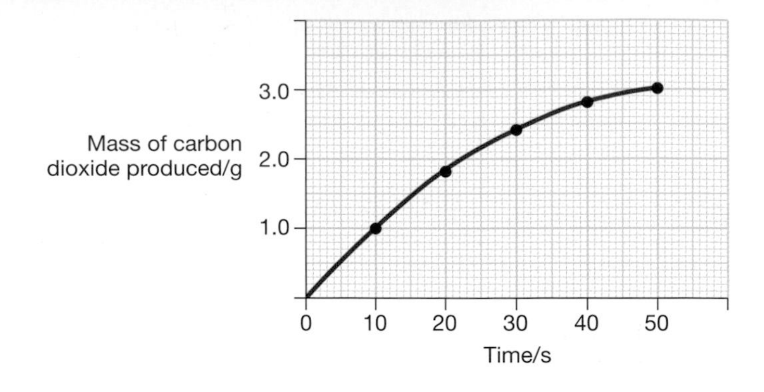

Q4 Heat a known volume of water in a beaker and note the temperature at which it boils using a thermometer. Repeat the experiment, using the same volume of water, but this time add a little salt to begin with. Again note the temperature of the boiling liquid.

Section 2.2

Q1 The temperature and the volume of the acid used should be kept the same.
Q2 The flour particles are very small.
Q3 The high temperature speeds up the rusting of the metal.

Section 2.3

Q1 To show that there is still 1 g of powder, filter it off, dry it and weigh it. To show that the powder is still manganese dioxide, add it to hydrogen peroxide and see if oxygen is released.
Q2 Select two zinc granules of equal mass and wind a piece of copper wire around one of them. Place the granules in equal volumes of dilute sulphuric acid of the same concentration and at the same temperature. If copper is a catalyst for the reaction, then bubbles of hydrogen gas will be given off faster from the granule with the copper wire attached.
Q3 Carrying out reactions at lower temperatures saves energy and reduces heating costs.
Q4 Bread.
Q5 Check your answer by referring to the chemical dictionary (Chapter 22). Examples of catalysts include manganese dioxide, nickel, iron, platinum, catalase and yeast. (There are many others.)

Chapter 3

Section 3.1

Q1 Find out if bismuth conducts electricity or not. If it conducts, then it is probably a metal. The following circuit could be used:

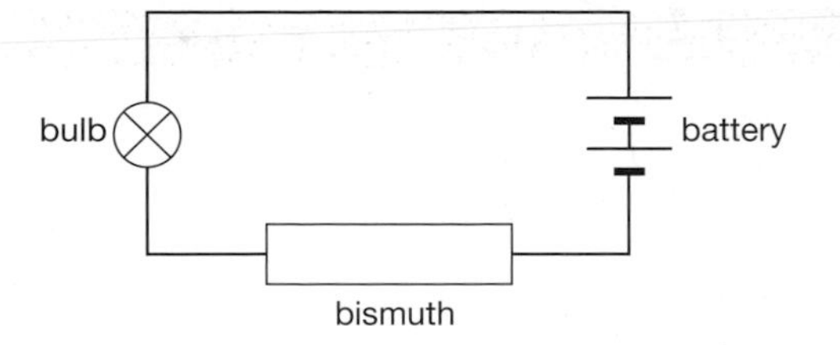

Q2 Various possible answers, for example (i) whether they are found as uncombined elements or not, (ii) according to colour.

Q3 **a)** metal;
b) metal;
c) non-metal;
d) metal.

Q4 **a)**, **b)** and **c)** – do not occur naturally, **d)** – occurs naturally.

Q5 Hydrogen, helium, nitrogen, oxygen, fluorine, chlorine, neon, argon, krypton, xenon and radon.

Section 3.2

Q1 Potassium oxide could be formed:

potassium + oxygen → potassium oxide

Q2 The alkali metals:
(i) react with water to produce the metal hydroxide and hydrogen (the metal hydroxide is an alkali),
(ii) react with oxygen to produce the metal oxide,
(iii) are soft and are easily cut, but the freshly cut surface quickly turns dull,
(iv) their compounds give bright colours when placed in a flame.

Q3 Pt and La are transition metals, but Ba and Pb are not.

Section 3.3

Q1 **a)** 1
b) 6
c) 8

Q2 **a)** 7
b) 6
c) 4

Q3 All of its elements, with the exception of helium, have 8 outer electrons.

Q4 Group 3.

Section 3.4

Q1 14

Q2 20

Q3

Element	Atomic number	No of protons	No of electrons
calcium	20	**20**	**20**
phosphorus	**15**	**15**	**15**
carbon	6	**6**	6
chlorine	**17**	17	**17**

Q4 **a)** 2,6
b) 2,3
c) 2,8,3
d) 2,8,8

Q5 **a)** nitrogen
b) fluorine
c) sodium
d) chlorine

Section 3.5

Q1 **a)** $^{4}_{2}He$
b) $^{32}_{16}S$

Q2 **a)** 16 protons, 16 neutrons, 18 electrons.
b) 20 protons, 20 neutrons, 18 electrons.
c) 7 protons, 8 neutrons, 10 electrons.

Q3 **a)** $^{23}_{11}Na^{+}$
b) $^{16}_{8}O^{2-}$

Q4 $^{18}_{8}O$

Q5 **a)** 12
b) 6
c) 8
d) 6

Chapter 4

Section 4.1

Q1 **a)** F-F
b) H-F

Q2 **a)** HF
b) CCl_4

Q3 4

Section 4.2

Q1 **a)** BF_3; **b)** CS_2; **c)** HI.

Q2 **a)** SO_2; **b)** NO; **c)** UF_6; **d)** P_2O_5.

Q3 **a)** Cu_2O; **b)** $SnCl_4$; **c)** FeO; **d)** Fe_2S_3.

Section 4.3

Q1 **a)** 2,7; **b)** 2,8.

Q2 2,8. This is the electron arrangement of noble gas neon.

Q3

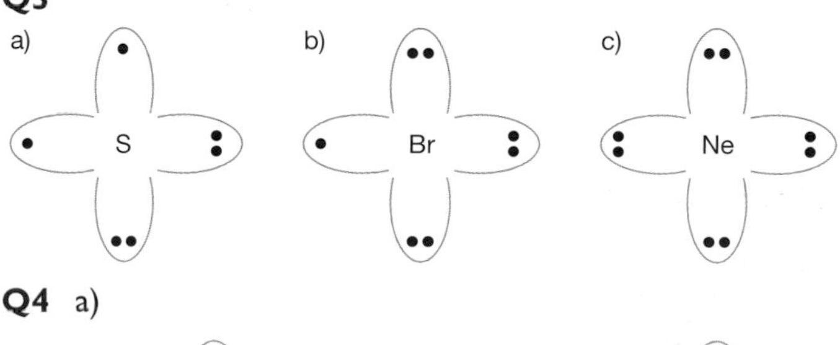

Q4 a)

b)

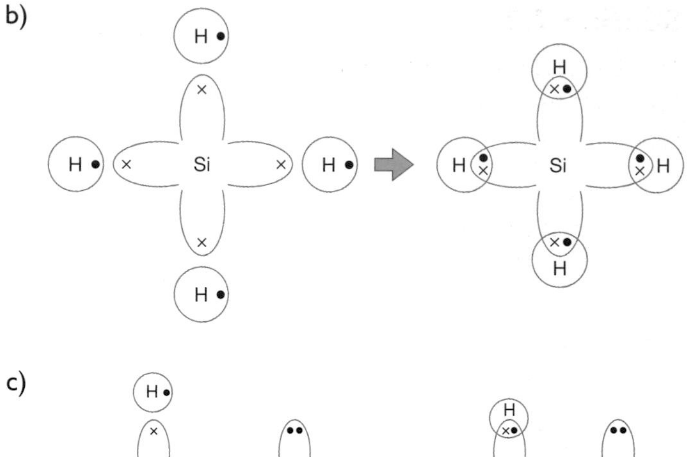

c)

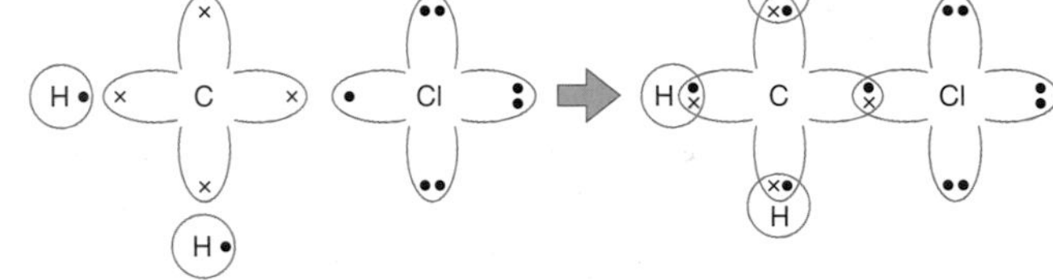

Section 4.4

Q1 a) b)

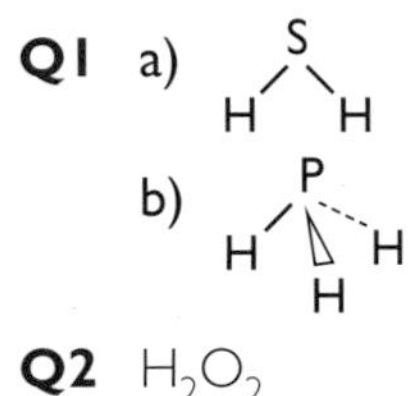

Q2 H_2O_2

Section 4.5

Q1 a) carbon monoxide + oxygen → carbon dioxide

$$2CO + O_2 \rightarrow 2CO_2$$

b) sodium + sulphur → sodium sulphide

$$2Na + S \rightarrow Na_2S$$

c) magnesium + hydrochloric acid → magnesium chloride + hydrogen

$$Mg + 2HCl \rightarrow MgCl_2 + H_2$$

Q2 a) $NH_3(g)$
b) $NH_3(l)$
c) $O_2(l)$
d) $CO_2(g)$
e) $CO_2(s)$

Q3 a) $NH_3(s)$
b) $NH_3(l)$

Chapter 5

Section 5.1

Q1 The damp cloth prevents air/oxygen from getting to the fat/oil and takes heat away from the flame.

Q2 Various possible answers, for example they were formed from material that was once living; the plants/animals were covered by layers of sediment; the decaying remains were subjected to pressure.

Section 5.2

Q1 Ethanol.

Section 5.3

Q1 $CH_4 + 2O_2 \rightarrow CO_2 + 2H_2O$

Q2 The oxygen present in the carbon dioxide and water could have been provided entirely from the air.

Q3 Various possible answers, for example glaciers and the polar ice-caps could melt; there could be a rise in sea level; low-lying areas could be flooded; normally fertile areas could become too hot for growing crops.

Q4 a) They poison the air (caused by carbon monoxide, nitrogen dioxide, etc.); they contribute to acid rain (for example due to nitrogen dioxide); they contribute to global warming (for example due to carbon dioxide). (You may be able to think of other answers.)

b) A 'lean burn' engine would reduce carbon dioxide levels by encouraging more complete combustion; using a smaller car would probably give a lower fuel consumption and this would reduce the levels of both nitrogen dioxide and carbon dioxide emitted; greater use of public transport would also reduce the pollution referred to in **a)**.

Chapter 6

Section 6.1

Q1 They are both used as fuels

Q2 a) ethane
b) butane
c) heptane

Q3 a)

```
  H H H
  | | |
H-C-C-C-H
  | | |
  H H H
```

b)

```
  H H H H H
  | | | | |
H-C-C-C-C-C-H
  | | | | |
  H H H H H
```

c)

```
  H H H H H H H H
  | | | | | | | |
H-C-C-C-C-C-C-C-C-H
  | | | | | | | |
  H H H H H H H H
```

Q4 a) $C_{10}H_{22}$
b) $C_{50}H_{102}$

Section 6.2

Q1 a) butene
b) pentene

Q2 a) (i) (ii)

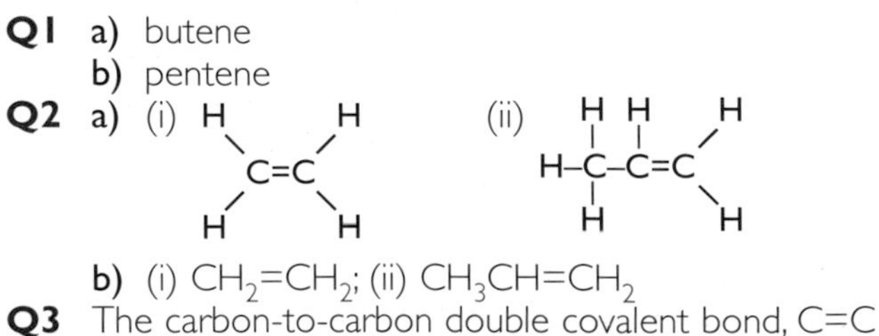

b) (i) $CH_2{=}CH_2$; (ii) $CH_3CH{=}CH_2$

Q3 The carbon-to-carbon double covalent bond, C=C

Q4 Melting points, flammability, viscosity (of liquid alkenes), etc. (Various answers possible.)

Q5 C_nH_{2n}

Q6 They can be represented by a general formula; they have at least one similar chemical property in that they all burn to give carbon dioxide and water.

Section 6.3

Q1

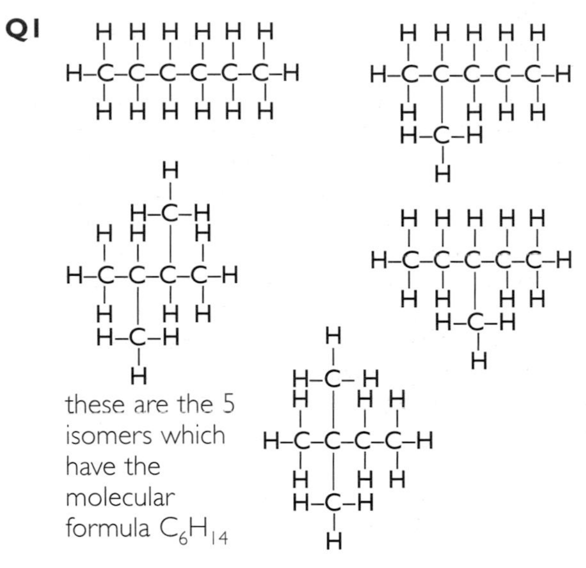

these are the 5 isomers which have the molecular formula C_6H_{14}

Q2 Ethene

Section 6.4

Q1 An alkene would decolorise bromine solution.
Q2 Hexene, C_6H_{12}.
Q3 Hexane.

Chapter 7

Section 7.1

Q1 a) Ca^{2+}
b) S^{2-}
c) Al^{3+}
d) Br^-
Q2 Cubic

Section 7.2

Q1 Examples include: acetone (modern name propanone), amyl acetate (pentyl ethanoate) and ethyl acetate (ethyl ethanoate).
Q2 a) silver chloride;
b) metal nitrates tend to be very soluble in water.
Q3 Metals and carbon as graphite conduct electricity. Non-metals, apart from carbon as graphite, do not. Refer to the periodic table on page 8 in the SQA Data Booklet for the names of elements.
Q4 a) conducts;
b) conducts;
c) does not conduct.

Section 7.3

Q1 Each carbon atom has four bonds.
Q2 For example, silicon.

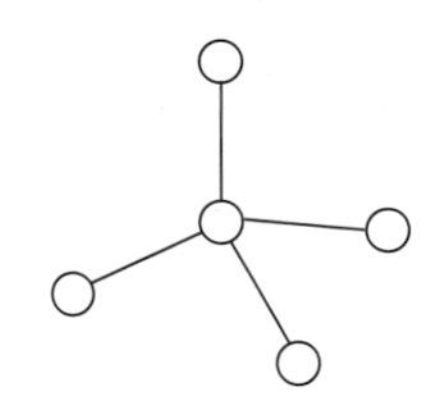

Q3 Silicon carbide, SiC. The structure is similar to that of diamond.

Section 7.4

Q1 Copper metal forms at the negative electrode because the copper ions present in the solution have a positive charge and are therefore attracted towards the electrode with the opposite charge.
Q2 Alternating current is not normally used because only with direct current do the electrodes keep the same charge.
Q3 At the negative electrode: $Pb^{2+} + 2e^- \rightarrow Pb$
At the positive electrode: $2Br^- \rightarrow Br_2 + 2e^-$

Section 7.5

Q1 a) LiF
b) CaI_2
c) Al_2O_3
d) Na_2S
Q2 a) Na_2CO_3
b) LiOH
c) NH_4Cl
Q3 a) $Mg(NO_3)_2$
b) $Pb(OH)_2$
c) $(NH_4)_2SO_4$
Q4 a) K^+I^-
b) $(Na^+)_2\ O^{2-}$
c) $(NH_4^+)_3PO_4^{3-}$

Section 7.6

Q1 The iron(II) ion, Fe^{2+}
Q2 The dichromate ion, $Cr_2O_7^{2-}$
Q3 a) The copper(II) ion, Cu^{2+}, is blue and is attracted towards the negative electrode.
b) The chromate ion, CrO_4^{2-}, is yellow and is attracted to the positive electrode.
c) Direct current must be used so that the electrodes keep the same charge, either positive all the time or negative all the time. If an alternating current is used the charge on each electrode keeps on changing rapidly from positive to negative and back again.
Q4 In salt A, the green ion is positively charged and could be the iron(II) ion, Fe^{2+}. In salt B, the orange ion is negatively charged and could be the dichromate ion, $Cr_2O_7^{2-}$. In salt C, the purple ion is negatively charged and could be the permanganate ion, MnO_4^-.

Chapter 8

Section 8.1

Q1

Metal oxides	Non-metal oxides
barium oxide lithium oxide calcium oxide	nitrogen dioxide

Q2 Phosphorus oxide.
Q3 One possibility – place one of the nails in tap water and the other in rain water. Use identical nails and conditions (for example, volume and temperature of water).

Section 8.2

Q1 Soft drinks and fruit juices tend to have low (acidic) pH values. Bicarbonate of soda, most soaps, some toothpastes and washing-up liquids have high (alkaline) pH values. (pH less than 7 = acidic, greater than 7 = alkaline.)
Q2 **a)** chloride
b) nitrate
c) sulphate
Q3 $K + O_2 \rightarrow K_2O$ ($4K + O_2 \rightarrow 2K_2O$ balanced)
$K_2O + H_2O \rightarrow KOH$ ($K_2O + H_2O \rightarrow 2KOH$ balanced)

Section 8.3

Q1 **a)** 103
b) 30
c) 63
Q2 **a)** 56 g
b) 148.5 g
c) 18 g
d) 4 g
Q3 **a)** (i) 470 g; (ii) 32 g;
b) (i) 0.1; (ii) 0.2; (iii) 0.3.
Q4 0.5 moles/litre

Section 8.4

Q1 0.5 litres (500 cm^3)
Q2 **a)** Kills fish by releasing poisonous aluminium into water; damages leaves so that trees cannot grow properly; washes nutrients which trees need from soil.
b)

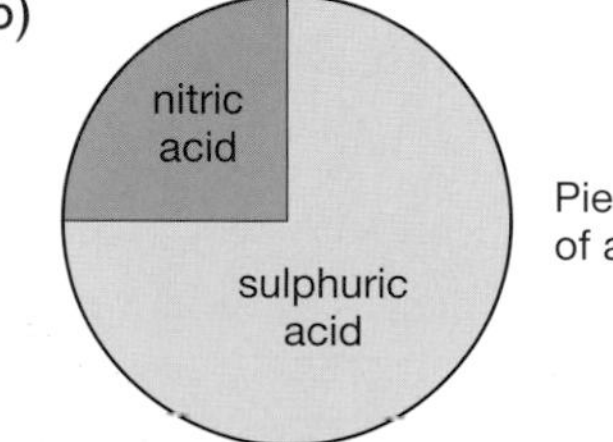

Pie chart to show the proportions of acids in acid rain

Chapter 9

Section 9.1

Q1 **a)** potassium hydroxide + nitric acid → potassium nitrate + water
b) calcium hydroxide + hydrochloric acid → calcium chloride + water
Q2 Sulphuric acid and sodium hydroxide.
Q3 **a)** sodium hydroxide + nitric acid → sodium nitrate + water
b) $NaOH + HNO_3 \rightarrow NaNO_3 + H_2O$
c) $Na^+ + OH^- + H^+ + NO_3^- \rightarrow Na^+ + NO_3^- + H_2O$
(Na^+ and NO_3^- are the spectator ions.)

Section 9.2

Q1 When mixed with air and a lighted taper is applied, hydrogen burns with a 'pop'.
Q2 aluminium + hydrochloric acid → aluminium chloride + hydrogen
Q3 **a)** → aluminium nitrate + water
b) → magnesium chloride + water
Q4 **a)** potassium carbonate + sulphuric acid → potassium sulphate + carbon dioxide + water
b) $2K^+ + CO_3^{2-} + 2H^+ + SO_4^{2-} \rightarrow 2K^+ + SO_4^{2-} + CO_2 + H_2O$

K^+ and SO_4^{2-} are spectator ions. Without these, equation becomes: $2H^+ + CO_3^{2-} \rightarrow H_2O + CO_2$

Section 9.3

Q1 **a)** nickel carbonate
b) silver chloride
Q2 **a)** $NaOH + HCl \rightarrow NaCl + H_2O$
b) $Na_2CO_3 + 2HCl \rightarrow 2NaCl + H_2O + CO_2$
Q3 **a)** A gas is given off in reactions (i) and (iv).
b) (i) $Mg + H_2SO_4 \rightarrow MgSO_4 + H_2$
(ii) $MgO + H_2SO_4 \rightarrow MgSO_4 + H_2O$
(iii) $Mg(OH)_2 + H_2SO_4 \rightarrow MgSO_4 + 2H_2O$
(iv) $MgCO_3 + H_2SO_4 \rightarrow MgSO_4 + H_2O + CO_2$

Section 9.4

Q1 **a)** 22.6 cm^3, 22.8 cm^3 and 22.7 cm^3
b) 22.7 cm^3
Q2 **a)** $NaOH + HCl \rightarrow NaCl + H_2O$
b) 0.08 mol/l

Chapter 10

Section 10.1

Q1 When the battery produces electricity, the zinc casing provides electrons and therefore gradually corrodes away.
Q2

Advantages	Disadvantages
portable low voltage, therefore safe	more expensive than mains uses up valuable chemicals

Section 10.2

Q1 a) As the reaction takes place zinc atoms turn into zinc ions in solution.
b) copper
c) The blue copper ions in solution gain electrons and turn into copper atoms; the copper metal is a brown solid.

Q2 a) yes
b) yes
c) no

Q3 On the left-hand side, the solution would change from yellow (due to Fe^{3+}) to green (due to Fe^{2+}). On the right-hand side, the solution would change from colourless (I^- ions are colourless) to reddish-brown (due to I_2).

Section 10.3

Q1 a) oxidation
b) reduction
c) reduction

Q2 B

Chapter 11

Section 11.1

Q1 They must be strong so they can be stretched without breaking and they must not corrode easily.

Q2 Various possible answers, for example recycling metals reduces pollution since less electrical energy is needed, so less sulphur dioxide and carbon dioxide are emitted from power stations burning fossil fuels; less mining would take place; less waste metal would need to be dumped.

Section 11.2

Q1 a) $4K + O_2 \rightarrow 2K_2O$
b) $4K + O_2 \rightarrow 2\,(K^+)_2O^{2-}$

Q2 a) $Ca + 2H_2O \rightarrow Ca(OH)_2 + H_2$
b) $Ca(s) + 2H_2O(l) \rightarrow Ca^{2+}(aq) + 2OH^-(aq) + H_2(g)$

Q3 a) $Mg + 2HCl \rightarrow MgCl_2 + H_2$
b) $Mg(s) + 2H^+(aq) + 2Cl^-(aq) \rightarrow Mg^{2+}(aq) + 2Cl^-(aq) + H_2(g)$

Q4 a) C A B **b)** C = K, Na, Li, Ca or Mg; A = Al, Zn, Fe, Sn or Pb; B = Cu, Hg, Ag or Au.

Section 11.3

Q1 a) $2HgO \rightarrow 2Hg + O_2$
b) $2Hg^{2+}O^{2-}(s) \rightarrow 2Hg(l) + O_2(g)$

Q2 As at August 1993, steel was made at, for example Scunthorpe, Sheffield, Llanwern and Port Talbot.

Q3 a) $Fe^{3+} + 3e^- \rightarrow Fe$
b) $Ag^+ + e^- \rightarrow Ag$

Section 11.4

Q1 $Fe = \dfrac{(3 \times 56)}{(3 \times 56) + (4 \times 16)} \times 100 = 72.4\%$

Q2 $Al = \dfrac{(2 \times 27)}{(2 \times 27) + (3 \times 16)} \times 100 = 52.9\%$

Q3

Element	Cu	O
mass/g	88.8	11.2
number of moles of atoms	$\frac{88.8}{63.5} = 1.40$	$\frac{11.1}{16} = 0.70$
simplest whole number ratio	$\frac{1.39}{0.70} = 2$	$\frac{0.70}{0.70} = 1$
empirical formula	Cu_2O	

Q4

Element	Mg	O
mass/g	0.1225	0.0800
number of moles of atoms	$\frac{0.1225}{24.5} = 0.005$	$\frac{0.0800}{16} = 0.005$
simplest whole number ratio	$\frac{0.005}{0.005} = 1$	$\frac{0.005}{0.005} = 1$
empirical formula	MgO	

Q5 PbO_2 (*Note*: mass of oxygen = 4.78 – 4.14 = 0.64 g)

Q6 Molecular formula of cyanogen is C_nN_n.
Formula mass of C_nN_n =
$(12 \times n) + (14 \times n) = 52$
$26n = 52$
$n = 2$
Thus molecular formula of cyanogen is C_2N_2.

Q7 $SnO_2 + 2H_2 \rightarrow Sn + 2H_2O$
1 mole ⟷ 1 mole
150.5 g ⟷ 118.5 g
7.55 g ⟷ $\dfrac{118.5 \times 7.55}{150.5} = 5.94\,g$

Chapter 12

Section 12.1

Q1 It is used as a control so that a fair comparison can be made. It shows that without the moist iron filings the water level at the mouth of the test tube does not change.

Q2 The presence of the sodium chloride increased the rate at which the iron nail rusted.

Section 12.2

Q1 It would cause the bodywork to rust (because electrons would flow from the iron in the bodywork to the positive terminal).
Q2 Zinc is higher than copper in the electrochemical series and therefore electrons flow from the zinc (which is sacrificed) to the copper (which is protected).
Q3 When iron is galvanised it is given a complete coating of zinc. The zinc therefore provides physical protection because it prevents air and water from reaching the iron. It also gives the iron sacrificial protection because zinc is above iron in the electrochemical series. Even if the zinc coating is broken, electrons, will still flow from the zinc (which is sacrificed) to the iron (which is protected).

Section 12.3

Q1 The other way – from the iron to the copper.
Q2 **b)** and **d)** (the other compounds are insoluble in water).
Q3 The steel bumpers would rust rapidly. The iron in the steel is above gold in the electrochemical series and therefore electrons would flow from the iron (sacrificed) to the gold (protected).

Chapter 13

Section 13.1

Q1 Light, strong, waterproof, re-useable, recyclable, etc.
Q2 Various answers possible, but probably efficient recycling and using less plastic are best. Not many plastic objects can be re-used, and biodegrading takes a long time.
Q3

Advantages	Disadvantages
most are currently cheap lightweight can be rigid or flexible do not corrode long-lasting	many are made from oil which will soon run out many burn, producing poisonous gases cause litter

(Other answers possible.)

Section 13.2

Q1 –
Q2 Carbon dioxide and water.
Q3 Hydrogen chloride, hydrogen cyanide and carbon monoxide.
Q4

Advantages	Disadvantages
glass is made from materials which will not run out; plastic is made from oil – a finite resource glass bottles can be reused	glass is heavier than plastic glass can shatter

(Other answers possible.)

Section 13.3

Q1 **a)** propene
b) phenylethene
Q2 The double bonds partly break leaving single electrons. The sharing of these electrons between monomer molecules produces a polymer molecule linked by single covalent bonds.

Section 13.4

Q1 **a)** The monomer would decolorise the bromine solution.
b) There would be no decolorisation (as only single carbon-to-carbon, bonds C–C, are now present).
Q2 **a)** 3
b) **c)**

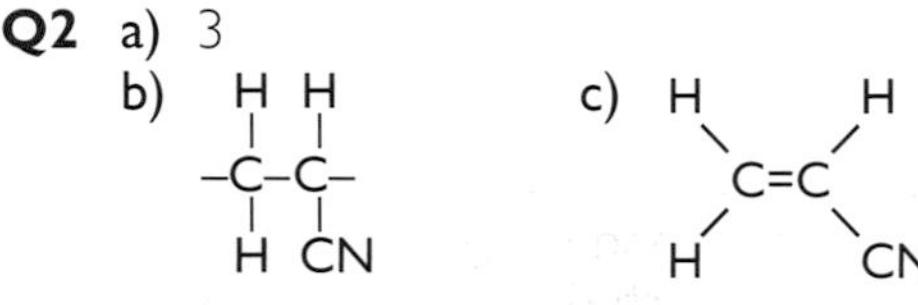

Section 13.5

Q1 thermoplastic
Q2 **a)**

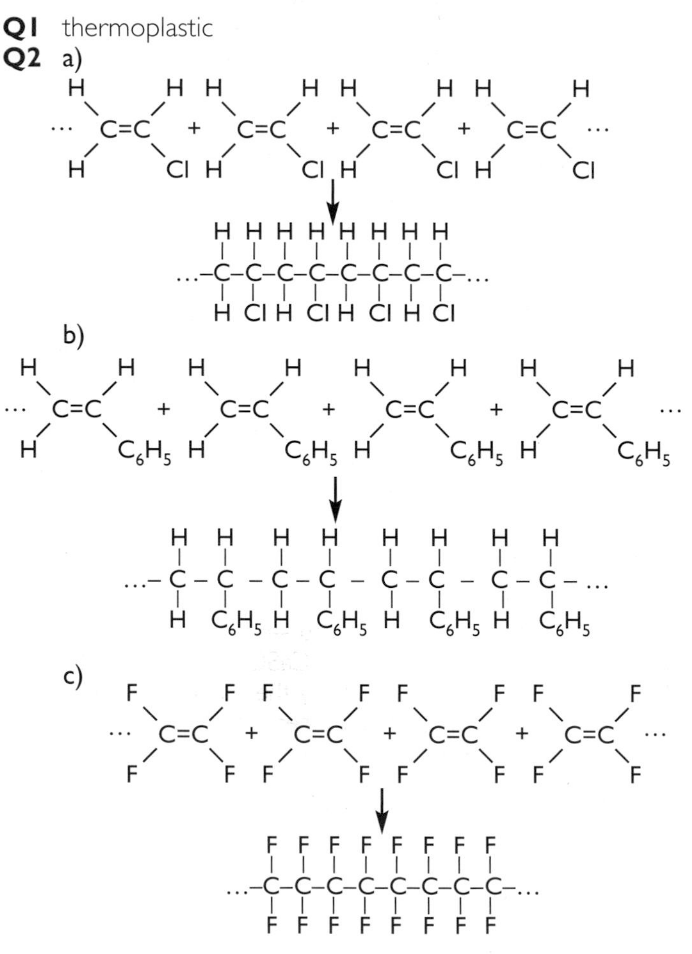

Chapter 14

Section 14.1

Q1 Bacon comes from pigs which are fed on cereals such as maize. Lettuce leaves and tomatoes come from plants. Bread is made from wheat. Butter comes from cattle which eat grass.

Q2 a) (i) NH_4^+; (ii) PO_4^{3-}; (iii) NO_3^-
b) (i) Unsuitable – 'insoluble'; (ii) suitable – 'very soluble'; (iii) suitable – 'very soluble'.

Q3

Advantages	Disadvantages
easy to store easy to apply are mostly odour-free greatly increase crop yields	increase pollution in water supplies

(Other answers possible.)

Section 14.2

Q1 Nitrates are very soluble in water.

Q2 During lightning, nitrogen and oxygen join together to form nitrogen dioxide which dissolves in rain to give nitric acid. Nitric acid contains the nitrate ion.

Q3 Legumes have root nodules which allow them to take nitrogen from the air and convert it into nitrates which can be used in the plants.

Section 14.3

Q1 a) Ammonia
b) Ammonia is useful because it can be easily converted into ionic compounds which can be used as fertilisers.

Q2 nitrogen + hydrogen $\rightleftharpoons$ ammonia

Q3 The reaction is reversible and does not go to completion. Liquefying the ammonia enables it to be removed from unreacted nitrogen and hydrogen, which can then be recycled through the reaction chamber.

Q4 Costs are reduced by making products more quickly. The use of a catalyst means a lower temperature can be used, thus reducing energy needs.

Section 14.4

Q1 a) ammonium sulphate
b) ammonia + sulphuric acid $\rightarrow$ ammonium sulphate

Q2 $Ca(OH)_2 + (NH_4)_2SO_4 \rightarrow CaSO_4 + 2NH_3 + 2H_2O$

Q3 Ammonia would be made by the Haber process. Nitrogen and hydrogen are heated to about 500°C under high pressure and in the presence of iron as a catalyst.

Q4 The reaction is exothermic.

Chapter 15

Section 15.1

Q1 a) water and carbon dioxide
b) glucose and oxygen

Q2 Various possible answers, for example general increase in the use of fossil fuels.

Q3 Global warming could result in climatic changes and a rise in sea levels. (Various answers are possible.)

Section 15.2

Q1 Test each with iodine solution – only starch reacts, giving a dark blue colour. Then heat with Benedict's or Fehling's solution – only glucose reacts, giving an orange colour.

Q2 Sucrose and maltose – molecular formula $C_{12}H_{22}O_{11}$.

Section 15.3

Q1 Initially, a dark blue colour would be obtained. This would gradually become less intense and eventually, when no more starch was present, there would be no colour change on testing with iodine solution.

Q2 It shows that maltose is obtained on hydrolysis of starch. This is because there is a spot on the chromatogram at the same height as one produced by pure maltose.

Section 15.4

Q1 A substance which speeds up a chemical reaction but is not itself used up.

Q2 $C_6H_{12}O_6 \rightarrow 2C_2H_5OH + 2CO_2$

Q3 37°C (approx.)

Q4 pH 2

Chapter 16

Section 16.1

Q1 a) $(Li^+)_2O^{2-}$
b) $Ca^{2+}S^{2-}$
c) $(Al^{3+})_2(O^{2-})_3$

Q2 a) $(Cu^+)_2O^{2-}$
b) $Fe^{2+}\ SO_4^{2-}$
c) $(NH_4^+)_2CO_3^{2-}$
d) $(Cr^{3+})_2(SO_4^{2-})_3$

Q3 a) $Cu^{2+}O^{2-}(s) + 2H^+(aq) + SO_4^{2-}(aq) \rightarrow Cu^{2+}(aq) + SO_4^{2-}(aq) + H_2O(l)$

Q4 a) $Zn(s) + 2H^+(aq) + 2Cl^-(aq) \rightarrow Zn^{2+}(aq) + 2Cl^-(aq) + H_2(g)$
b) $Cl^-(aq)$ are spectator ions
c) $Zn(s) \rightarrow Zn^{2+}(aq) + 2e^-$ (Zn(s) is oxidised)
$2e^- + 2H^+(aq) \rightarrow H_2(g)$ ($H^+(aq)$ is reduced)